Proceedings in Life Sciences

Radiation and Cellular Control Processes

Edited by J. Kiefer

With 176 Figures

Springer-Verlag
Berlin Heidelberg New York 1976

Professor Dr. JÜRGEN KIEFER
Strahlenzentrum der
Justus Liebig Universität
6300 Gießen/FRG

Proceedings of a conference at the Strahlenzentrum der Justus Liebig Universität, Gießen, Germany, October 6–10, 1975; organized in cooperation with "Deutsche Gesellschaft für Biophysik"; cosponsored by the Commission of the European Communities, Directorate General Research, Science and Education, Biology Division.

ISBN-13: 978-3-642-66457-1 e-ISBN-13: 978-3-642-66455-7
DOI: 10.1007/978-3-642-66455-7

Library of Congress Cataloging in Publication Data. Main entry under title: Radiation and cellular control processes. (Proceedings in life sciences) "Proceedings of a conference held at the 'Strahlenzentrum der Justus Liebig-Universität', Giessen, Germany, from October 6 to October 10, 1975." Bibliography: p. Includes index. 1. Cellular control mechanisms--Congresses. 2. Radiobiology--Congresses. I. Kiefer, Jürgen, 1936-. II. Giessen. Universität. Strahlenzentrum. QH604.R3. 574.1'915. 76-41874.

Softcover reprint of the hardcover 1st edition 1976

Offsetprinting: Beltz Offsetdruck, Hemsbach/Bergstr.;

Preface

This book contains the proceedings of a conference held at the Strahlenzentrum der Justus Liebig-Universität, Gießen, Germany, from October 6 to October 10, 1975. About 40 scientists, mostly from European countries, came together to discuss specific aspects of modern radiation biology. The majority of the papers deal with "simple eukaryotic systems" such as yeast, algae, slime moulds etc. It was felt that these study objects offer good prospects to tackle fundamental questions. This scope, however, was never intended to be exclusive, and the reader will see from the table of contents that a variety of different systems was discussed. Therefore we thought the general title would be appropriate. The rationale that led to the planning of the symposium and the titles of the sessions is outlined in the Introduction. We hope that this volume gathers some useful information, not only for radiation biologists, but also for colleagues engaged either in fundamental cell research or in the application in the field of radiation therapy and protection.

The conference would not have been possible without the generous support of the European Communities and the cosponsorship of the Deutsche Gesellschaft für Biophysik. If it was to a certain extent successful, then the merit goes to the very active engagement of all participants. I should like to thank my coworkers for their devoted help in the organisation, particularly Dr. Eckart Schneider, Dr. Elmar Gocke, Maria Hlawica, and Eva-Maria Peter for secretarial help.

Gießen, Autumn 1976 JÜRGEN KIEFER

Contents

List of Participants and Contributors

ABBONDANDOLO, A., Laboratorio di Mutagenesi e Differenziamento del C.N.R., Via Cisanello 147, Pisa / Italy

ALPER, T., Gray Laboratory, Mount Vernon Hospital, Northwood, Middlesex HA6 2RN / Great Britain

ANDERS, A., Fakultät für Physik der Universität Bielefeld und Zentralinstitut für Biochemie und Biophysik der Freien Universität Berlin, 1000 Berlin / FRG

AVERBECK, D., Fondation Curie, Institut du Radium, Paris / France

BEER, J.Z., Instytut Badan Jadrowych, Oddzial Zeran, Warszawa / Poland

BERTSCHE, U., Gesellschaft für Strahlen- und Umweltforschung, Paul-Ehrlich-Straße 20, 6000 Frankfurt/M. / FRG

BOYLE, J.M., Paterson Laboratories, Christie Hospital and Holt Radium Institute, Wilmslow Road, Manchester M20 9BX / Great Britain

BRENDEL, M., Arbeitsgruppe Mikrobengenetik im Fachbereich Biologie, J.W. Goethe-Universität, Robert-Mayer-Straße 7-9, 6000 Frankfurt/M. / FRG

BRYANT, P.E., MRC Cyclotron Unit, Hammersmith Hospital, Ducane Road, London W12 OHS / Great Britain

COQUERELLE, Th., Krebsforschungszentrum, Institut für Strahlenbiologie, 7400 Karlsruhe / FRG

DÖHREN, Fr. R. v., Institut für Biochemie, Fachbereich Chemie, Universität Mainz, 6500 Mainz / FRG

DORANGE, J.L., Station de Technologie des Produits Végétaux I.N.R.A., B.P. 1540, 7 Rue Sully, 21034 Dijon Cedex / France

EBERT, M., Paterson Laboratories, Christie Hospital and Holt Radium Institute, Wilmslow Road, Manchester M20 9BX / Great Britain

ECKSTEIN, H., Universität Hamburg, Institut für Physiologische Chemie, Martinistraße 52, 2000 Hamburg 20 / FRG

ESSER, G., Battelle Institut, 6000 Frankfurt/M. / FRG

FABER, G., Strahlenzentrum der Justus Liebig-Universität, 6300 Gießen / FRG

FINGERHUT, R., Strahlenzentrum der Justus-Liebig-Universität, 6300 Gießen / FRG

FRANKENBERG, D., Gesellschaft für Strahlen- und Umweltforschung, Abteilung Biophysikalische Strahlenforschung, Paul-Ehrlich-Straße 20, 6000 Frankfurt/M. / FRG

FRANKENBERG-SCHWAGER, M., Gesellschaft für Strahlen- und Umweltforschung, Abteilung Biophysikalische Strahlenforschung, Paul-Ehrlich-Straße 20, 6000 Frankfurt/M. / FRG

FRIESS, E., Strahlenzentrum der Justus Liebig-Universität, 6300 Gießen / FRG

GOCKE, E., Strahlenzentrum der Justus Liebig-Universität, Institut für Biophysik, 6300 Gießen / FRG

HARTMANN, G., Deutsches Krebsforschungszentrum, Institut für Nuklearmedizin, 6900 Heidelberg / FRG

HÖLTZ, G.W., Gesellschaft für Strahlen- und Umweltforschung München, Abteilung für biophysikalische Strahlenforschung, Paul-Ehrlich-Straße 20, 6000 Frankfurt/M. / FRG

HOWARD, A., Paterson Laboratories, Christie Hospital and Holt Radium Institute, Manchester M20 9BX / Great Britain

INGLIS, R.J., Biophysics Laboratories, Physics Department, Portsmouth Polytechnic, Gun House, Hampshire Terrace, Portsmouth / Great Britain

JAIN, V.K., Gesellschaft für Strahlen- und Umweltforschung, Paul-Ehrlich-Straße 20, 6000 Frankfurt/M. / FRG

JENTSCH, K.D., Max-Planck-Institut für Biophysikalische Chemie, Postfach 968, 3400 Göttingen / FRG

KIEFER, J., Strahlenzentrum der Justus Liebig-Universität, 6300 Gießen / FRG

KINSELLA, A.R., (not present, paper given by J.M.BOYLE) Paterson Laboratories, Christie Hospital and Holt Radium Institute,Manchester M20 9BX /Great Britain

KONINGS, A.W.T., Laboratory of Radiopathology, Groningen / Netherlands

KOWALSKI, S., Forschungsinstitut für Mikrobiologie, Institut für Gärungsgewerbe, Seestraße 13-15, 1000 Berlin 65 / FRG

LAMPRECHT, I., Zentralinstitut für Biochemie und Biophysik der Freien Universität Berlin, 1000 Berlin/FRG

LASKE, B., Strahlenzentrum der Justus-Liebig-Universität, 6300 Gießen / FRG

LASKOWSKI, W., Zentralinstitut für Biochemie und Biophysik der Freien Universität Berlin, 1000 Berlin/FRG

LEHMANN, A.R., MRC Cell Mutation Unit, University of Sussex, Falmer, Brighton BN1 9QG / Great Britain

MOORE, P.D., National Institute for Medical Research, Mill Hill, London NW7 1AA / Great Britain

MOUSTACCHI, E., Fondation Curie-Institut du Radium, Bâtiment 110, Orsay 91 / France

OTTO, F., Institut für Aerobiologie, Grafschaft / FRG

PARRY, J.M., Department of Genetics, University College of Swansea, Swansea SA2 8PP / Great Britain

PARTKE, H.-J., Institut für Biophysik, 5300 Bonn / FRG

PEARLMAN, R.E., Dept. of Biology, York University, 4700 Keele Street, Downsview, Ontario M3J 1P3 / Canada

REINHARD, R.D., Gesellschaft für Strahlen- und Umweltforschung München, Abteilung für Biophysikalische Strahlenforschung, 6000 Frankfurt/M. / FRG

RETÈL, J., Biochemisch Laboratorium, Vrije Universiteit, De Boelelaan 1085, Amsterdam / Netherlands

RINK, H., Institute of Biophysics, Section Radiobiology, Annabergerweg 15, 5300 Bonn-Venusberg / FRG

SACHSENMAIER, W., Institut für Biochemie und Experimentelle Krebsforschung der Universität Innsbruck, Fritz-Pregl-Straße 3/VII, 6020 Innsbruck / Austria

SCHNEIDER, E., Strahlenzentrum der Justus Liebig-Universität, 6300 Gießen / FRG

SCHRADER-REICHARDT, U., Radiologische Klinik, Abteilung Strahlenphysik und Strahlenbiologie, 3400 Göttingen / FRG

WALICKA, M., (not present, paper given by J.Z.BEER) Dept. of Radiobiology and Health Protection, Institute of Nuclear Research, 03-195 Warszawa-Zerân/ Poland

WALLER, H., Strahlenzentrum der Justus Liebig-Universität, 6300 Gießen / FRG

WANGENHEIM, K.-H. v., Institut für Botanik und Mikrobiologie, Kernforschungsanlage Jülich GmbH, 5170 Jülich / FRG

WEBER, K., Strahlenzentrum der Justus Liebig-Universität, 6300 Gießen / FRG

WESTERGAARD, O., Dept. of Molecular Biology, University of Aarhus, 8000 Aarhus C / Denmark

WIENHARD, I., Strahlenzentrum der Justus Liebig-Universität, Institut für Biophysik, 6300 Gießen / FRG

WINTERSBERGER, E., Physiologisch-Chemisches Institut der Universität Würzburg, Koellikerstraße 2, 8700 Würzburg / FRG

WINTERSBERGER, U., Institut für Krebsforschung der Universität Wien, Borschkegasse 8a, 1090 Wien / Austria

YASUI, A., Fakultät für Physik der Universität Bielefeld und Zentralinstitut für Biochemie und Biophysik der Freien Universität Berlin, 1000 Berlin / FRG

Introduction - On the Rationale of this Conference

J. Kiefer

Historically, radiation biology has one of its roots in the practical problems of radiation protection and radiation therapy, although it was realized at early stages that radiation action on cells poses intriguing fundamental questions. Their solution requires the cooperation of physicists and biologists. The target theory of radiation action as developed by Dessauer, Crowther, Lea, and Zimmer constitutes one of the still fascinating examples of a theoretical attempt to link physical and biologic aspects. It was learned at later stages that radiation is an effective tool for probing the biologic object. The action spectra for mutation induction and cell killing indicated the involvement of nucleic acids, quite some time before molecular biology proved their importance by different means. The more recent discoveries of the cell's capability to repair damage in the genetic material, which are of utmost importance in the understanding of living structures, are almost exclusively due to radiation biology.

We know that the cell is a delicately controlled system that has to exchange substances and energy with its environment to maintain a steady state, far off from thermodynamic equilibrium. This balance may be upset by irradiation. The study of metabolic key processes may provide a means to gain more insight into the interwoven network that guarantees cellular stability.

Long-term stability, survival, however, can only be achieved if damage inflicted can be removed in some way. Repair and recovery processes form therefore a centerpiece of all attempts in the understanding of cellular organisation and biologic evolution.

Cell multiplication, the replication of genetic material, and its inheritance are essential for species survival and evolution. On the other hand, it requires elaborate interplay and synchronisation of many biochemical processes. A dividing cell in "balanced growth" magnifies the integration of many different pathways that can easily be disturbed by comparatively low doses of irradiation that hardly affect survival, thus allowing studies under almost "physiologic" conditions. To be honest, undisturbed cell division is still one of the breathtaking miracles of biology. Radiation biology may help to unravel it.

A meeting devoted to the discussion of the questions sketched above would be doomed to be completely fruitless if it were confined to a closed shop of radiation biologists. Therefore it was attempted to foster the exchange of thoughts between many disciplines. This conference was purposely aimed at fundamental questions, which is not very fashionable in these times when application is asked for and the merit of scientific work is judged by its obvious gain for society.

Certainly, this is not quite at hand for our enterprise at first sight. It will turn out, however, that dealing with fundamental problems, apparently far off from direct application, might ultimately be more useful than those investigations that like to label themselves as "applied", "directly applicable" and very often lead only to shortcomings. Reasonable applications are only possible if the fundamentals are understood.

To be more specific, it is hoped that by conferences of this kind the radiation therapist may gain some information to help him with his very important task.

Chapter I
Radiation and Metabolic Key Processes

1. Radiation Action on Biochemical Key Processes: An Introduction

J. M. BOYLE

A. Key Targets and Cell Survival

In a short review of this kind it is necessary for the author to be selective, and I have chosen to review recent work on the effects of radiation on transcription and translation in *Escherichia coli* and yeast. The resultant effects on respiration and cell permeability will be reviewed from the point of view of controlled derepression of operons induced by radiation. Previous reviews (Rothstein, 1959; Okada, 1970; Dertinger and Jung, 1970; Pollard, 1970) contain summaries of information up to 1970.

As recognised by Pollard and Vogler (1961), the central dogma (DNA → RNA → protein) determines that radiation damage to DNA results in amplification of the damage in a temporal sequence. It is generally agreed that protein synthesis is necessary for recovery from irradiation. The case for RNA synthesis is less clear: some workers indicate that it is necessary whilst others find the opposite. Respiration may be important in determining the sparing effect of split-dose recovery (Howard, 1968). Hence the effects of radiation on the molecular biology of the cell are significant to cell survival, as measured by colony forming ability. Key targets will therefore show dose responses similar to those of cell survival. Table 1 lists some survival parameters for various strains of *E. coli* and yeast relevant to the studies to be discussed.

Table 1. X ray survival of c.f.a.

Cells	Minimal			Nutrient		Reference
	D_{37}	D_0	n	D_0	n	
A. *E. coli*						
K12 (λ)	11					Pollard and Davis, 1970
$15T^-L^-$	3					Pollard and Davis, 1970
		3.5	1.6	7.8	4	Pollard and Randall, 1973
$15T^-$JG151	1.8					Pollard and Davis, 1970
		2.9	1.5	2	1	Pollard and Randall, 1973
B/r	5.5					Pollard and Davis, 1970
B	4					Pollard and Davis, 1970
		4.5	1	5.5	2	Pollard and Randall, 1973
B_{s-1}	2.2					Pollard and Davis, 1970
		1.8	1	1.6	1	Pollard and Randall, 1973
B. Yeast						
S. cereviseae strain 211	~88					Kiefer, 1971
	47					Kiefer, 1974

B. Transcription

I. Total RNA Synthesis

Early studies of the in vivo incorporation of RNA precursors in yeast showed that 50-100 krads stimulated RNA synthesis by 20% (Rothstein, 1959). In *E. coli* transcription was either resistant (Frampton, 1964) or sensitive (Pollard et al., 1965; Pollard and Barone, 1966) to radiation. The conditions under which transcription was sensitive, i.e., irradiation of cells in mineral salts medium at densities less than 1×10^8 per ml in the presence of air, permitted the formation of H_2O_2 and also affected the electron transport system (Frey and Pollard, 1968). When the effects of peroxide were minimised by the presence in the medium of 10^{-7} *M* catalase at the time of irradiation, the rate of uracil incorporation at 5 min postirradiation showed a bimodal dose response (Fig. 1) (Pollard and Davis, 1970). This finding was interpreted as implying that there are two modes of RNA synthesis, one sensitive to radiation and the other insensitive. Since operon transcription is sometimes, but not always, more sensitive to radiation than total RNA synthesis (Pollard and Davis, 1970; Pollard and Lydersen, 1972), it was suggested that cofactors of RNA polymerase might be involved.

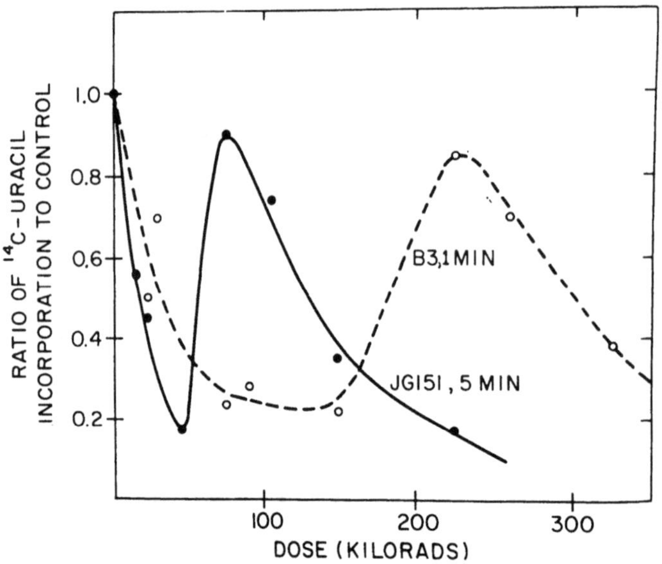

Fig. 1. Bimodal dose-response of uracil incorporation. *E. coli* strains JG 151 and B_3 were exposed to γ-irradiation from ^{60}Co source and then incubated. At 1 min (B_3) or 5 min (JG 151) after irradiation, 2-4 min pulse of ^{14}C uracil was given followed by chase with excess cold uracil. Incorporation of radioactive label rapidly reached a plateau, and these plateau values relative to that obtained with unirradiated controls are plotted as function of radiation dose. (Reproduced from Pollard and Davis, 1970)

II. RNA Polymerase

This idea has received support from in vitro studies with *E. coli* RNA polymerase (Sumegi et al., 1972). Transcription of phage T4 DNA requires σ factor whereas transcription of calf thymus DNA is independent of σ. Irradiation of holoenzyme (core polymerase + σ) resulted in an exponential dose response for transcription from calf thymus DNA and in a biphasic response for transcription from T4 DNA. Irradiation of core enzyme and σ factor separately showed that the biphasic response was due to the sensitivity of the σ factor. Responses were dependent upon the enzyme concentration, but D_{37} values for the inactivation of σ were of the order of 5 krad (Table 2) and suggested it may play a key role in cellular inactivation. The observed inactivation of σ was associated with the irreversible oxidation of sulfhydryl groups that could be protected by cystamine.

Table 2. *E. coli* RNA polymerase in vitro

Enzyme	Function assayed	D_{37} (krads)	Reference
Core	RNA synthesis on	~40*	Sumegi et al., 1972
Sigma	T4 template	< 5*	
Holoenzyme	binding to homopolymers	>140	Strniste et al., 1973a

*Enzyme concentration dependent

Further in vitro studies using synthetic homopolymer templates for assaying polymerase activity showed that the binding of holoenzyme to the template is relatively resistant (Strniste et al., 1973a). Initiation of RNA synthesis from homopolymers is rate limiting and may require σ. When initiation is bypassed by adding to the assay system an oligomer complementary to the template, the sensitivity of the holoenzyme is abolished (Strniste et al., 1973b), confirming that σ is a radio-sensitive target. Irradiation of RNA polymerase was also shown to induce a linear dose response for misreading that is the same for *E. coli* strains B/r and B_{s-1} and that occurs at the rate of 1-10% per 25 krads depending on the homopolymer template.

These results suggest that transcription, particularly σ-initiated transcription, is a key radiation target. The relative number of core and σ molecules present in a cell at the time of irradiation should thus be expected to influence the cellular response to radiation.

III. RNA Species Made after Irradiation

In general the effect of ionising radiation on bacteria is to cause a decrease in total RNA synthesis with an associated increase in intracellular accumulation of ribonucleotides (Frampton, 1964; Pollard and Davis, 1970). Sucrose gradient centrifugation of RNA pulse-chased after irradiation permits the separation of stable RNA species, or ribosomal subunits, depending on the conditions of extraction.

Using this technique Fujita and Miyakoshi (1973) made a detailed study of the effect of γ-irradiation on *E. coli* K 12 rec^+ and rec^- bacteria, with D_{37} values for colony forming ability (c.f.a.) of

10 and 1.3 krads, respectively. Total uridine uptake was very resistant to radiation (D_{37} ~ 400 krad) and was the same in both strains despite the rapid DNA degradation of the irradiated rec^- cell. Figure 2 shows sucrose gradient profiles obtained after extraction of the pulse-chased RNA. Measurement of the areas under each peak (rec^+ strain) or the peak heights (rec^- strain) allowed an estimation of the dose response for the rate of synthesis of each RNA species. In each case, synthesis was depressed exponentially with dose, and to approximately the same extent in both rec^+ and rec^- bacteria.

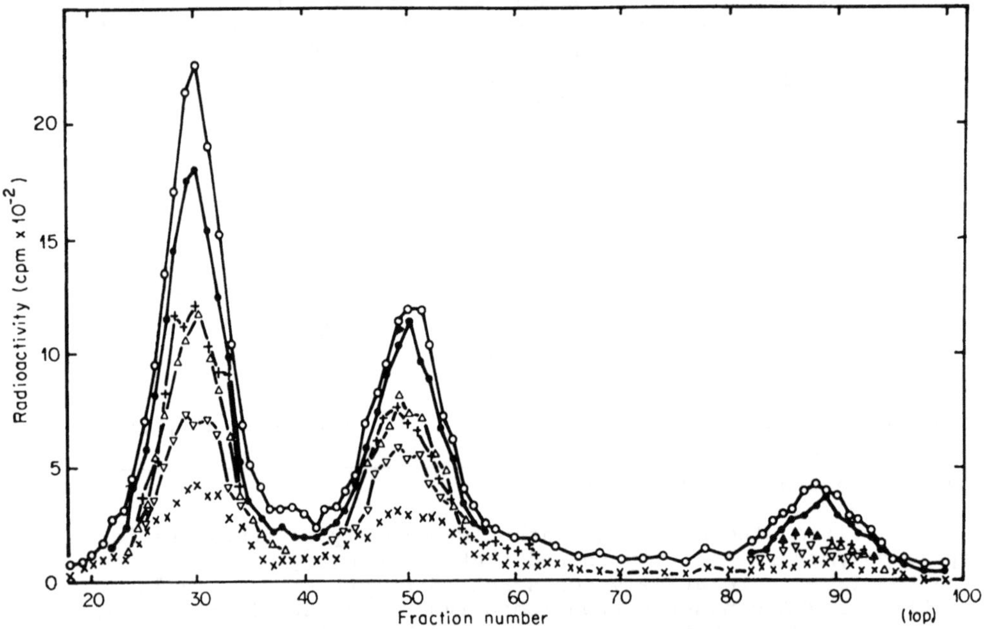

Fig. 2. Sedimentation patterns of RNA synthesized after irradiation. *E. coli* K 12 strain JC 1557 was irradiated with 70 (●), 140 (+), 210 (Δ), 280 (∇), and 350 (x) krads from a ^{60}Co source. Unirradiated (o) and irradiated cells were labelled after irradiation by 10 min incubation with ^3H-uridine (0.2 μCi/ml) in the presence of unlabelled uridine (0.08 μg/ml). RNA was extracted and centrifuged through 5-20% sucrose gradients. Radioactivity profiles of the gradients show major peaks corresponding to 23-S, 16-S, and 4-S RNA. (Reproduced from Fujita and Miyakoshi, 1973)

Table 3 shows that the D_{37} values for 16-S and 23-S rRNA are similar and greater than that of 4-S tRNA. However, these values are an order of magnitude greater than those reported by Pollard and Lyderson (1972). Bacterial strain differences have been suggested to account for this difference. Both groups showed that even after high radiation doses, the size of the RNA species is not reduced (Fig. 2). This suggests that the radiation-induced reduction in rRNA synthesis is due to inhibition of initiation (alternatively, prematurely terminated RNA molecules must degrade rapidly and fairly completely in vivo). Hence the in vivo observations do not support the in vitro observations of Hagen and co-workers (1970) that reduction in transcription of irradiated DNA was due to the production of as many, but smaller, RNA molecules as from unirradiated DNA. Rather the in vivo situation lends support to the interpretation of Weiss and Wheeler (1967) that irradiated DNA causes increased binding of RNA polymerase at new sites, which do not act as initiation sites.

Table 3. Synthesis of RNA in vivo

E. coli strain	RNA species	D_{37} (krads)	Reference
15T⁻JG151	23-S	35	Pollard and Lyderson, 1972
	16-S	50	
K12JC1557	23-S	465	Fujita and Miyakoshi, 1973
	16-S	445	
	4-S	295	

After 360 krad, 23.5-S and 17-S RNA species accumulated, which are precursors of 50-S and 30-S ribosomal subunits (Fujita and Miyakoshi, 1973). Hence, at doses outside the range of cell viability, radiation may cause slowing of the enzymatic maturation of RNA, perhaps indirectly as a result of general metabolic disturbance in a dying cell.

In yeast also, total RNA synthesis was reduced after ultraviolet (UV), x-ray, and α particle doses that reduced cell survival to 50%. However, Koch and Kiefer (1972) showed that when corrected for changes in total cell number, related to radiation-induced division delay, synthesis per cell of 26-S, 17-S, and 4-5-S RNA was actually increased after UV and x-rays, but reduced after densely ionising α particles.

C. Ribosomes and Translation

Petranovic et al. (1971) made a careful reassessment of the inactivation by direct action of γ-rays on the capacity of *E. coli* ribosomes to make polyphenylalanine from a poly U message. Ribosomes from strain MRE 600 (lacking ribosome-bound ribonuclease) were irradiated at -80°C in the presence of protectors. Transcriptional capacity of 70-S ribosomes was inactivated exponentially with a D_{37} of 2.2 Mrads (Table 4). This implies that 1 in 8 hits inactivates, which is about the same rate as strand scission of rRNA (Kucan et al., 1971).

Table 4. Effect of irradiating ribosomes on in vitro translation

Source material	Radiation action	D_{37} (Mrads)	Reference
E. coli MRE 600			
70-S	direct	2.2	Petranovic et al., 1971
30-S	direct	4.8	Kucan et al., 1971
E. coli B			
70-S			
A. Centrifugation and washing	indirect	0.080	Ekert and Pattein, 1970
B. Amm. sulphate precipitation	indirect	0.025	Ekert and Pattein, 1970

Measurements of indirect inactivation of ribosomes are dependent on the amino acid used in the translation assay and also on the method of isolation of ribosomes. Within these restrictions Ekert and Pattein (1970) obtained D_{37} values for indirect inactivation of *E. coli* strain B ribosomes in the range of 25-80 krads. Binding of mRNA, amino acid, and tRNA to ribosomes was four times more resistant.

Loss of translational ability is associated with an increased lability of 70-S ribosomes to dissociate into 30-S and 50-S subunits. The dissociation was clearly demonstrable after 25 krads and could be reversed by 10 m*M* Mg^{++} (Tamaoki et al., 1968). Similarly, x-irradiation induced dissociation of yeast 80-S ribosomes to 60-S + 40-S subunits and their subsequent degradation on removal of Mg^{++} (Ghosh, 1970). The author concluded that x-rays affected the salt linkages responsible for the formation of 80-S ribosomes and the maintenance of structural compactness of the subunits. It would seem possible that in local intracellular environments such changes might be responsible for the polar effects observed for the translation of the lac operon in *E. coli* (Pollard and Davis, 1970).

In two in vivo studies, synthesis of ribosomes after radiation paralled the formation of rRNA (Pollard and Lydersen, 1972; Fujita and Miyakoshi, 1973). Evidently the supply of ribosomal proteins is not rate limiting in ribosome synthesis. In both studies sedimentation of 50-S and 30-S ribosomes made after irradiation was normal, although after 360 krad their rate of formation was slowed.

D. Enyzme Induction

Both in vitro and in vivo experiments suggest that translation is relatively radiation resistant in comparison with cell viability and transcription requiring σ factor for initiation. This being so, measurements involving enzyme synthesis induced after radiation may be interpreted as reflecting largely the action of radiation on transcription, whereas measurements of enzymes induced before irradiation are interpreted as reflecting the action of radiation on mRNA stability or on translation.

A comparison of the effect of radiation on enzyme synthesis in *E. coli* cells induced before or after irradiation led Pollard and Davis (1970) to conclude: (1) the process of derepression is not radiation sensitive; (2) the RNA made after high doses of radiation (Fig. 1) is not involved in enzyme induction; (3) transcription of operons was more sensitive than total protein synthesis (confirming Novelli et al., 1961); (4) the sensitivity of operons varies over a wide range. The latter is true for different enzymes of a given bacterial strain and also for the same enzyme in different bacteria. In the case of *E. coli* strain B_3 irradiated under nitrogen, the sensitivity of colony formation and that of enzyme induction (tryptophanase and β-galactosidase) were not markedly different. Thus under some conditions failure to induce operons could lead to cell death.

E. Respiration

The effects of radiation on energy metabolism have received more attention in yeast than in bacteria. When stationary phase cells are irradiated, respiration of both yeast and bacteria is very resistant. For example, Barron et al. (1953) showed that respiration in yeast was unimpaired by 380 krad. Billen et al. (1953) showed that respiration of glucose by *E. coli* B/r was normal for 40 min then showed a dose-dependent inhibition: respiration still continued at a reduced rate 3 h after 90 krads, a dose that yielded a 0.002% survival (Billen et al., 1953).

These early observations suggested that the respiratory machinery per se was radiation resistant and that the delay in respiratory inhibition was a secondary response to radiation. Kiefer (1974) investigated the effects of x-rays and UV on aerobic respiration of *S. cereviseae* actively growing on a glucose carbon source. For 4 h after irradiation there was no effect: then a dose-dependent depression of respiration occurred from which the cells subsequently recovered. When respiration was measured per cell, then as with RNA synthesis (see Sect. B, III), respiration during the first 6 h after radiation was actually greater than that of unirradiated controls. This effect increased from 0-60 krads, above which dose respiration dropped towards control levels. The apparent correlation between RNA synthesis and respiration does not appear to hold for irradiated *Candida utilis* buds. Svihla et al. (1971) found that 24-48 h after 200 krad (0.1% survival), buds became deficient in respiration and DNA synthesis while RNA metabolism, assimulation of N_2, and ATP synthesis were little affected.

A long series of careful experiments by Swenson and co-workers have established the hypothesis that in *E. coli*, the control of respiration is modified by an inhibitor protein synthesised after de novo transcription and translation of the irradiated genome.

In this system cells were grown on glycerol, a nonfermentable substrate, as the sole carbon source. Swenson et al. (1971) showed that γ-irradiation of *E. coli* B/r caused a delayed inhibition of respiration, the severity and extent of which were dose-dependent (Fig. 3).

The complete cessation of respiration, seen after 30 krads was prevented by postirradiation addition of rifampin, chloramphenicol, or 5-fluorouracil (FU), which is assumed to cause miscoding errors in RNA. Unlike the uncoupling of respiration and growth seen in yeast by Kiefer, respiration and growth of *E. coli* B/r normally remain coupled after x-rays. However, addition of FU after irradiation prevented cessation of respiration and uncoupled growth and viability. This latter result strongly contrasted with the 20- to 30-fold increase in viability produced by FU within 60 min after UV irradiation. An interpretation of this difference was that FU prevented synthesis of proteins required for recovery of cell division after x-rays that are not required after UV.

Significantly *E. coli* B_{s-1}, which rapidly degrades its DNA and is exr^- (lex^-), did not cease respiring after irradiation, and respiration and growth became uncoupled.

F. Coordinated Responses and the Cell Membrane

I. Hypotheses

The requirements for the formation of the postulated inhibitor of respiration are transcription and translation of a damaged genome in cells that are rec A^+ and exr^+ (lex^+). The same requirements are needed for induction of prophage λ, UV reactivation of λ, UV mutagenesis in *E. coli* and λ, filamentation of lon^- mutants, and induction of an inhibitor of DNA degradation after UV or X-irradiation (Grady and Pollard, 1968; Pollard and Randall, 1973; Pollard et al., 1974; Tolun et al., 1974). These phenomena appear to reflect a coordinated response of cellular control systems in *E. coli*, as has been advanced by Witkin (1974) and Radman (1974).

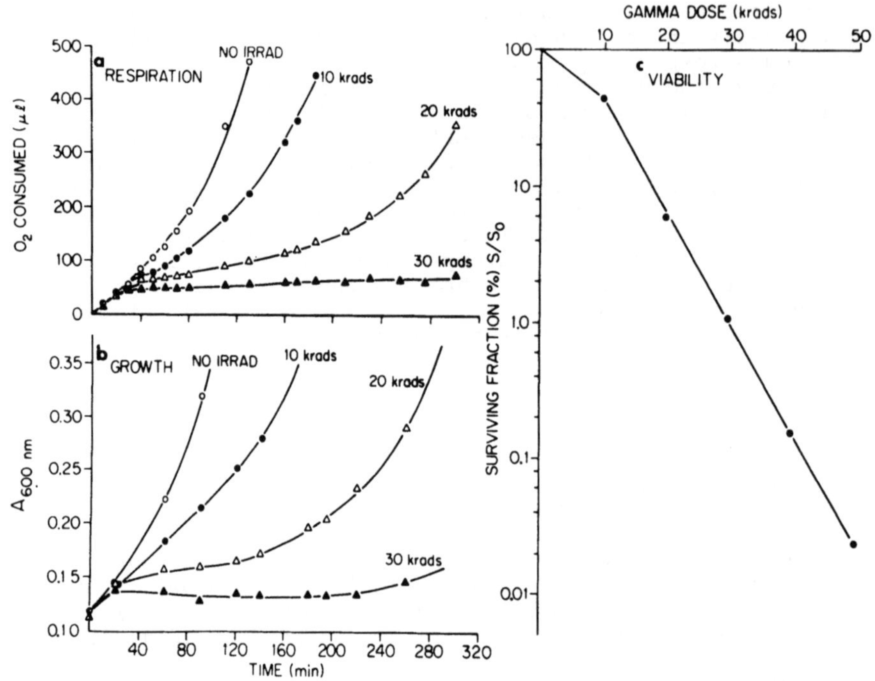

Fig. 3. Effects of γ-irradiation on (a) respiration, (b) growth, and (c) viability of *E. coli* B/r. In panels a and b the doses (krads) are indicated by the respective curves. In panel c the cells were irradiated with various doses of γ-rays and plated immediately afterwards; the results are expressed as per cent of viable (surviving) cells (S/S_0). The medium was M63gly. (Reproduced from Swenson et al., 1971)

Recently, Swenson (in press) has reviewed the physiologic effects of far-ultraviolet irradiation in *E. coli* and has formulated a general hypothesis for radiation cell death in repair competent cells. The irreversible physiologic change (IPC) hypothesis states that following DNA damage, a coordinated sequence of repair and recovery steps that would normally assure cell survival, is interrupted by an irreversible physiologic change(s). The essence of the hypothesis is that IPC can cause death even when DNA repair has been completed.

II. Cell Membranes

The effect of radiation on cell membranes, particularly the bacterial cell membrane, has received considerable attention. Together with DNA, this structure shares the important property of being present as a single copy whose integrity is essential for normal cell metabolism. Although many effects of radiation on membranes have been demonstrated, their cause-effect relationship with cell survival has been difficult to interpret. Frequently the argument can be made that the effects observed result from metabolic events of a cell already committed to die.

Despite these reservations, it is of some interest to review the situation regarding membranes, in light of the hypotheses outlined above. Myers and Johnson (1974) measured radiation-induced loss of K^+ from *E. coli* strains of B/r and B_{s-1}, to test whether the cessation of respiration observed by Swenson et al. (1971) was a result of direct radiation damage to the cell membrane. Unfortunately, these authors used broth-grown cells for their experiments, so the comparison is not a fair one, although the potassium effluxes were measured after radiation doses (75 krad) that depressed respiration. Only 10% loss of potassium was observed with cells grown after irradiation in nutrient broth containing 6 m*M* K^+. Incubation in phosphate buffer, in the absence of potassium, caused considerable loss of K^+ from the cells, with an approximate D_{37} of 50 krads for both B/r and B_{s-1}. Potassium losses were greatest under conditions where respiration was inhibited most, i.e., in the absence of extracellular K^+. Potassium retentivity was 5-10 times more resistant than cell survival.

These results were in agreement with earlier observations in yeast. Bair and Stannard (1955) had shown that postirradiation respiration was protected by high K^+ concentrations in the growth medium. Rothstein (1959) suggested that the increased permeability of yeast cells resulting from small doses of radiation led to loss of K^+, which, under some conditions, led secondarily to inhibition of metabolism. Bruce (1958) showed that the retentivity of K^+ after x-rays under oxygenated conditions, was fivefold more resistant than cell survival.

Correlations of the kind attempted by Myers and Johnson are complicated by a differential effect of molecular size and structure on the movement of molecules through the membrane (Spoerl et al., 1964). Radiation may induce almost immediate alterations in permeability to ions and small molecules, whereas leakage of macromolecules may be delayed (Pollard and Weller, 1968). The dose response and kinetics of leakage of β-galactosidase from *E. coli* K12 (λ) was similar to the induction of prophage λ, except that maximum leakage of enzyme occurred after 14 krad whereas the maximum induction of λ occurred after 4 krad. Both values are within the range of D_{37} values for *E. coli* K12 cells. Consistent with the hypothesis of coordinated induction of radiation responses, strain B_{s-1} (exr^-) was among the least leaky of the strains tested.

Another line of evidence correlating induction phenomenon with the cell membrane comes from the work of Swenson and Schenley (1974). These authors demonstrated that Triton X-100 causes lysis of nonrespiring cells that can be measured by loss of turbidity. Figure 4 shows the effect of Triton X-100 added at the beginning of postirradiation incubation of *E. coli* B/r growing on glycerol medium. Comparison with Figure 3 shows that Triton X-100 causes lysis at the same time, if not slightly before, respiration ceases. Since the action of this detergent is selectively to dissolve cell membranes, it is apparent that cessation of respiration in this system is related to membrane changes. This conclusion is supported by observations that membrane-bound NAD is lost from the cell during cessation of respiration (Swenson and Schenley, 1970).

Thus evidence is accumulating that associates membrane changes with radiation-inducible events. This area would seem deserving of further study since conceivably the membrane could provide the clue to the coordination of radiation-induced effects.

Apart from the relationship between induced respiratory responses and the membrane, transcription itself has been shown to be sensitive to inhibition by H_2O_2 formed by the action of radiation on growth media that is in intimate contact with the membrane (Pollard et al., 1965). In bacteria the membrane contains nucleases, and DNA is bound to the membrane at the mesosome. Radiation-induced filamentation in lon^- cells occurs as a result of failure of septation.

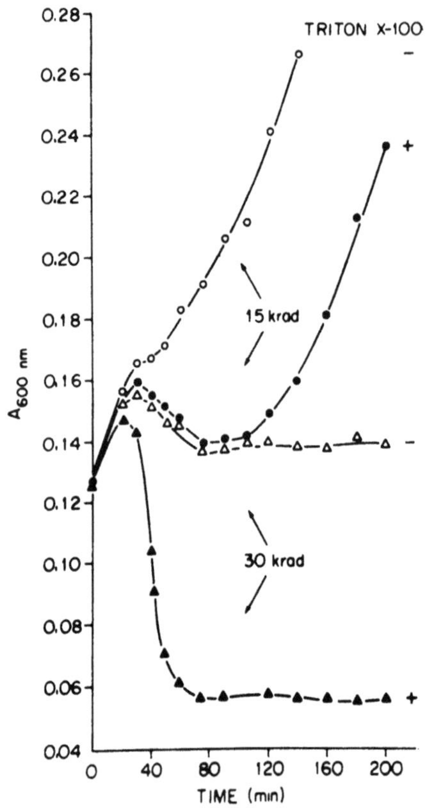

Fig. 4. Effects of Triton X-100 on the turbidity of γ-irradiated *E. coli* B/r. Triton X-100 was added immediately after irradiation (at zero time). The presence (+) or absence (-) of the detergent is indicated to the right of each curve. (Reproduced from Swenson and Schenley, 1974)

The development of methods that will allow greater resolution of the timing of early events (during the first generation time) following irradiation appear to be necessary, as well as the development of methods that would allow the separation of populations of radiation-modified cells from normal populations. A step in this direction has been taken by Swenson and co-workers with the exploitation of the action of Triton X-100 to provide a means of separating viable and nonviable populations (Fisher et al., 1975).

G. Summary

In conclusion, current evidence indicates that the σ factor involved in *E. coli* operon transcription is a radiation sensitive target. Translation is relatively radiation resistant, and hence observations on the sensitivity of enzyme induction appear to reflect the sensitivity of operon transcription.

At the cellular level, in vivo modifications of transcription are suggested by a set of inducible functions that occur after transcription and translation of irradiated DNA, requiring the activity of genes rec A and lex (exr). Whether these "inductions" are always entirely new expressions, or whether in some cases they reflect an altered balance of control, is not clear. The model for our thinking is the induction of prophage λ, where induction does represent new genetic expression. In the case of respiratory control, the switching off of respiration could result from excess production of an inhibitor protein(s) that is normally present within the cell.

Whether the sensitivities of σ factor in vitro and enzyme induction in vivo are related is not known. Both implicate transcription as being a key process affected by radiation and which, perhaps by different means, may contribute to cell lethality in cells that are DNA-repair proficient.

In yeast there is far less evidence from which to draw a coherent picture. Comparing the results of Swenson and Kiefer, a major difference between the results in bacteria and yeast appears to be the uncoupling of cell division from respiration that occurs in yeast but not in *E. coli*. Such differences provide constant reminders that *E. coli* cells are not after all, eukaryotes. It is hoped that the participants at this meeting will provide new data to show more clearly how specific differences between pro- and eukaryote organisation affect cellular response to radiation.

Acknowledgements. It is a pleasure to acknowledge the enthusiastic encouragement and generosity in communicating information before publication given to the author by Dr. Paul A. Swenson.

Financial support for this review was provided by block grants from the Medical Research Council and the Cancer Research Campaign.

References

Bair, W.J., Stannard, J.N.: Role of electrolytes and starvation in altering apparent radiosensitivity of bakers' yeast. J. Gen. Physiol. 38, 493-504 (1955)

Barron, L.S., Spiegelman, S., Quaster, H.: Enzyme formation in non-viable cells. J. Gen. Physiol. 36, 631-641 (1953)

Billen, D., Stapleton, G.E., Hollaender, A.: The effect of X-radiation on the respiration of *Escherichia coli*. J. Bacteriol. 65, 131-135 (1953)

Bruce, A.K.: Response of potassium retentivity and survival of yeast to far-ultraviolet, near-ultraviolet and visible, and X-radiation. J. Gen. Physiol. 41, 693-702 (1958)

Dertinger, H., Jung, H.: Molecular Radiation Biology. New York-Heidelberg-Berlin: Springer, 1970, pp. 163-173

Ekert, B., Pattein, A.: Effets des rayons gamma sur les ribosomes d'*E. coli*. Int. J. Radiat. Biol. 17, 47-59 (1970)

Fisher, W.D., Schenley, R.L., Swenson, P.A.: American Society for Photobiology, Abstracts 3, 91 (1975)

Frampton, E.W.: Synthesis of ribonucleic acid by X-irradiated bacteria. J. Bacteriol. 87, 1369-1376 (1964)

Frey, H.E., Pollard, E.C.: The action of γ-ray-irradiated media on bacteria: relation to the electron transport system. Radiat. Res. 36, 59-67 (1968)

Fujita, H., Miyakoshi, M.: Effect of γ-irradiation on the synthesis of ribosomal RNA and ribosomal subparticles in *Escherichia coli*. Radiat. Res. 56, 186-200 (1973)

Ghosh, N.: Effect of X-irradiation on yeast ribosomes in vitro. Int. J. Radiat. Biol. 18, 567-575 (1970)

Grady, L.J., Pollard, E.C.: Ionizing radiation-initiated degradation of deoxyribonucleic acid in bacteria. A possible role for defective prophage. Radiat. Res. 36, 68-86 (1968)

Hagen, U., Ullrich, M., Petersen, E.E., Werner, E., Kroger, H.: Enzymatic RNA synthesis on irradiated DNA. Biochim. Biophys. Acta (Amst.) 199, 115-125 (1970)
Howard, A.: The oxygen requirement for recovery in split-dose experiments with Oedogonium. Int. J. Radiat. Biol. 14, 341-350 (1968)
Kiefer, J.: The importance of cellular energy metabolism for the sparing effect of dose fractionation with electrons and ultraviolet light. Int. J. Radiat. Biol. 20, 325-336 (1971)
Kiefer, J.: Energy metabolism in yeast cells after UV and X-irradiation. Int. J. Radiat. Biol. 26, 167-179 (1974)
Koch, H., Kiefer, J.: RNA-synthesis in irradiated diploid yeast at 50 per cent survival level. Int. J. Radiat. Biol. 21, 223-234 (1972)
Kucan, Z., Herak, J.N., Pecevsky-Kucan, I.: Functional inactivation and appearance of breaks in RNA chains caused by gamma irradiation of *Escherichia coli* ribosomes. Biophys. J. 11, 237-251 (1971)
Myers, D.K., Johnson, L.D.: Loss of potassium from *E. coli* after X-irradiation. Int. J. Radiat. Biol. 26, 395-397 (1974)
Novelli, G.D., Kameyama, T., Eisenstadt, J.M.: The nature of the system catalysing the synthesis of induced β-galactosidase. Cold Spring Harbor Symp. Quant. Biol. 26, 133-143 (1961)
Okada, S.: Molecular events in irradiated bacteria. In: Radiation Biochemistry (Altman, K.I., Gerber, G.B., Okada, S., eds.). New York-London: Academic Press, 1970, Vol. I, pp. 156-189
Petranovic, D., Pecevsky-Kucan, I., Kucan, Z.: A comparison of the direct effect of γ-rays on *Escherichia coli* ribosomes and bacteriophage f2. Radiat. Res. 46, 621-630 (1971)
Pollard, E.C.: The effect of ionizing radiation on the molecular biology of *Escherichia coli*. Current Topics Radiat. Res. 6, 53-127 (1970)
Pollard, E.C., Barone, T.C.: The effect of ionizing radiation on genetic transcription: aspects of the mechanism. Radiat. Res. Suppl. 6, 194-200 (1966)
Pollard, E.C., Billen, D., Achey, P.M.: Kinetic studies of the effect of restoration of a required amino acid on post-irradiation DNA degradation and cell permeability. Int. J. Radiat. Biol. 26, 107-119 (1974)
Pollard, E.C., Davis, S.: The action of ionizing radiation on transcription (and translation) in several strains of *Escherichia coli*. Radiat. Res. 41, 375-399 (1970)
Pollard, E.C., Ebert, M.J., Miller, C.M., Kolacz, K., Barone, T.F.: Ionizing radiation: effect of irradiated medium on synthetic processes. Science 147, 1045-1047 (1965)
Pollard, E.C., Lydersen, B.K.: The action of ionizing radiation on the synthesis of RNA and two inducible enzymes in *Escherichia coli*. Radiat. Res. 50, 293-300 (1972)
Pollard, E.C., Randall, E.P.: Studies on the inducible inhibitor of radiation-induced DNA degradation of *Escherichia coli*. Radiat. Res. 55, 265-279 (1973)
Pollard, E.C., Vogler, C.: Radiation action on some metabolic processes in *Escherichia coli*. Radiat. Res. 15, 109-119 (1961)
Pollard, E.C., Weller, P.K.: Post irradiation permeability of *Escherichia coli* cells. Radiat. Res. 35, 722-739 (1968)
Radman, M.: Phenomenology of an inducible mutagenic DNA repair pathway in *Escherichia coli*: SOS repair hypothesis. In: Molecular and Environmental Aspects of Mutagenesis (Prakash, L., Sherman, F., Miller, M.W., Lawrence, C.W., Taber, H.W., eds.). Springfield: C.C. Thomas, 1974, pp. 128-142
Rothstein, A.: Biochemical and physiological changes in irradiated yeast. Radiat. Res. Suppl. 1, 357-371 (1959)
Spoerl, E., Niedzielski, R.J., Biernat, R.W.: Permeability changes in yeast cells after X-irradiation and starvation. Radiat. Res. 21, 86-90 (1964)

Strniste, G.F., Smith, D.A., Hayes, F.N.: X-ray inactivation of the *Escherichia coli* deoxyribonucleic acid dependent ribonucleic acid polymerase in aqueous solution. I. Studies on binding and substrate concentration dependence in the inactivation of enzymatic activity. Biochemistry 12, 596-602 (1973a)
Strniste, G.F., Smith, D.A., Hayes, F.N.: X-ray inactivation of the *Escherichia coli* deoxyribonucleic acid dependent ribonucleic acid polymerase in aqueous solution. II. Studies on initiation and fidelity of transcription. Biochemistry 12, 603-608 (1973b)
Sumegi, J., Sanner, T., Pihl, A.: Inactivation of DNA-dependent RNA polymerase from *Escherichia coli* by X-rays in solution. Biochim. Biophys. Acta (Amst.) 262, 145-153 (1972)
Svihla, G., Schlenk, F., Dainko, J.L.: Growth and metabolism of yeast cells after lethal X-irradiation. Int. J. Radiat. Biol. 20, 461-474 (1971)
Swenson, P.A.: Physiological responses of *Escherichia coli* to far-ultraviolet radiation. Photochem. Photobiol. Reviews, in press
Swenson, P.A., Schenley, R.L.: Role of pyridine nucleotides in the control of respiration in ultraviolet-irradiated *Escherichia coli* B/r cells. J. Bacteriol. 104, 1230-1235 (1970)
Swenson, P.A., Schenley, R.L.: Evidence relating cessation of respiration, cell envelope changes and death in ultraviolet-irradiated *Escherichia coli* B/r cells. J. Bacteriol. 117, 551-559 (1974)
Swenson, P.A., Schenley, R.L., Boyle, J.M.: Interference with respiratory control by ionizing radiations in *Escherichia coli* B/r. Int. J. Radiat. Biol. 20, 213-223 (1971)
Tamaoki, T., Miyazawa, F., Nakamura, K., Scrimger, J.W.: Effect of γ-irradiation on *Escherichia coli* ribosomes: reversible dissociation. Biochim. Biophys. Acta (Amst.) 157, 655-657 (1968)
Tolun, A., Christensen, R., Pollard, E.C.: Repair of radiation-induced strand breaks as related to the inducible inhibitor of post-irradiation DNA degradation. Biophys. J. 14, 691-696 (1974)
Witkin, E.M.: Thermal enhancement of ultraviolet mutability in tif-1 uvrA derivative of *Escherichia coli* B/r: Evidence for an inducible error-prone system of DNA repair ('SOS repair'). Proc. Nat. Acad. Sci. USA 71, 1930-1934 (1974)
Weiss, J.J., Wheeler, C.M.: Some studies on γ-irradiated DNA as primer in RNA synthesis. Biochim. Biophys. Acta (Amst.) 145, 68-75 (1967)

2. On the Transcriptional Organization of the Ribosomal RNA Genes of Yeast

J. Retèl, T. J. Stoof, and R. J. Planta

A. Introduction

It has been well established by our previous studies that the 26-S and 17-S ribosomal RNA (rRNA) genes in yeast form part of a large common transcriptional unit: the rRNA synthesis in yeast starts with the formation of a 42-S precursor molecule, which contains the sequences of 26-S and 17-S rRNA as well as nonribosomal segments, the latter comprising about 40% of the 42-S precursor sequence (Retèl and Planta, 1970; Planta et al., 1972; van den Bos and Planta, 1973). This excess RNA is removed in a number of discrete steps during the subsequent maturation process. Approximately 140 42-S transcriptional units are present in yeast per haploid genome (Retèl and Planta, 1968; Schweizer et al., 1969). They are grouped together in a rather limited number of clusters (Retèl and Planta, 1972; Cramer et al., 1972; Goldberg et al., 1972). It is possible to isolate by Hg^{2+}-Cs_2SO_4 gradient centrifugation a high-molecular-weight DNA fraction, representing 10-12% of the total nuclear DNA, which contains all transcriptional units for rRNA (Fig. 1). About 60% of this so-called ribosomal DNA (rDNA) consists of sequences coding for 42-S precursor rRNA (Retèl and Planta, 1972; Retèl and van Keulen, 1975). The precursor cistrons are interspersed with (A+T)-rich sequences, which make up most of the remaining 40% of the rDNA. In contrast to the precursor cistrons, which show very little or no base sequence divergency, these intervening spacer sequences display considerable base sequence heterogeneity (Retèl and van Keulen, 1975).

Apart from the 42-S precursor and spacer sequences, the rDNA fraction appears to contain all the 5-S rRNA genes, which have about the same degree of redundancy as the genes coding for the two high-molecular-weight rRNA species (Retèl and Planta, 1972; see also Fig. 1). This finding strongly suggests that the 5-S rRNA genes in yeast are located very close to those specifying 26-S and 17-S rRNA. Indications for a close physical linkage between the genes for the three rRNA species were recently obtained also by Rubin and Sulston (1973) and by Aarstad and Øyen (1975).

In order to investigate whether the 5-S gene is part of the same transcriptional unit as the genes coding for 26-S and 17-S rRNA, the technique of ultraviolet (UV) irradiation mapping was used. This method has been successfully applied to the study of the transcriptional organization of genes in both prokaryotic and eukaryotic systems (van den Bos et al., 1971; Bräutigam and Sauerbier, 1973, 1974; Hercules and Sauerbier, 1973, 1974; Hackett and Sauerbier, 1974, 1975). In this paper the effect of UV irradiation on the synthesis of the various rRNA components of yeast is described and compared with the results we obtained with bacterial cells, in which the rRNA genes are arranged behind a common promoter in the order 5'-16S-23S-5S-3' (for a review see Pace, 1973). The results indicate that the 5-S rRNA genes in yeast are not part of the transcriptional unit containing the genes of 26-S and 17-S rRNA, but are transcribed separately.

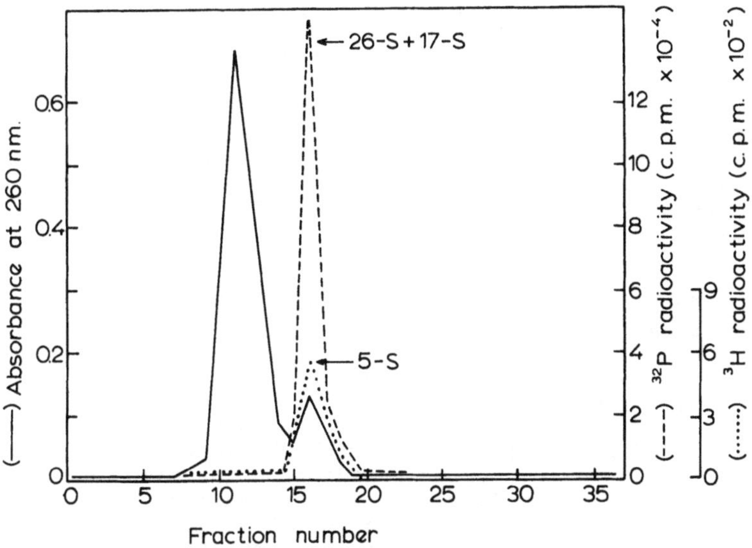

Fig. 1. Preparative Hg^{2+}-Cs_2SO_4 gradient centrifugation of high-molecular-weight nuclear yeast DNA. Preparative Hg^{2+}-Cs_2SO_4 gradient centrifugation of high-molecular-weight nuclear yeast DNA (40 x 10^6) was performed at 39,000 rev/min for 60 h at 20°C in no. 50 fixed-angle Spinco rotor as described previously (Retèl and Planta, 1972). Gradient fractions were hybridized with ^{32}P-labelled 26-S plus 17-S rRNA or ^{3}H-labelled 5-S rRNA as previously described (Retèl and Planta, 1972)

B. Methods

I. Analysis of the Effect of UV Irradiation on the Synthesis of the rRNA Components in Yeast

Yeast cells (*Saccharomyces carlsbergensis*, strain S 74) were cultured in a standard medium to which ^{3}H-uridine (The Radiochemical Centre, Amersham, England; spec. act. 29 Ci/mmol) was added at a final concentration of 1 μCi/ml, until the midlogarithmic growth phase as described previously (Retèl and Planta, 1967). Then 100-ml samples were transferred to petri dishes (12 cm internal diameter). The samples were irradiated for various times with UV light from a Philips TUV-30 ultraviolet lamp (Philips, Eindhoven, The Netherlands), placed at a distance of 90 cm. During the irradiation the suspension was continuously stirred and the temperature maintained at 29°C. Subsequently the cultures were transferred into 200-ml centrifuge tubes containing 100 μCi ^{32}P-orthophosphate (Philips-Duphar, Petten, The Netherlands; carrier-free). After incubation for 30 min at 29°C in the dark (to prevent photoreactivation), isotope incorporation was stopped by pouring the cultures on crushed ice. Cells were collected by centrifugation and washed three times with ice-cold water. Total RNA was extracted from the cells as described previously (van den Bos et al., 1971) and fractionated by sedimentation through a linear 5-20% sucrose gradient (Retèl and Planta, 1967). The 26-S and 17-S rRNA fractions, and those containing the small RNA components (4-5-S) were pooled, the RNA precipitated by the addition of 3 volumes of ethanol, and redissolved in

0.05 *M* Tris-HCl (pH 7.0). The high-molecular-weight RNA components were analyzed by electrophoresis on 2.6% polyacrylamide gels according to the method of Loening (1969). The small RNA species were fractionated by electrophoresis on 10% polyacrylamide gels using the discontinuous buffer system described by Richards et al. (1965). The gels were cut in 1-mm slices and assayed for ^{32}P and ^{3}H radioactivity as described previously (van den Bos et al., 1971). Subsequently the specific radioactivities (defined as the $^{32}P/^{3}H$ ratio) of the various RNA components from the irradiated cultures and the nonirradiated control were determined.

II. Analysis of the Effect of UV Irradiation on the Synthesis of the rRNA Components in *Bacillus licheniformis*

Bacillus licheniformis cells (strain S 244) were cultured at 37°C until the early logarithmic growth phase as described previously (van Dijk-Salkinoja et al., 1970). Fifty-ml samples were transferred to petri dishes and irradiated for various times with UV light as described previously for yeast cells. During irradiation the samples were continuously stirred and the temperature was maintained at 37°C. After irradiation the samples were transferred to 200-ml flasks containing 50 µCi ^{3}H-uridine (spec. act. 29 Ci/mmol). Incubation was carried out for 15 min at 37°C in the dark and then stopped by pouring the culture on crushed ice. Before collecting the cells by centrifugation, each sample was mixed with 50 ml of a culture that was uniformly labelled with ^{14}C-uracil (The Radiochemical Centre, Amersham, England; spec. act. 50 mCi/mmol; ^{14}C-uracil was added to the culture in a final concentration of 0.25 µCi/ml). The cells were collected, washed two times with ice-cold standard buffer, and converted into protoplasts by lysozyme treatment as described previously (van Dijk-Salkinoja et al., 1970). From the protoplasts, total RNA was extracted at 0°C as described by Retèl and Planta (1967). Fractionation of the various RNA components and determination of the specific radioactivities (defined as the $^{3}H/^{14}C$ ratio) were performed as described for yeast.

C. Results

UV irradiation mapping is based on the finding that lesions induced by UV light cause premature termination of transcription at the site of damage in the DNA, resulting in the release of the incomplete transcription product and the RNA polymerase from the template (Michalke and Bremer, 1969; Sauerbier et al., 1970; Bräutigam and Sauerbier, 1973, 1974; Hercules and Sauerbier, 1973, 1974; Hackett and Sauerbier, 1975). Thus, it can be expected that the decrease in the rate of synthesis of a given RNA molecule upon UV irradiation is determined by: (1) the length of the corresponding gene; (2) the distance of that gene from the promoter site in the case that the RNA forms part of a larger or a polycistronic transcription product; (3) the base composition and base sequence of the DNA transcribed. The base compositions of the three rRNA species in *Bacillus licheniformis* and thus of the DNA sequences coding for these RNA's are very similar (Stoof, 1974). The same holds true for the various rRNA species in yeast (Retèl and Planta, 1967). In addition, the segments coding for the nonribosomal excess RNA of the yeast 42-S precursor have about the same G+C content as the 26-S and 17-S rRNA sequences (Retèl and van Keulen, 1975). Thus, the effect of UV irradiation on the synthesis of the various rRNA species both in *Bacillus* and yeast will be largely determined by the size of their respective genes and the location of these genes relative to the common promotor site.

The influence of UV irradiation on the synthesis of the rRNA components of yeast was determined by irradiating yeast cells for various periods of time and then incubating the cells with ^{32}P-orthophosphate for 30 min. Prior to irradiation the cells were uniformly labelled with ^{3}H-uridine. Total RNA was extracted from the cells and the large rRNA components (26-S and 17-S) and the small RNA species (4-5-S) were analyzed by electrophoresis on 2.6% and 10% polyacrylamide gels, respectively. The $^{32}P/^{3}H$ ratios of the various RNA components from the irradiated cultures were determined and compared to those of the non-irradiated control. The influence of UV irradiation on the synthesis of the rRNA components of *Bacillus* was studied in a similar way (see Methods). The results for yeast and *Bacillus* are summarized in Figures 2 and 3, respectively. Both in yeast and *Bacillus* synthesis of the larger of the two high-molecular-weight rRNA components (26-S, 23-S) is much more affected by irradiation than the synthesis of the smaller one (17-S, 16-S). Since in both cases the transcription of the genes coding for the two high-molecular-weight rRNA species is directed by a common promotor, this result indicates that both in *Bacillus* and yeast the smaller rRNA gene is closer to the promoter site than the larger one. Arrangements of the two rRNA genes in the reverse order, i.e., when the 17-S (16-S) gene in the ribosomal transcription unit is preceded by the 26-S (23-S) gene, will always result in higher UV sensitivity of the synthesis of the smaller rRNA component as compared to the larger rRNA component.

On the other hand, the effect of UV irradiation on the synthesis of 5-S rRNA is totally different for the two types of cells. In *Bacillus*, 5-S rRNA synthesis is inactivated even more rapidly than that of 23-S rRNA (Fig. 3). From Figure 3 it can be calculated that the synthesis of 5-S RNA is 1.1 times more sensitive to UV light than the formation of 23-S rRNA. This finding is in agreement with the location of the 5-S rRNA gene behind the 23-S gene at the 3'-end of the common ribosomal transcription unit (Pace, 1973). The 5-S rRNA synthesis in yeast is much less affected by UV irradiation: its inactivation rate is comparable to that of tRNA (Fig. 2). This result indicates that the 5-S rRNA genes are not part of the same transcriptional unit containing the genes of 26-S and 17-S rRNA, but are transcribed separately.

D. Discussion

The results described in this paper indicate that in yeast the 17-S rRNA gene is closer to the promotor of the 42-S precursor cistron than the 26-S rRNA gene, in agreement with our previous findings (van den Bos et al., 1971). That the two high-molecular-weight rRNA species in yeast are transcribed in the order 5'-17S-26S-3' was recently confirmed by a study of the labelling kinetics of the mature and precursor rRNA species (Trapman and Planta, 1975). So, the topographic arrangement of the two high-molecular-weight rRNA species of yeast relative to the promotor appears to be very similar to that of bacteria. However, this concurrency apparently does not apply to the genetic organization of the 5-S rRNA genes. It is well established that in bacteria the 5-S rRNA genes are situated at the 3'-end of the approx. six ribosomal transcription units, which also contain the 23-S and 16-S rRNA genes (Pace, 1973; Ginsburg and Steitz, 1975). The UV inactivation data recently obtained by Hackett and Sauerbier (1974) as

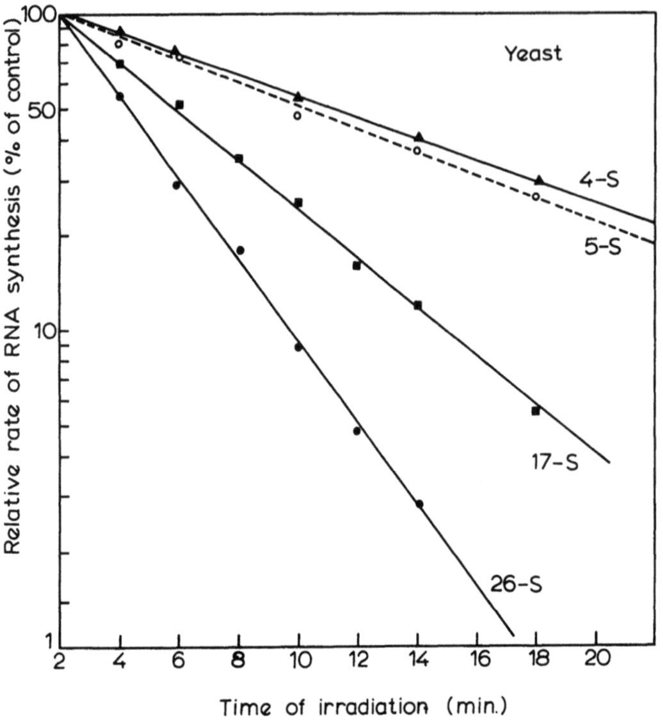

Fig. 2. Effect of UV irradiation on synthesis of 26-S, 17-S, 5-S rRNA and tRNA of yeast. Synthesis of various RNA components after various times of irradiation with UV light was measured as described in Methods. Synthesis of RNA components for each UV dose was expressed as percentage of RNA synthesized by nonirradiated control and has been plotted on semilogarithmic scale as function of time of irradiation

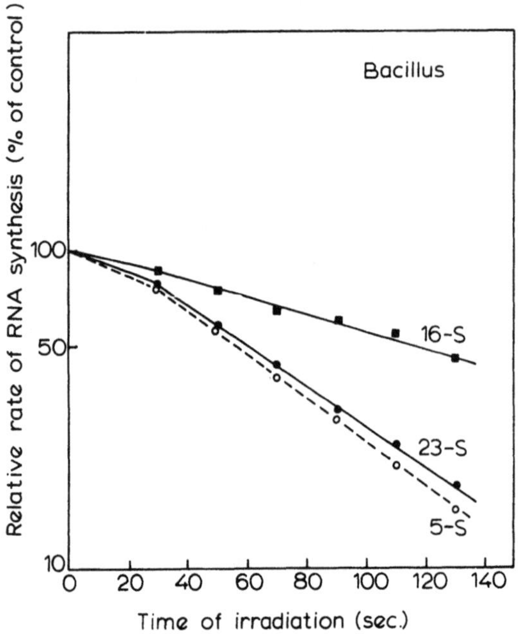

Fig. 3. Effect of UV irradiation on synthesis of 23-S, 16-S, and 5-S rRNA of *Bacillus licheniformis*. Synthesis of various RNA components after various times of irradiation with UV light was measured as described in Methods. Results are plotted as described in legend to Fig. 2

well as the results described in this paper are in accordance with that arrangement. On the other hand, the inactivation kinetics of the 5-S rRNA synthesis in yeast, which are quite different from those in bacteria (Figs. 2 and 3), suggest that the yeast 5-S RNA genes are arranged in separate transcriptional units. This situation would be in agreement with the results of Hindley and Page (1972), indicating that 5-S RNA is a primary transcription product, and with those of Udem and Warner (1972), showing that 5-S rRNA can be synthesized independently from the common precursor of 26-S and 17-S rRNA. Furthermore, comparison of the labelling kinetics of 5-S rRNA with those of the mature 26-S, 17-S and precursor rRNA species also indicates that the 5-S genes are transcribed separately (Trapman and Planta, 1975). Taking into account these data together with the indications that the genes for 5-S RNA and for 17-S and 26-S rRNA are physically linked (Fig. 1; Retèl and Planta, 1972; Rubin and Sulston, 1973; Aarstad and Øyen, 1975), the assumption seems to be justified that the 5-S RNA genes in yeast are organized on the rDNA in separate clusters close to the 42-S precursor cistrons.

Acknowledgements. The present investigations were supported in part by the Netherlands Foundation for Chemical Research (S.O.N.) with financial aid from the Netherlands Organization for the Advancement of Pure Research (Z.W.O.). The authors are indebted to Dr. H.A. Raué for helpful critical comment on this manuscript and to Mr. S. Simonian and Mr. C.D.J. de Lange for their skilled technical assistance.

References

Aarstad, K., Øyen, T.B.: On the distribution of 5-S RNA cistrons on the genome of *Saccharomyces cerevisiae*. FEBS Letters 51, 227-231 (1975)

Bos, R.C., van den, Planta, R.J.: Structural comparison of 37-S and 32-S ribosomal precursor RNA in yeast with the mature ribosomal RNA components. Biochim. Biophys. Acta (Amst.) 294, 464-471 (1973)

Bos, R.C., van den, Retèl, J., Planta, R.J.: The size and the location of the ribosomal RNA segments in ribosomal precursor RNA of yeast. Biochim. Biophys. Acta (Amst.) 232, 494-508 (1971)

Bräutigam, A., Sauerbier, W.: Transcription unit mapping in bacteriophage T7. I. In vivo transcription by *Escherichia coli* RNA polymerase. J. Virol. 12, 882-886 (1973)

Bräutigam, A., Sauerbier, W.: Transcription unit mapping in bacteriophage T7. II. Proportionality of number of gene copies, mRNA, and gene product. J. Virol. 13, 1110-1117 (1974)

Cramer, J.H., Bhargava, M.M., Halvorson, H.O.: Isolation and characterization of γ-DNA of *Saccharomyces cerevisiae*. J. Molec. Biol. 71, 11-20 (1972)

Dijk-Salkinoja, M.S., van, Stoof, T.J., Planta, R.J.: The distribution of polysomes and ribosomal subunits in exponential-phase cells of *Bacillus licheniformis*. Eur. J. Biochem. 12, 474-482 (1970)

Ginsburg, D., Steitz, J.A.: The 30-S ribosomal precursor RNA from *Escherichia coli*. A primary transcript containing 23-S, 16-S, and 5-S sequences. J. Biol. Chem. 250, 5647-5654 (1975)

Goldberg, S., Øyen, T., Idriss, J.M., Halvorson, H.O.: Use of disomic strains to study the arrangement of ribosomal cistrons in *Saccharomyces cerevisiae*. Molec. Gen. Genet. 116, 139-157 (1972)

Hackett, P.B., Sauerbier, W.: Radiological mapping of the ribosomal RNA transcription unit in *E. coli*. Nature (Lond.) 251, 639-641 (1974)

Hackett, P.B., Sauerbier, W.: The transcriptional organization of the ribosomal RNA genes in mouse L cells. J. Molec. Biol. 91, 235-256 (1975)
Hercules, K., Sauerbier, W.: Transcription units in bacteriophage T4. J. Virol. 12, 872-881 (1973)
Hercules, K., Sauerbier, W.: Transcription units in bacteriophage T4. J. Virol. 13, 341-348 (1974)
Hindley, J., Page, S.M.: Nucleotide sequence of yeast 5-S ribosomal RNA. FEBS Letters 26, 157-160 (1972)
Loening, U.E.: The determination of the molecular weight of ribonucleic acid by polyacrylamide-gel electrophoresis. The effects of changes in conformation. Biochem. J. 113, 131-138 (1969)
Michalke, H., Bremer, H.: RNA synthesis in *Escherichia coli* after irradiation with ultraviolet light. J. Molec. Biol. 41, 1-23 (1969)
Pace, N.: Structure and synthesis of the ribosomal ribonucleic acid of prokaryotes. Bacteriol. Rev. 37, 562-603 (1973)
Planta, R.J., Bos, R.C., van den, Klootwijk, J.: Biosynthesis of ribosomal RNA in eukaryotic cells. In: Ribosomes: Structure, Function and Biogenesis (Bloemendal, H., Planta, R.J., eds.). Amsterdam: North-Holland, 1972, pp. 183-196
Retèl, J., Keulen, H., van: Characterization of yeast ribosomal DNA. Eur. J. Biochem. 58, 51-57 (1975)
Retèl, J., Planta, R.J.: Ribosomal precursor RNA in *Saccharomyces carlsbergensis*. Eur. J. Biochem. 3, 248-258 (1967)
Retèl, J., Planta, R.J.: The investigation of the ribosomal RNA sites in yeast DNA by the hybridization technique. Biochim. Biophys. Acta (Amst.) 169, 416-429 (1968)
Retèl, J., Planta, R.J.: On the mechanism of the biosynthesis of ribosomal RNA in yeast. Biochim. Biophys. Acta (Amst.) 224, 458-468 (1970)
Retèl, J., Planta, R.J.: Nuclear satellite DNA's of yeast. Biochim. Biophys. Acta (Amst.) 281, 299-309 (1972)
Richards, E.G., Coll, J.A., Gratzer, W.B.: Disc electrophoresis of ribonucleic acid in polyacrylamide gels. Anal. Biochem. 12, 452-471 (1965)
Rubin, G.M., Sulston, J.E.: Physical linkage of the 5-S cistrons to the 18-S and 28-S ribosomal RNA cistrons in *Saccharomyces cerevisiae*. J. Molec. Biol. 79, 521-530 (1973)
Sauerbier, W., Milette, R.L., Hackett, P.B.: The effects of ultraviolet irradiation on the transcription of T4 DNA. Biochim. Biophys. Acta (Amst.) 209, 368-386 (1970)
Schweizer, E., MacKechnie, C., Halvorson, H.O.: The redundancy of ribosomal and transfer RNA genes in *Saccharomyces cerevisiae*. J. Molec. Biol. 40, 261-277 (1969)
Stoof, T.J.: De biosynthese van ribosomaal RNA in *Bacillus licheniformis*. Ph.D. thesis, Vrije Universiteit: Amsterdam (1974)
Trapman, J., Planta, R.J.: Detailed analysis of the ribosomal RNA synthesis in yeast. Biochim. Biophys. Acta (Amst.) 414, 115-125 (1975)
Udem, S.A., Warner, J.R.: Ribosomal RNA synthesis in *Saccharomyces cerevisiae*. J. Molec. Biol. 65, 227-242 (1972)

3. Control of DNA Polymerase Activity in Synchronized Growing Yeast after X-Irradiation

H. ECKSTEIN

With regard to the time course of metabolic processes in a growing cell, two main groups can be established: many processes, whose function it is to maintain the viability of the growing cell, e.g., energy metabolism, occur continually during the cell cycle. They exhibit a nearly constant activity rate, except for a doubling of their rate once during each cell division. Because there is a constant demand for enzymes belonging to this type, they are always present in the cell. In extracts from synchronized growing yeast cells, their activity rate remains essentially constant during most of the cell cycle and is doubled rather rapidly in a specific phase of the cycle (Eckstein et al., 1966; Gorman et al., 1964).

Those processes that are clearly bound to cell proliferation, such as DNA replication, are restricted to certain phases of the cell cycle. Several investigators have shown that some of the enzymes belonging to this second type are present in the cell only when they are needed (Sylvén et al., 1959; Brent et al., 1965; Eckstein et al., 1967; Sachsenmaier et al., 1967; Golombek et al., 1974). Apparently, the growing cell possesses distinct regulatory mechanisms, which guarantee that temporarily needed enzymes indeed are available in the cell only temporarily. The details of these control mechanisms are as yet unknown. We have focussed our attention upon one of the most essential processes in a growing cell, namely DNA replication and DNA polymerase activity.

We demonstrated earlier (Eckstein et al., 1966; Eckstein et al., 1967) that in synchronized growing yeast cells DNA is augmented stepwise, exhibiting a well-defined S phase. The activity of DNA polymerase from cell extracts fluctuates during the cell cycle: maximum activity is found at the beginning of each S phase; it decreases to minor values at the end of DNA synthesis, and it starts to increase again shortly before the next DNA replication cycle. In parallel to the increase in cell number, maximum DNA polymerase activity increases, too, with each cell generation (Eckstein et al., 1967; Golombek et al., 1974). Fluctuations of certain DNA polymerase activities recently were reported also in mammalian cell systems (Lindsay et al., 1970; Chang et al., 1973; Spadari and Weissbach, 1974a).

Considering both the increase in activity to maximum values and the augmentation of activity due to cell growth, one could guess that increase and decrease of DNA polymerase activity are governed in principle by degradation and de novo synthesis of the enzyme. Our results with x-irradiated yeast, however, led to a different conclusion.

Irradiation of synchronized yeast cells with 50 kR of x-rays (200 kV) destroys the proliferation ability of the cells without essential lesions in cell growth. Of many metabolic processes, only DNA synthesis is severely affected. It exhibits a pronounced delay in x-irradiated synchronized growing yeast cells (Hilz and Eckstein, 1964). This delay cannot be explained either by radiation-induced breakdown of precursor synthesis (Eckstein et al., 1974) or by a general lesion of DNA polymerase activity (Eckstein et al., 1967). It has been shown that extensive repair processes are performed by the yeast cell before DNA replication begins again (Eckstein, 1974).

DNA polymerase activity, though not destroyed in principle, exhibits an interesting behavior after x-irradiation of the synchronized yeast cells: the fluctuation pattern of enzyme activity remains nearly unaffected - it persists with almost the same rhythm as in the nonirradiated cells. Augmentation of maximum activity due to cell growth, i.e., de novo synthesis of enzyme, is, however, prevented as long as DNA augmentation is delayed (Eckstein et al., 1967). This is in clear contrast to many other enzymes, especially those involved in energy metabolism, whose activity and stepwise augmentation are damaged, at most, only slightly (Eckstein et al., 1966). Concerning DNA polymerase activity, we have to distinguish therefore between radiation-sensitive de novo synthesis of the enzyme and, independently, fluctuations of enzyme activity. These fluctuations of course also could be based on turnover of the enzyme protein. In this case, however, it would be difficult to explain why only this part of de novo synthesis remains undisturbed after x-irradiation, whereas enzyme duplication is delayed, in contrast to many other enzymes. This all the more, since Halvorson et al. (1964) pointed out that growing yeast cells possess no protein turnover. The recent finding that inhibition of protein synthesis by cycloheximide does not prevent the fluctuations (Golombek et al., 1974) finally excludes an explanation of enzyme fluctuations in terms of protein turnover.

The temporary appearance of DNA polymerase activity during the yeast cell cycle therefore seems to be caused rather by an activation - inactivation process. This could be based on a modification of the enzyme protein itself or on the influence of some effector(s), or both. The best known example of interference with protein structure is the change between phosphorylase A and B, mediated by protein kinases and cyclic AMP. Golombek et al. (1974) supposed a similar mechanism to be responsible for the alterations in DNA polymerase activity in the growing yeast cell. As far as activation is concerned, several results with bacteria and HeLa cells point rather to participation of distinct proteins and ribopolynucleotides (Spadari and Weissbach, 1974a; Highfield and Dewey, 1972; Stalker et al., 1974; Seki and Mueller, 1975; Hanna and Carl, 1975; Seki et al., 1975; Probst et al., 1975). In recent studies, we looked for possible control of DNA polymerase activity by inactivating effectors. In a previous investigation (Eckstein et al., 1967), we found no indication of effectors of DNA polymerase activity from cell extracts in vitro. However, the action of effectors in vivo cannot be ruled out completely by those experiments. In the course of our recent studies on deoxynucleotide synthesis, we observed a distinct inhibition of DNA polymerase activity by acid-soluble material from yeast. The following is a report of some of our data.

Preparation of material and other technical details are described elsewhere (Eckstein et al., 1974; Eckstein and Ahnefeld, in preparation). Estimation of DNA polymerase activity is based on the assay described by Lindberg and Skoog (1970) for deoxynucleotide estimation. This assay has proved useful also for analysis of DNA precursor contents in extracts from synchronized growing yeast cells. As far as dATP and dTTP are concerned, the values correlate with the amount of "concentrate" per assay, and with cell growth. Difficulties arise, however, in the estimation of dCTP. As shown in Figure 1, the incorporation of radioactivity from labelled dGTP decreases sharply with increasing amounts of yeast concentrate in the assay.

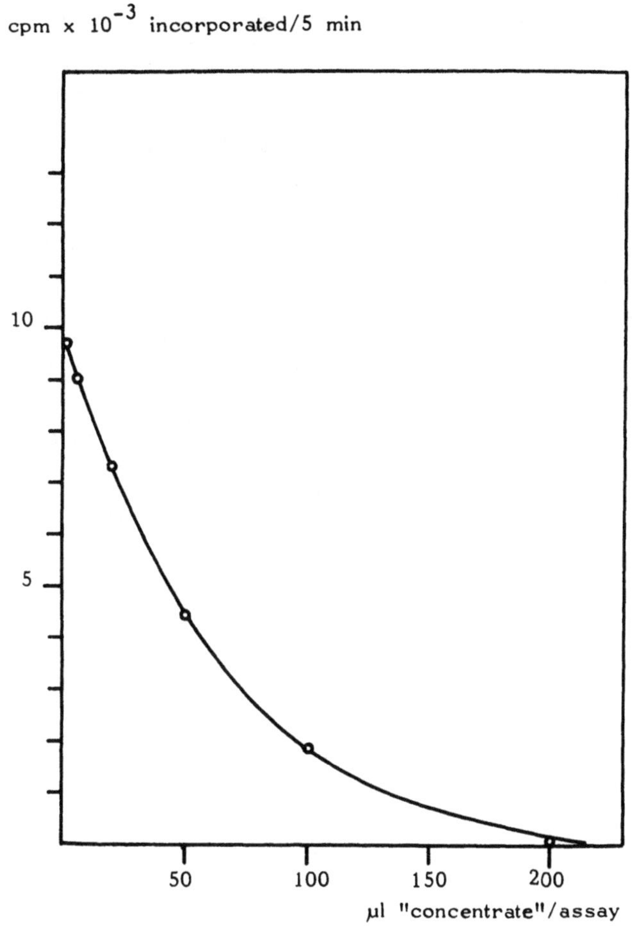

Fig. 1. Poly-d(G-C)-synthesis: Decrease of incorporation of radioactivity from ^{3}H-dGTP with increasing amounts of yeast concentrate. Enzyme assay was performed with 8-^{3}H-dGTP in excess, a limiting concentration of dCTP, and DNA polymerase I from *E. coli*, as described by Eckstein et al. (1974) and in Table 1

Apparently, the concentrates prepared from TCA extracts from growing yeast cells contain a substance that strongly inhibits the activity of highly purified DNA polymerase in vitro. This inhibitor acts only when poly-d(I-C) or poly-d(G-C) is used as a primer-template. It does not react with poly-d(A-T). This result points to a very differentiated interference of the inhibitor with the DNA polymerase reaction. An explanation of the phenomenon in terms of one of the known low-molecular-weight, random inhibitors, such as pyrophosphate, mono- or divalent ions, etc., can be excluded, since the inhibitor is not dialyzable. Several additional observations on the nature of the inhibitor are summarized in Table 1.

When concentrate is applied to a DEAE cellulose column pre-equilibrated with 0.01 *M* $MgCl_2$ in 0.1 *M* Tris-HCl buffer, pH 7.6, several ultraviolet (UV)-absorbing fractions are found by elution with a linearly increasing salt gradient (Fig. 2). Inhibitor activity is eluted together with UV-absorbing material, mainly between 0 and 0.14 *M* NaCl. A minor part,

Table 1. Inhibitor activity in yeast concentrates after various treatments. Commercially available highly purified DNA polymerase I from *E. coli* MRE 600 (Kornberg-Polymerase, Boehringer, Mannheim, Grade I) was assayed in total volume of 0.4 ml with 5 nmoles dGTP supplemented with 500 nCi 8-^{3}H-dGTP (The Radiochemical Centre, Amersham), 40 nmoles dCTP, 2 x 10^{-3} OD_{260}-units poly-d(I-C) (Boehringer, Mannheim), 500 nmoles mercaptoethanol, 4 µmoles $MgCl_2$, 40 µmoles Tris-HCl buffer, pH 7.6, 100 µg bovine serum albumin (Behringwerke, Marburg), and 50 µl concentrate pretreated as indicated. Reaction was started by addition of 10 units of enzyme and by incubation at 37°C. Every 5 min, 50 µl aliquots were taken, applied to filter paper discs and handled as usual (Eckstein et al., 1974). Radioactivity incorporated into acid-insoluble material was counted in a TriCarb liquid scintillation counter. Concentrate was prepared as described in the text and pretreated as follows: dialysis overnight against the 2500-fold volume of 0.01 *M* $MgCl_2$ in 0.1 *M* Tris-HCl buffer, pH 7.6. Heat treatment 15 min in boiling water bath. Digestion with trypsin (from bovine pancreas, 9000 u/mg, Boehringer) or pronase (pronase P from *Streptomyces griseus*, 4000 DMC-units/mg, Serva, Heidelberg): 1.0 ml concentrate were incubated with 10 mg of enzyme, free of nucleases, for 60 min at 37°C, the reaction was stopped by heating assay 30 min at 100°C. Digestion with DNase I: 1.0 ml concentrate were incubated with 1.0 mg enzyme (DNase I from bovine pancreas, 2000 units/mg Serva, Heidelberg) for 60 min at 37°C. Enzyme was inactivated by heating assay 30 min at 100°C

Concentrate treatment	cpm incorporated in 5 min/assay	%
Control	1184	100
Untreated	735	62.1
Dialysis	806	68.0
Heat	768	64.8
Trypsin	656	55.4
Pronase P	768	64.8
DNase I	2150	182.0

Fig. 2. Ion exchange chromatography of dialyzed concentrate. All operations were carried out at 4°C. Concentrate was prepared and dialyzed as described in text and in Table 1. 6.0 ml dialyzed concentrate were applied to column, 12 mm in diameter, 25 cm high, which was pre-equilibrated with 0.01 *M* $MgCl_2$ in 0.1 *M* Tris-HCl buffer, pH 7.6. The column was washed with 25 ml Tris-Mg^{++} buffer. Elution was performed with a linearly increasing NaCl gradient, from 0 to 0.5 *M*. Absorption of eluate at 280 nm was automatically recorded, samples of 60 drops were collected. Fractions no. 1-5, 6-11, 12-18, 19-22, 23-28, 29-35 were pooled, dialyzed against water, evaporated to dryness, and redissolved with 6.0 ml Tris-Mg^{++} buffer. 50 and 100 µl, respectively, of the solutions were applied to *E. coli* DNA polymerase assays, as described in Table 1

Fig. 3. Influence of precursor concentration on inhibition of poly-d(G-C)-synthesis by yeast concentrate. Poly-d(G-C)-synthesis was performed with DNA polymerase I from *E. coli* and with 100 µl concentrate. "cpm incorporated/5 min" are mean values from 30 min of incubation time, they refer to 50 µl aliquots from the assay. (a) 12.5 x 10^{-6} *M* dGTP; (b) 125 x 10^{-6} *M* dGTP

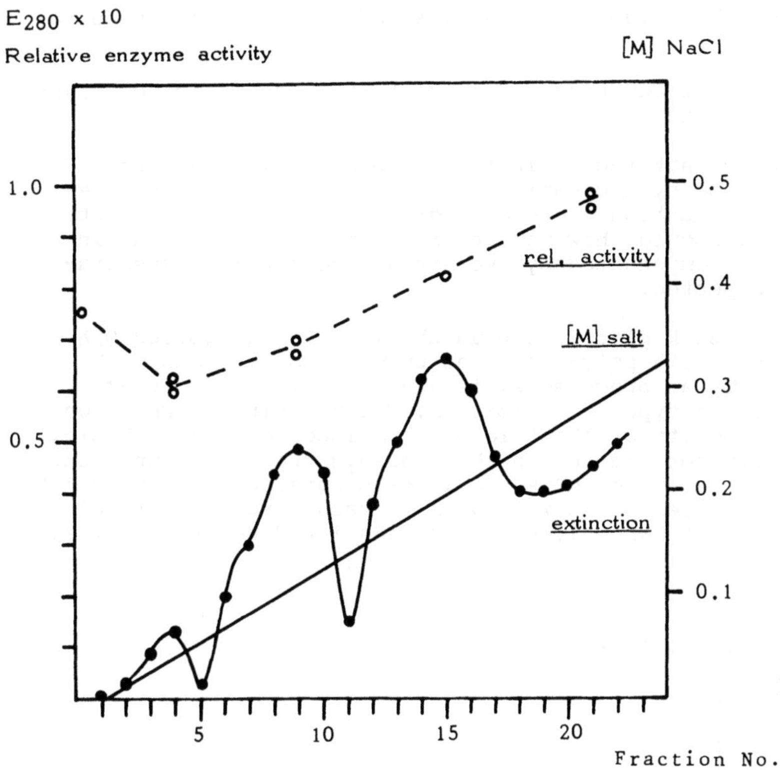

Fig. 2

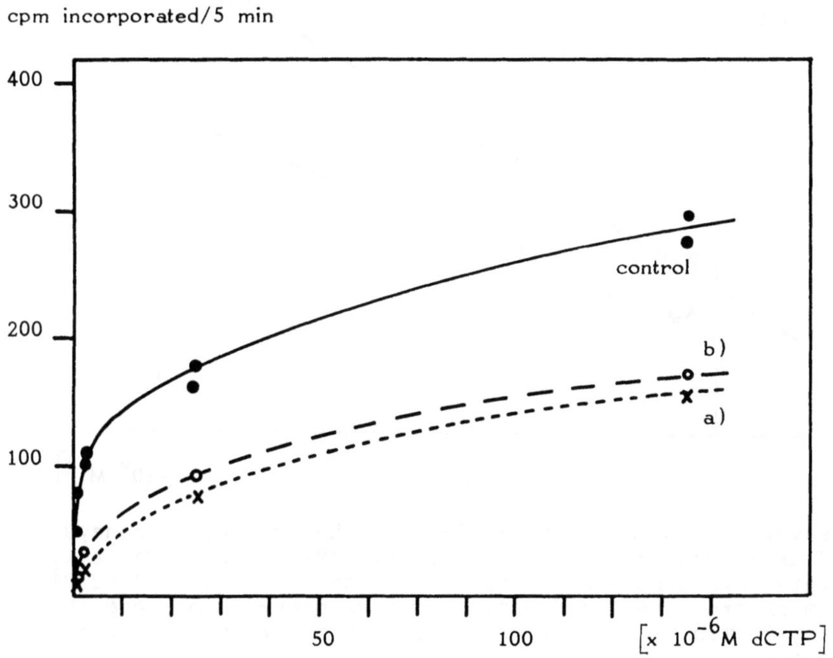

Fig. 3

Legends see opposite page

however, comes from the column between 0.14 and 0.25 *M* salt concentration.

The results from Table 1 and Figure 2 suggest that the inhibitor might be an oligodeoxynucleotide.

Figure 3 demonstrates that the inhibition of DNA polymerase I from *E. coli* cannot be overcome by increasing the concentrations of the precursors. A competitive action of some acid-soluble cell constituents is thus excluded. It should, however, be mentioned here that the inhibitory activity is compensated by increasing the enzyme concentration, as would be expected.

A typical Lineweaver-Burk plot of the uninhibited and inhibited DNA polymerase reactions, with poly-d(I-C) as primer-template and with dCTP and labelled dGTP as precursors, is given in Figure 4. It exhibits a noncompetitive type of inhibition. The uninhibited reaction reaches a maximum velocity of about 10,000 cpm incorporated in 5 min. This value is clearly not reached in those assays that are supplemented with yeast concentrate. For dCTP, K_m is calculated as 10.5 x 10^{-6} *M* in both cases. This value agrees well with K_m values recently found with other DNA-dependent DNA polymerases (Spadari and Weissbach, 1974a).

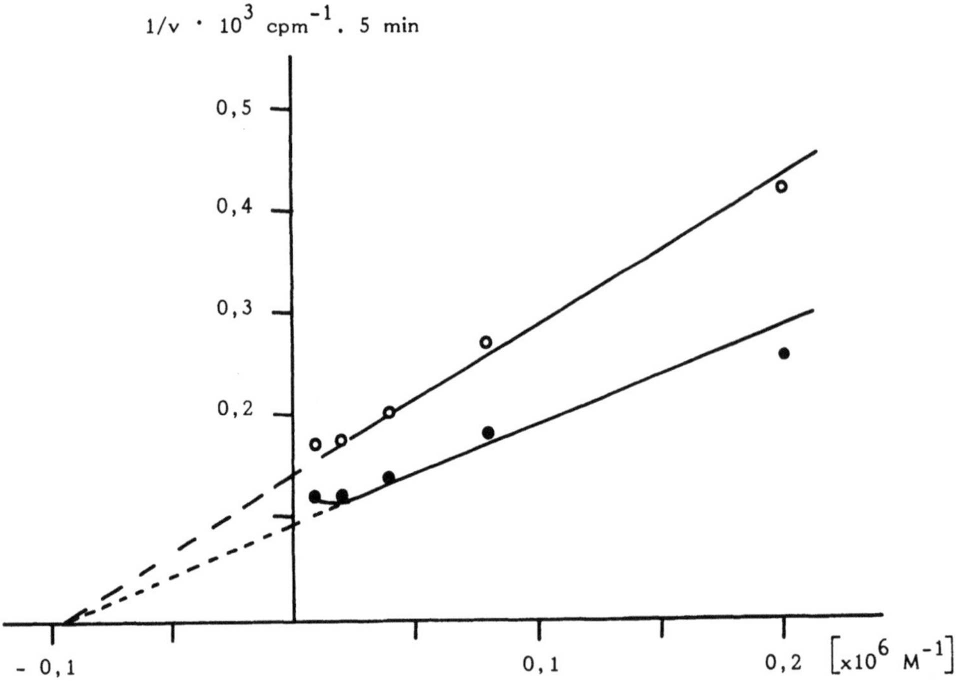

Fig. 4. Lineweaver-Burk plot of poly-d(G-C)-synthesis. Enyzme assays were performed as described in Eckstein et al. (1974) and in Table 1. 50 µl concentrate were supplemented to each assay. *Abscissa*: 1/[dCTP]. *Ordinate*: 1/cpm incorporated in 5 min. Cpm incorporated/5 min are the mean values from 30 min of incubation time; they refer to 50 µl aliquots from the assays

Let me present now two findings that possibly elucidate the biologic relevance of the inhibitor.

Figure 5 shows the time course of DNA synthesis in vitro, with "activated" salmon sperm DNA as a primer, all four deoxynucleoside triphosphates as precursors, and with purified DNA polymerase I from *E. coli* . It is obvious that yeast concentrate, when added to the assay, again effectively inhibits the enzyme activity.

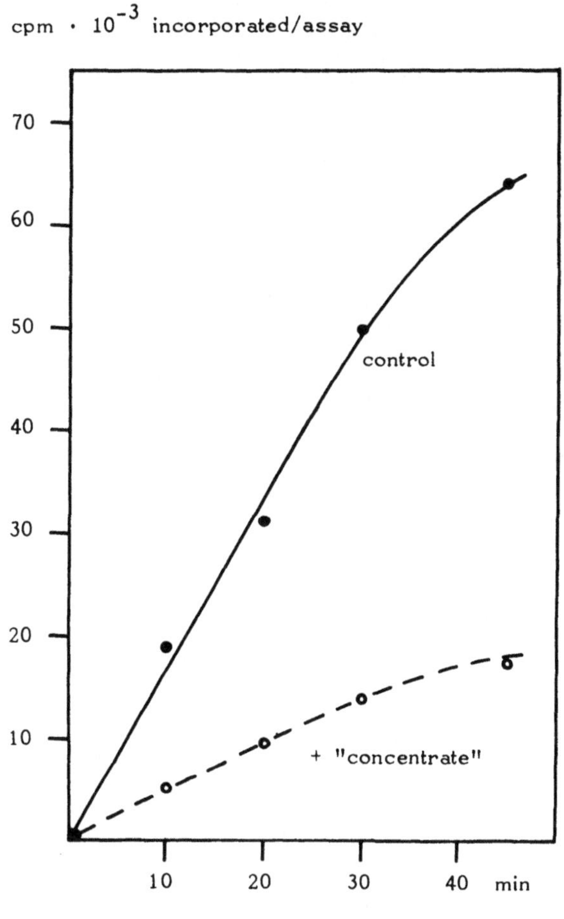

Fig. 5. Inhibition of DNA synthesis in vitro by yeast concentrate. DNA synthesis was performed in an assay with highly purified DNA polymerase I from *E. coli*, as described in Table 2, supplemented with 100 µl concentrate

With synchronized growing yeast cells, the acid-soluble inhibitor shows fluctuating behavior during the cell cycle (Figure 6). When the activity of DNA polymerase from the synchronized growing yeast cells has reached its maximum, the inhibitor starts to increase to particularly high values. At the same time, when the inhibitor has reached its maximum, DNA polymerase activity drops to low values, i.e., at the end of DNA synthesis. After that, the inhibitor also decreases sharply and reaches a minimum at the beginning of the next S phase.

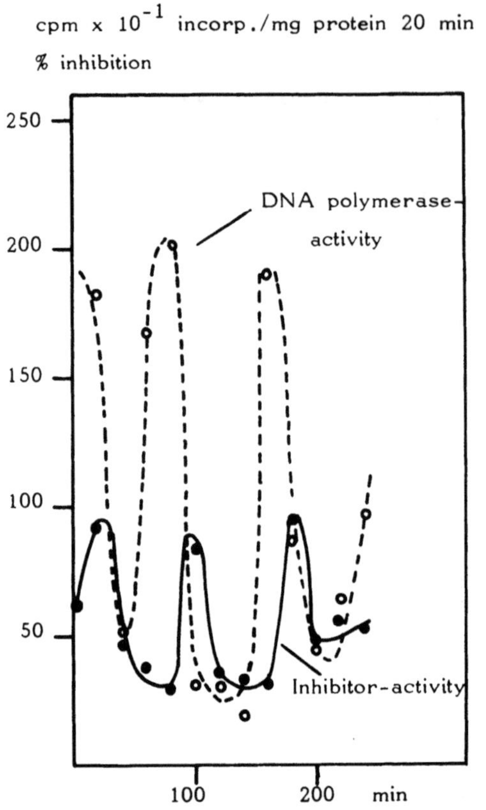

Fig. 6. DNA polymerase activity and inhibitor activity in synchronized growing yeast cells. Baker's yeast was synchronized and grown, as described earlier (Eckstein et al., 1967). Aliquots were taken from the culture at indicated times. Concentrates were prepared from 50 ml samples (0.25 x 10^9 cells at zero time of incubation), as described by Eckstein et al. (1974) and in Table 1. Test system with poly-d(I-C) and *E. coli* DNA polymerase I was supplemented with 50 µl each of the concentrates. DNA polymerase activity was estimated from crude extracts prepared from the cells from 50 ml samples, according to Eckstein et al. (1967). Protein was estimated from crude extracts by the biuret reaction

A third piece of evidence, which emphasizes a possible biologic function of this inhibitor, comes from experiments with partially purified DNA polymerase from yeast. As is shown in Table 2, DNA synthesis by partially purified DNA polymerase activity from growing yeast cells is inhibited remarkably when concentrate is added to the assay. This concerns not only the soluble enzyme fraction, but also the insoluble DNA polymerase activity, which apparently consists of two distinguishable enzymes (Wintersberger and Wintersberger, 1970). Inhibition of yeast DNA polymerase apparently is more pronounced than that of highly purified DNA polymerase I from *E. coli*. This could be a further indication of the biologic significance of the inhibitor in yeast.

As yet nothing is known about the biologic role of the different DNA polymerase activities described from growing yeast cells. Thus, the degree to which the inhibitor affects those functions of yeast DNA polymerase that are directly concerned with DNA replication in vivo remains to be determined.

Taken together, our findings nevertheless support the view that DNA synthesis in synchronized growing yeast cells might be regulated at least in part by a process of activation and inactivation of DNA polymerase. Some acid-soluble cell constituent(s), presumably an oligodeoxynucleotide, seems to play an essential role in this respect.

Table 2. DNA polymerase activity from growing yeast in assays supplemented with concentrate. Yeast DNA polymerase was partially purified according to Wintersberger and Wintersberger (1970) from homogenates from yeast cells grown to late log phase. *E. coli* DNA polymerase I was the highly purified preparation commercially available from Boehringer, Mannheim. Yeast concentrate was prepared as described in text; it contained 523 pmoles endogenous dTTP/100 µl. Enzyme assay contained in total volume of 0.40 ml : 5.0 nmoles dTTP supplemented with 5.0 µCi (methyl-^{3}H)-dTTP (The Radiochemical Centre, Amersham); 50.0 nmoles each of dATP, dCTP, and dGTP; 1 µmole ATP; 500 nmoles mercaptoethanol, 50 µg "activated" salmon sperm DNA; 4 µmoles $MgCl_2$; 40 µmoles Tris-HCl buffer, pH 7.6. Reaction was started by adding 5 µl each of yeast DNA polymerase preparations, or 0.025 µl of *E. coli* enzyme (50 units), respectively, and by incubation at 37°C. Every 5 min 50 µl aliquots were taken, applied to filter paper discs, and handled in the usual way (Eckstein et al., 1974). Radioactivity incorporated into acid-insoluble material was counted in a BF Betascint 5000 liquid scintillation counter. Salmon sperm DNA was activated by incubating in total volume of 0.56 ml, 0.5 mg DNA with 5 µmoles $MgCl_2$, and with 100 ng DNase I (from bovine pancreas, Serva, 2000 units/mg) at 37°C; after 30 min, 10 µmoles Na_2EDTA were added. Assay was heated at 90°C for 15 min and then allowed to cool slowly at room temperature

DNA polymerase from	cpm incorporated/5 min·assay		
	control	+ 100 µl concentrate	%
Yeast, 115,000 x *g* - supernatant	1095	563	51.9
Yeast, 115,000 x *g* - residue	789	438	57.3
Escherichia coli	609	429	71.8

References

Brent, T.P., Butler, J.A.V., Crathorn, R.A.: Variations in phosphokinase activities during the cell cycle in synchronous populations of HeLa cells. Nature (Lond.) 207, 176-177 (1965)

Chang, L.M.S., Brown, M., Bollum, F.J.: Induction of DNA polymerase in mouse L-cells. J. Molec. Biol. 74, 1-8 (1973)

Eckstein, H.: DNA turnover in growing yeast after x-irradiation. Z. Naturforsch. 29c, 48-53 (1974)

Eckstein, H., Ahnefeld, S.: An acid-soluble inhibitor of DNA-polymerases from growing yeast cells. In preparation

Eckstein, H., Ahnefeld, S., Albiez-Loges, K.: Acid-soluble deoxynucleotides and DNA synthesis in growing yeast after x-irradiation. II. Synthesis of deoxynucleoside tri- and monophosphates in synchronized and asynchronously growing cells. Z. Naturforsch. 29c, 272-282 (1974)

Eckstein, H., Paduch, V., Hilz, H.: Teilungssynchronisierte Hefezellen. II. Enzymsynthese nach Hemmung der Zellteilung durch Röntgenstrahlen. Biochem. Z. 344, 435-445 (1966)

Eckstein, H., Paduch, V., Hilz, H.: Synchronized yeast cells. III. DNA synthesis and DNA polymerase after inhibition of cell division by x-rays. Eur. J. Biochem. 3, 224-231 (1967)

Golombek, J., Wolf, W., Wintersberger, E.: DNA synthesis and DNA-polymerase activity in synchronized yeast cells. Molec. Gen. Genet. 132, 137-145 (1974)

Gorman, J., Tauro, P., La Berge, M., Halvorson, H.O.: Timing of enzyme synthesis during synchronous division in yeast. Biochem. Biophys. Res. Comm. 15, 43-49 (1964)
Halvorson, H.O., Gorman, J., Tauro, P., Epstein, R., La Berge, M.: Control of enzyme synthesis in synchronous cultures of yeast. Fed. Proc. 23, 1002-1008 (1964)
Hanna, M.H., Carl, P.L.: Reinitiation of deoxyribonucleic acid synthesis by deoxyribonucleic acid initiation mutants of *Escherichia coli*: Role of ribonucleic acid synthesis, protein synthesis, and cell division. J. Bacteriol. 121, 219-226 (1975)
Highfield, D.P., Dewey, W.C.: Inhibition of DNA synthesis in synchronized chinese hamster cells treated in G-1 or early S-phase with cycloheximide or puromycin. Exp. Cell Res. 75, 314-320 (1972)
Hilz, H., Eckstein, H.: Teilungssynchronisierte Hefezellen. I. Unterschiedliche Wirkungen von Röntgenstrahlen und cytostatischen Verbindungen auf Stoffwechsel und Zellteilung. Biochem. Z. 340, 351-382 (1964)
Lindberg, U., Skoog, L.: A method for the determination of dATP and dTTP in picomole amounts. Analyt. Biochem. 34, 152-160 (1970)
Lindsay, J.G., Berryman, S., Adams, R.L.P.: Characteristics of deoxyribonucleic acid polymerase activity in nuclear and supernatant fractions of cultured mouse cells. Biochem. J. 119, 839-848 (1970)
Probst, G.S., Stalker, D.M., Mosbaugh, D.W., Meyer, R.R.: Stimulation of DNA polymerase by factors isolated from Novikoff Hepatoma. Proc. Nat. Acad. Sci. USA 72, 1171-1174 (1975)
Sachsenmaier, W., Bollig, T., Bohnert, E., Fournier, D., von, Gürtler, R.E., Rupff, H.: Thymidine kinase induction correlated with synchronous mitosis in Physarum polycephalum. Abstr. 4th FEBS Meeting, Oslo, p. 57 (1967)
Seki, S., LeMahieu, M., Mueller, G.C.: A permeable cell system for studying DNA replication in synchronized HeLa cells. Biochim. Biophys. Acta (Amst.) 378, 333-343 (1975)
Seki, S., Mueller, G.C.: A requirement for RNA, protein and DNA synthesis in the establishment of DNA replicase activity in synchronized HeLa cells. Biochim. Biophys. Acta (Amst.) 378, 354-362 (1975)
Spadari, S., Weissbach, A.: The interrelation between DNA synthesis and various DNA polymerase activities in synchronized HeLa cells. J. Molec. Biol. 86, 11-20 (1974a)
Spadari, S., Weissbach, A.: HeLa cell R-deoxyribonucleic acid polymerases. J. Biol. Chem. 249, 5809-5815 (1974b)
Sylvên, B., Tobias, C.A., Malmgreen, H., Ottoson, R., Thorell, B.: Cyclic variations in the peptidase and catheptic activities of yeast cultures synchronized with respect to cell multiplication. Exp. Cell Res. 16, 75-87 (1959)
Stalker, D.M., Probst, G.S., Mosbaugh, D.W., Kunkel, T.A., Meyer, R.R.: Purification and properties of stimulatory proteins for Novikoff Hepatoma DNA polymerase. J. Cell Biol. 63, 332a (1974)
Wintersberger, U., Wintersberger, E.: Studies on deoxyribonucleic acid polymerases from yeast. 1. Partial purification and properties of two DNA polymerases from mitochondria-free cell extracts. Eur. J. Biochem. 13, 11-19 (1970)

4. DNA-Specific Labelling by Deoxyribonucleoside 5'-Monophosphates in *Saccharomyces cerevisiae*

M. BRENDEL, W. W. FÄTH, and R. TOPER

A. Introduction

DNA-specific labelling in *Saccharomyces cerevisiae* in vivo is possible with 5'-dTMP only (Jannsen et al., 1968; Brendel and Haynes, 1973; Fäth and Brendel, 1974). Thymine and thymidine cannot be used for DNA-specific labelling, because thymidine kinase apparently is absent in this organism (Grivell and Jackson, 1968). The introduction of a genetic (Fäth et al., 1974; Brendel and Fäth, 1974) or the use of a phenotypic block of thymidylate biosynthesis (Jannsen et al., 1973; Laskowski and Lehmann-Brauns, 1973) yielded a selective system for the isolation of 5'-dTMP low-requiring mutants, some of which show utilization (incorporation into DNA) of exogenously offered 5'-dTMP of between 25 and 50 per cent (Fäth and Brendel, 1975).

These 5'-dTMP low-requiring strains cease nucleic acid synthesis and exhibit rapid loss of viability when 5'-dTMP is offered at 10^{-3} *M*, a concentration not affecting wild-type or 5'-dTMP high-requiring strains (Langjahr et al., 1975). Because the same phenomenon is also observed with all the other ribo- and deoxyribonucleoside 5'-monophosphates in 5'-dTMP low-requiring strains, these were employed in our study to achieve DNA-specific labelling by deoxyribonucleoside 5'-monophosphates other than 5'-dTMP in *Saccharomyces cerevisiae*.

B. Materials and Methods

Strains. The yeast strains employed are given in Table 1. The two 5'-dTMP low-requiring strains are derived from the haploid strain 211-1aM, T-series (Laskowski and Lehmann-Brauns, 1973). Both strains have lost their mating type, apparently due to their 5'-dTMP low-requirement.

Table 1. Characterization of yeast strains

Strain	Genotype	5'-dTMP requirement
211-1aM	α *ilv2 rho*$^+$	none$^+$
T6-425	(α) *ilv2 typ1 rho*$^-$	none^{++}
T6-425*tmp1-51*	(α) *ilv2 typ1 tmp1-51 rho*$^-$	3 μg/ml

In the presence of aminopterin and sulfanilamide the strains require ($^+$) 60 to 70 μg 5'-dTMP/ml and ($^{++}$) 3 μg 5'-dTMP/ml

Standard Growth Conditions. Strains were grown in liquid medium N with appropriate supplement of nucleotides (Fäth and Brendel, 1974). Nature and concentration of the supplement(s) are always given in parantheses. For all other details like microscopic counting etc. see Brendel and Langjahr (1974).

Nucleic Acid Synthesis. Procedures of labelling RNA and DNA were essentially as described by Brendel and Langjahr (1974).

Nucleotides and Bases. Radioactive compounds were purchased from Amersham Searle Ltd., all other chemicals from Merck AG.

Assay of Radioactivity. This was done according to Brendel and Haynes (1973) and Fäth and Brendel (1974).

Abbreviations. 5'-dNMP: 2'-deoxyribonucleoside 5'-monophosphate; 5'-NMP: ribonucleoside 5'-monophosphate; Ade: adenine; dAdo: deoxyadenosine; Ura: uracil; Gua: guanine; Guo, dGuo: guanosine, deoxyguanosine.

C. Results and Discussion

At concentrations between 10^{-3} and 10^{-4} *M* the five nucleotides 5'-dTMP, 5'-dUMP, 5'-dCMP, 5'-dAMP and 5'-dGMP inhibit growth of 5'-dTMP low-requiring mutants of *Saccharomyces cerevisiae* but not that of the wild-type strain. This phenomenon has been described for 5'-dTMP (Langjahr et al., 1975) and will be demonstrated here with the purine monophosphates 5'-dGMP and 5'-GMP. An exponentially growing culture of strain T6-425 is severely inhibited by both nucleotides at exogenous concentrations higher than 5 x 10^{-5} *M* whereas all possible building blocks of the nucleotides do not have any effect on cellular growth (Fig. 1b). 5'-dTMP high-requiring (after inhibition of thymidylate synthesis) strain 211-laM is not influenced by 5'-dGMP and 5'-GMP at concentrations up to 10^{-3} *M* (Fig. 1a).

Microscopic observation of cells of an exponentially growing culture of strain T6-425 after transfer to medium N (5 x 10^{-4} *M* 5'-dGMP) reveals the apparent "freezing" of the cells in their respective place in the cell cycle. Morphological changes are absent and the fractions of budded and unbudded cells remain of the same order of magnitude whereas the cell number remains nearly constant over 24 h of incubation (Table 2). There is some residual and slowly occurring macromolecular synthesis as the fraction of unbudded cells slowly increases, in parallel with a minor increase (as compared to the control) in optical density of the culture (Table 3). Thus the cells show the same qualitative response to 5 x 10^{-4} *M* 5'-dGMP as they do toward the same concentration of 5'-dTMP (Langjahr et al., 1975).

While the cell number of strain T6-425 in medium N (5 x 10^{-4} *M* 5'-dGMP) remains approximately constant the titer of colony forming units declines slowly but steadily (Fig. 2a). The arrest of cellular growth is due to the cessation of nucleic acid synthesis (Fig. 2b). Growth inhibition is reversible, in that RNA and DNA synthesis resume after the excessive 5'-dGMP has been removed (data not shown). Again we observe phenomena similar to those already witnessed with 5 x 10^{-4} *M* 5'-dTMP (Langjahr et al., 1975).

The fact that cells of strain T6-425 and T6-425*tmp1-51* are sensitive to ribo- and deoxyribonucleoside 5'-monophosphates but not to their

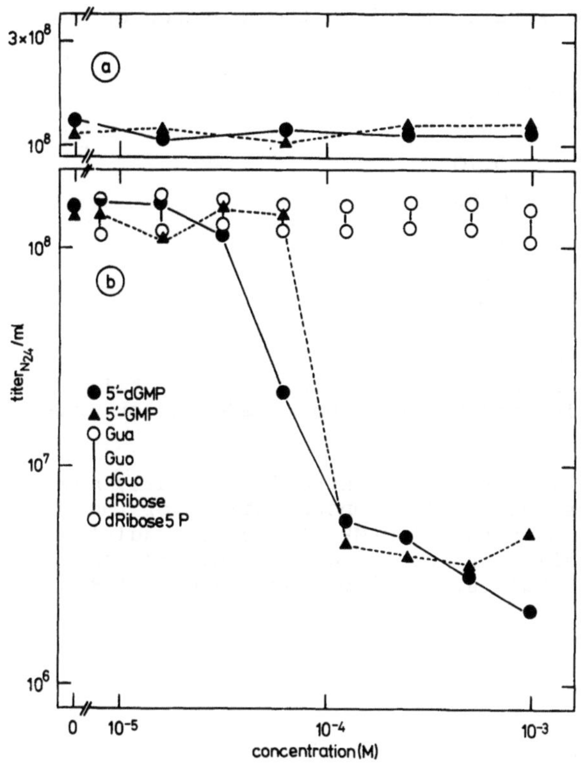

Fig. 1 a and b. Concentration-dependent inhibition of growth in yeast by 5'-dGMP, 5'-6MP and their possible cleavage products. (a) strain 211-1aM; (b) strain T6-425. Media were inoculated at titer of 2 x 10^6/ml with cells of exponentially growing culture

Table 2. Budding pattern of exponentially growing cells of strain T6-425 after addition of 5 x 10^{-4} *M* 5'-dGMP

Time (h)	Number of cells according to bud size: unbudded	tiny bud	bud	large bud	Titer/ml x 10^{-7}
0	43 (26)[a]	51 (54)	50 (80)	62 (50)	1.1
1	(60)	(49)	(51)	(56)	
3	96 (70)	22 (37)	43 (56)	47 (39)	1.2
6	120 (91)	32 (39)	46 (51)	43 (32)	1.0
24	134 (111)	30 (34)	35 (54)	43 (50)	1.1

[a]Data of two independent experiments are given

respective nucleosides and bases was taken as an indication that there is uptake of the monophosphates themselves. In this case we expected to achieve DNA-specific labelling with all 5'-dNMPs.

The results of such a 5'-dNMP labelling are shown in Table 4. Obviously all nucleotides are taken up and incorporated into nucleic acid. Efficiency of uptake, with the exception of 5'-dUMP, is about one third that of 5'-dTMP. In DNA specificity (alkali stability), the 5'-dNMPs exhibit significant differences. However, even 5'-dUMP, which

Table 3. Synthesis of biomass of strain T6-425 in the presence of 5×10^{-4} *M* 5'-dGMP

Time (h)	Optical[a] density in medium N + 5'-dGMP	N
0	0.70	0.70
3	0.81	0.98
6	0.88	1.45
22	0.96	8.8

[a]OD_{540} nm

yields the lowest percentage of alkali stable radioactivity, still contributes less to RNA synthesis than the free bases adenine or uracil (Table 4, bottom).

The apparent utilization of 5'-dNMPs for RNA synthesis obviously may be totally prevented in the presence of high concentrations of the expected cleavage products, as shown for 5'-dAMP (Table 4, middle). Here we obtain DNA-specific labelling as can be derived by the comparison with the percentage of alkali stable label achieved by the DNA-specific precursor 5'-dTMP.

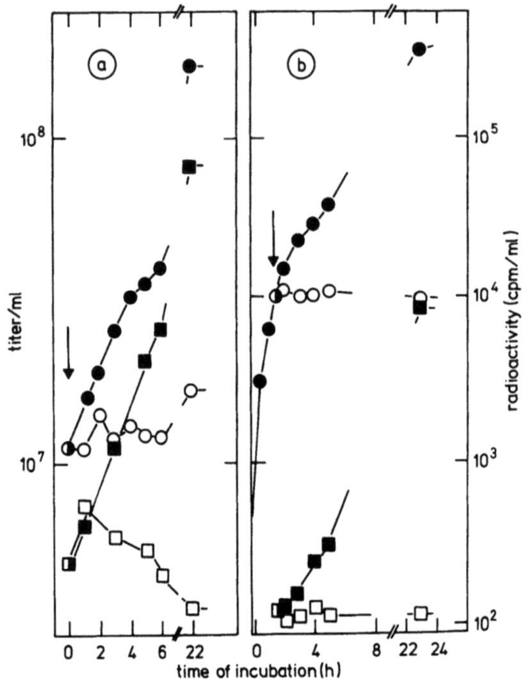

Fig. 2 a and b. Inhibition of growth of strain T6-425 by 5×10^{-4} *M* 5'-dGMP (open symbols). Filled symbols represent controls (no nucleotide). Arrow marks the time of addition of 5'-dGMP. (a) o----o, cell number as determined by microscopic count; □---□, viability as determined by plating. (b) o----o, total nucleic acid (TCA precipitable counts); □---□, DNA (TCA precipitable counts after alkaline treatment)

Table 4. DNA labelling by deoxyribonucleoside 5'-monophosphates in strain T6-425*tmp1-51*

Medium	TCA precipitable radioactivity (cpm)	TCA precipitable radioactivity after alkali treatment (cpm)	% alkali stable radioactivity
N^+(5 μCi ^{3}H·dTMP/ml)	18,890	22,695	120[a]
N^+(3.3 μg Na_2·dUMP/ml; 5 μCi ^{3}H·dUMP/ml)	838	65	7.8
N^+(2.5 μg dCMP/ml; 5 μCi ^{3}H·dCMP/ml)	5,700	923	16.2
N^+(3.2 μg Na_2·dGMP/ml; 5 μCi ^{3}H·dGMP/ml)	7,700	1,500	19.5
N^+(2.7 μg dAMP/ml; 5 μCi ^{3}H·dAMP/ml)	7,900	5,700	72.1
N^+(2.7 μg dAMP/ml; 10^{-2} *M* Ade; dAdo; 5 μCi ^{3}H·dAMP/ml)	1,180	1,300	110[a]
N^+(20 μg Ade/ml; 1 μCi ^{14}C·Ade/ml)	17,589	777	4.4
N^+(20 μg Ura/ml; 1 μCi ^{14}C·Ura/ml)	25,951	1,027	4.0

Medium N^+ contains 3.0 μg Na_2·dTMP/ml

[a]Values of over 100% are found because alkali treatment reduces quenching

D. Summary

Growth of 5'-dTMP low-requiring strains is inhibited by exogenous 5'-dGMP and 5'-GMP at concentrations higher than 5 x 10^{-4} *M*. Synthesis of nucleic acids ceases and cells remain fixed in their respective stage in the cell cycle. At concentrations lower than 10^{-5} *M*, deoxyribonucleoside 5'-monophosphates may be employed for radioactive labelling, the label being preferentially used for DNA synthesis. Affinity to DNA of the 5'-dNMPs is in the order of 5'-dAMP > 5'dGMP > 5'-dCMP > 5'-dUMP. DNA-specific labelling is achieved with 5'-dAMP when the medium is supplemented with adenine and deoxyadenosine.

Acknowledgements. This work was supported by a grant from the Deutsche Forschungsgemeinschaft to the first author. We want to thank Ms. Gabi Jung for excellent technical assistance.

References

Brendel, M., Fäth, W.W.: Isolation and characterization of mutants of *Saccharomyces cerevisiae* auxotrophic and conditionally auxotrophic for 5'-dTMP. Z. Naturforsch. 29c, 733-738 (1974)

Brendel, M., Haynes, R.H.: Exogenous thymidine-5'-monophosphate as a precursor for DNA synthesis in yeast. Molec. Gen. Genet. 126, 337-348 (1973)

Brendel, M., Langjahr, U.G.: "Thymineless death" in a strain of *Saccharomyces cerevisiae* auxotrophic for deoxythymidine-5'-monophosphate. Molec. Gen. Genet. 131, 351-358 (1974)

Fäth, W.W., Brendel, M.: Specific DNA-labelling by exogenous thymidine-5'-monophosphate in *Saccharomyces cerevisiae*. Molec. Gen. Genet. 131, 57-67 (1974)

Fäth, W.W., Brendel, M.: An improved assay of UV-induced thymine-containing dimers in *Saccharomyces cerevisiae*. Z. Naturforsch. 30c, 804-810 (1975)

Fäth, W.W., Brendel, M., Laskowski, W., Lehmann-Brauns, E.: Economizing DNA-specific labelling by deoxythymidine-5'-monophosphate in *Saccharomyces cerevisiae*. Molec. Gen. Genet. 132, 335-345 (1974)

Grivell, A.R., Jackson, J.F.: Thymidine kinase: evidence for its absence in *Neurospora crassa* and some other microorganisms, and the relevance of this to the specific labelling of deoxyribonucleic acid. J. Gen. Microbiol. 54, 307-317 (1968)

Jannsen, S., Lochmann, E.-R., Laskowski, W.: DNS-Synthese nach Röntgenbestrahlung bei homozygoten Hefestämmen verschiedenen Ploidiegrades. Z. Naturforsch. 23b, 1500-1507 (1968)

Jannsen, S., Witte, I., Megnet, R.: Mutants for the specific labelling of DNA in *Saccharomyces cerevisiae*. Biochim. Biophys. Acta (Amst.) 299, 681-685 (1973)

Langjahr, U.G., Hartmann, E.-M., Brendel, M.: Nucleic acid metabolism in yeast. I. Inhibition of RNA and DNA synthesis by high concentrations of exogenous deoxythymidine 5'-monophosphate in 5'-dTMP low-requiring strains. Molec. Gen. Genet. 143, 113-118 (1975)

Laskowski, W., Lehmann-Brauns, E.: Mutants of *Saccharomyces* able to grow after inhibition of thymidine phosphate synthesis. Molec. Gen. Genet. 125, 275-277 (1973)

5. Studies on DNA Synthesis in *Saccharomyces cerevisiae*

U. WINTERSBERGER

A. Introduction

Extensive studies in recent years have established that DNA replication is a complex reaction, requiring a variety of enzymes and protein factors that are involved in the initiation and elongation of deoxynucleotide chains. The identification of the components of the DNA synthesizing apparatus as well as the clarification of their specific functions has now become of vital interest in biochemistry. Progress in this area of research can be expected from the use of conditionally lethal mutants exhibiting temperature sensitive DNA synthesis. Provided the availability of adequate test methods, such mutants prove most valuable for the isolation and characterization of constituents of the replication machinery. This was recently demonstrated by the isolation of several proteins involved in the replication of bacterial and phage DNA (Kornberg, 1974).

With few exceptions, conditionally lethal mutants of higher cells are isolated with great difficulty. One such exception is the simple eukaryote, *Saccharomyces cerevisiae*, which is as readily amenable to genetic analysis as are many prokaryotes. For this reason it was hoped that this organism would serve as a useful model for studies on the biochemistry of DNA replication in eukaryotes. Evidence for a surprising similarity of the general mechanism of DNA replication in yeast and in higher organisms is provided by electron-microscopic examination of replicating molecules of yeast DNA: each DNA molecule shows several replication bubbles, also observed in higher eukaryotes, which means that each chromosome - although on the average smaller than the chromosome of *Escherichia coli* - consists of more than one replication unit. This is in contrast to the bacterial chromosome, which represents a singe replicon (Newlon et al., 1973).

One of the greatest obstacles in the work on the control of DNA replication and transcription in lower eukaryotes, such as yeast or *Neurospora crassa*, was the difficulty in finding adequate methods for cell fractionation, in particular for the isolation of intact, pure nuclei. This is mainly due to the rigid cell wall and the fragile nuclear membrane of these organisms. So far, successful isolation of intact nuclei from yeast requires the prior conversion of cells into spheroplasts and the lysis of these spheroplasts in the presence of substances, such as Ficoll or polyvinylpyrrolidone, which stabilize the structure of the nuclei. Such methods have been developed recently, and various properties of isolated nuclei and of chromatin prepared therefrom were determined (Wintersberger et al., 1973).

In our work on the structure and function of the genetic material in the yeast *S. cerevisiae*, we were interested in the DNA polymerases present in isolated nuclei and in the development of a DNA replicating system that could be used to purify and characterize components involved in DNA synthesis. The following is a summary of our results in this area.

B. Yeast DNA Polymerases

Isolated, intact nuclei from *S. cerevisiae* in many respects resemble those of higher eukaryotes. Although there seem to be certain characteristic differences in the chromosome structures of yeast and of higher organisms (Wintersberger et al., 1975), important similarities between the chromatin components were found (Wintersberger et al., 1973). Among these components, the enzymes involved in the replication of nuclear DNA are expected. In contrast to this expectation, however, it is generally observed in higher cells that after cell fractionation a large part of the DNA polymerizing activity is present in the cytoplasm. The same situation appears to be true also for other proteins that function in the initiation and continuation of DNA replication. Apparently, these proteins are only loosely bound in the nucleus and are efficiently extracted during the preparation and purification of nuclei. This fact has to be considered if efforts are directed toward the development of nuclear DNA replication systems (see below). In yeast the situation is by no means exceptional: when the distribution of DNA polymerase activity between the cell fractions was determined, isolated nuclei contained not more than about 15% of the total cellular DNA polymerase activity while 82% was found in the cytoplasm and 3% in the mitochondria (Wintersberger, U., 1974). DNA polymerase activity from cytoplasm as well as from purified nuclei was characterized. These experiments revealed that the enzyme activity quantitatively extracted from nuclei is separable on a DEAE-cellulose column into two fractions, eluting at salt concentrations identical with those found earlier for the yeast DNA polymerases A and B isolated from the cytoplasm (Wintersberger and Wintersberger, 1970a; Wintersberger, E., 1974; Wintersberger, U., 1974). Other properties of the enzymes isolated from nuclei also are identical with those of the cytoplasmic DNA polymerases A and B. These include, for instance, the primer-template preference, the inhibition by salt, and the sensitivity to SH-blocking compounds (Table 1). The molecular weight of these enzymes (determined by sedimentation analyses in sucrose gradients in the presence of high salt concentrations) is around 170,000 daltons and thus corresponds to that of the 6-8-S DNA polymerase found in the cytoplasm of higher eukaryotes (Wintersberger, U., 1974).

Because nuclei of all higher cells studied so far were found to contain a small, 3.4-S DNA polymerase (see Loeb, 1974, for review), it was of interest to see whether this enzyme has only escaped our detection or whether it was altogether absent in *S. cerevisiae*. Experiments were therefore designed to find this enzyme even if it were present only in small quantities compared to the 6-8-S polymerases. Since the 6-8-S and the 3.4-S DNA polymerases are most readily distinguished by their different sedimentation rates in sucrose gradients, we used this method to analyze nuclear extracts for the presence of the low-molecular-weight DNA polymerase. Much to our surprise we could not detect this enzyme in yeast, despite the use of a variety of conditions for extraction of nuclei and analysis of the extracts (Wintersberger, U., 1974).

This result indicates that yeast cells may lack the small nuclear DNA polymerase, an observation that is of interest with regard to the function of this enzyme in higher eukaryotes. In lower eukaryotes, therefore, there could be another enzyme that differs from the 3.4-S polymerase of mammalian nuclei in size but that plays a similar role. A candidate for such an enzyme in yeast is DNA polymerase B. We have evidence that this enzyme, like the 3.4-S DNA polymerase, is not reg-

Table 1. Properties of yeast DNA polymerases

	Nuclear of cytoplasmic DNA polymerases		Mitochondrial DNA polymerase
	Enzyme A	Enzyme B	
Template-primers			
Activated DNA	100 %	100 %	100%
poly d(AT)	228 %	108 %	212%
poly(dA)poly(dT)	98 %	144 %	
poly(dA)(dT)$_{10}$	79 %	2070 %	450%
poly(rA)(dT)$_{10}$	1.5%	1.7%	3%
Mg^{++} optimum (m*M*)	10	20	50
Elution from DEAE-cellulose (*M* NaCl)	0.18	0.26	0.36
Effect of NaCl			
Without NaCl	100 %	100 %	100%
0.05 *M* NaCl	130 %	190 %	110%
0.10 *M* NaCl	100 %	140 %	90%
0.20 *M* NaCl	10 %	10 %	25%
0.40 *M* NaCl	5 %	5 %	10%
Effect of p-chloromercuribenzoate			
0.02 *M*	18 %	38 %	70%
0.04 *M*	10 %	25 %	45%
0.08 *M*	5 %	5 %	10%
Effect of cytosine-arabinoside triphosphate			
0.05 m*M*	30 %	60 %	38%
0.10 m*M*	20 %	50 %	28%
Molecular weight			
from sedimentation in sucrose gradients at high ionic strength	170,000	160,000	120,000

ulated in the cell cycle, quite in contrast to yeast DNA polymerase A, the activity of which fluctuates in the cell cycle and which exhibits maximal activity at the beginning of the S phase, like the 6-8-S polymerase of animal cells (Golombek et al., 1974; Wintersberger, E., 1974). We would like to suggest, therefore, that yeast DNA polymerase A functions in DNA replication as the enzyme responsible for the synthesis of replicative intermediates, while DNA polymerase B might catalyze gap-filling reactions and in addition be involved in DNA repair.

In addition to the nuclear or cytoplasmic DNA polymerases A and B, a third distinct enzyme is present in yeast mitochondria. This enzyme was purified by chromatography on DEAE-cellulose and DNA-carboxymethylcellulose (Potuzak and Wintersberger, U., in preparation) and its properties were found to be clearly different from those of the two main cellular activities (Wintersberger and Wintersberger, 1970b; Wintersberger and Blutsch, unpublished). Some of these properties are included in Table 1 for comparison. So far we have no evidence for the presence in yeast mitochondria of a second DNA polymerase beside the one we have isolated and characterized. It is likely that this single polymerase catalyzes the reactions in the replication of mitochondrial DNA for which in nuclei two enzymes are required. The mitochondrial DNA replication apparatus is therefore autonomous with respect to the DNA polymerase. Little is known about other components that function in mitochondrial DNA synthesis. Experiments do indicate, however, that nuclei and mitochondria share some proteins involved in DNA replication: mutants, temperature-sensitive in the elongation step of nuclear DNA replication, exhibit also temperature-sensitive synthesis of mitochondrial DNA (Wintersberger et al., 1974).

C. Synthesis of DNA in Isolated Yeast Nuclei

In the past ten years a number of experiments have been described in which isolated nuclei were used with the aim of unraveling the mechanism of DNA replication in higher cells (see for example Winnacker et al., 1971; Hershey et al., 1973; Schiebel and Schneck, 1974; Benz and Strominger, 1975). In all these systems, radioactive deoxyribonucleoside triphosphates were used as substrates, and their incorporation into DNA was tested under a variety of conditions. With few exceptions, these experiments do not allow one to distinguish beyond doubt a replicative from a repair reaction. Furthermore, there are great variations in the requirements for optimal incorporation among different nuclear in vitro systems.

The use of mutants, temperature sensitive in the initiation or elongation reaction of DNA synthesis (Hartwell et al., 1974), should provide a means to test for the replicative versus repair capacity of such an in vitro system. At a more advanced stage, a nuclear replicating system from yeast could be the basis of a complementation assay that would be necessary for the purification and characterization of proteins required for DNA replication.

With a method for the preparation of intact yeast nuclei at hand, we initiated experiments to set up a nuclear in vitro DNA synthesizing system from *S. cerevisiae*; the following is a summary of the current state of our efforts.

After numerous unsuccessful trials in which extensively purified but inactive nuclei were obtained, conditions were worked out that were optimal for the isolation of nuclei active in the incorporation of deoxynucleoside triphosphates. These conditions require (1) that harvest of cells and spheroplast preparation are carried out at room temperature and (2) that nuclei are isolated from spheroplasts as rapidly as possible, avoiding lengthy centrifugation, which are necessary to obtain optimal yields of pure nuclei. In other words, it was necessary to find a compromise between reasonable purity and active DNA synthesizing capacity of the isolated nuclei. This is not surprising in view of the fact that many components of the replication machinery rapidly leak out of the nuclei during an extensive purification. Figure 1a shows an electron-microscopic picture of a nuclear fraction as it is used in the DNA synthesis experiments; Figure 1b shows a nucleus at higher magnification. The inner struc-

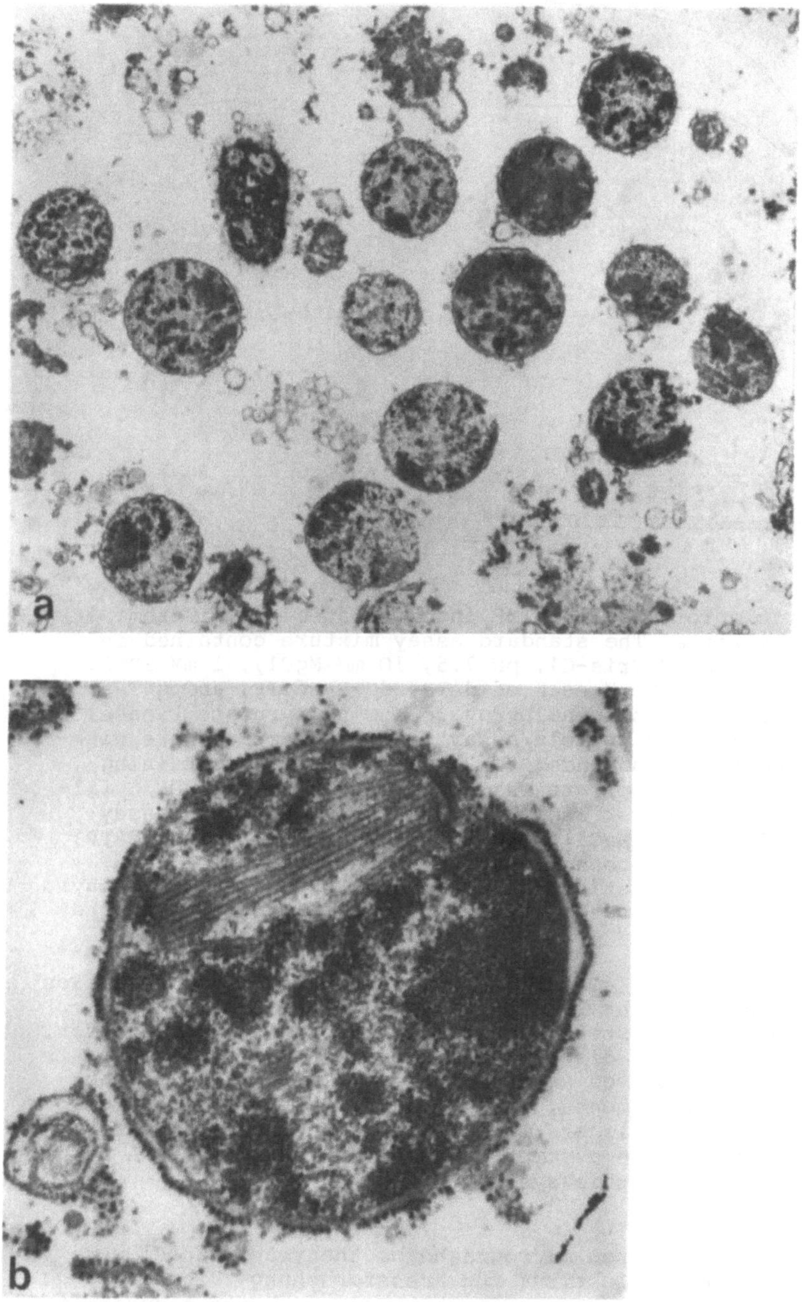

Fig. 1 a and b. Electron micrographs of isolated nuclei from *Saccharomyces cerevisiae*. (a) general view, X7,700; (b) single nucleus, X55,000

tural elements such as microtubuli and spindle plaques are well preserved. These nuclei were incubated with the four deoxyribonucleoside-triphosphates, and the incorporation of labelled TMP was measured under various conditions. From Figure 2 it can be seen that incorporation is dependent on the presence of Mg ions and all four deoxytriphosphates and that it is stimulated by an ATP regenerating system. Additional

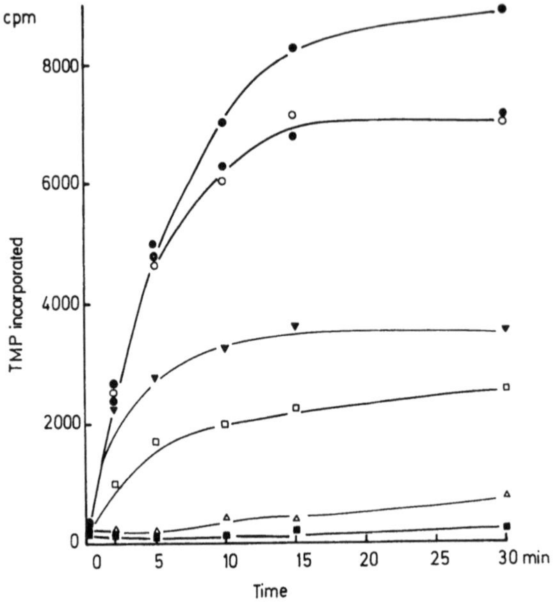

Fig. 2. Incorporation of labelled TMP into isolated nuclei from *Saccharomyces cerevisiae*. The standard assay mixture contained in a total of 625 µl: 0.05 *M* Tris-Cl, pH 7.5, 10 m*M* $MgCl_2$, 1 m*M* EDTA, 2 m*M* β-mercaptoethanol, 50 µ*M* each of dATP, dCTP, dGTP, and dTTP, 5 µC of ^{3}H-TTP, 625 µg phosphoenolpyruvate, 50 µg pyruvate kinase, and suspended nuclei (1 mg protein/assay). 100 µl samples were withdrawn at indicated times and added to 2 ml cold 5% TCA containing 1% pyrophosphate. The precipitate was collected on Whatman GF/C filters, washed, and counted for radioactivity. ○——○ standard assay; ●——● plus 1 m*M* ATP, 40 m*M* NaCl; ■——■ minus $MgCl_2$; △——△ minus dATP, dCTP, dGTP; □——□ minus phosphoenolpyruvate, minus pyruvate kinase; ▼——▼ nuclei after preincubation for 30 min at 36°C (standard assay); ◉——◉ nuclei after storage for 3 weeks in liquid nitrogen (standard assay)

ATP (1 m*M*) as well as NaCl (40 m*M*) does not enhance the incorporation rate. Nuclei that are kept at 36°C rapidly loose their ability to synthesize DNA while those stored in liquid nitrogen stay active for several weeks. A convincing discrimination between DNA replication and DNA repair in the isolated yeast nuclei is not possible at the present state of our experiments, though the fact that nuclei from logarithmically growing cells are more active than those from stationary phase cells supports the assumption that replication of DNA rather than repair is measured in this system.

Acknowledgements. The electron micrographs of the yeast nuclei were kindly taken by M. Binder, Institut für Krebsforschung, Vienna. I am grateful to Erhard Wintersberger for his help with the manuscript.

References

Benz, W.C., Strominger, J.L.: Viral and cellular DNA synthesis in nuclei from human lymphocytes transformed by Epstein-Barr virus. Proc. Nat. Acad. Sci. USA 72, 2413-2417 (1975)

Golombek, J., Wolf, W., Wintersberger, E.: DNA synthesis activity in synchronized yeast cells. Molec. Gen. Genet. 132, 137-145 (1974)

Hartwell, L.H., Culotti, J., Pringle, J.R., Reid, B.J.: Genetic control of the cell division cycle in yeast. Science 183, 46-51 (1974)

Hershy, V.W., Stieber, J.F., Mueller, G.C.: DNA synthesis in isolated HeLa nuclei. A system for continuation of replication in vivo. Eur. J. Biochem. 34, 383-394 (1973)

Kornberg, A.: DNA Synthesis. San Francisco: W.H. Freeman and Company, 1974

Loeb, L.: Eukaryotic DNA polymerases. In: The Enzymes. New York: Academic Press, 1974, Vol. 10, pp. 173-209

Newlon, C.S., Petes, T.D., Hereford, L.M., Fangman, W.L.: Replication of yeast chromosomal DNA. Nature (Lond.) 247, 32-35 (1973)

Schiebel, W., Schneck, U.: DNA replication continued in isolated nuclei of synchronously growing *Physarum polycephalum*. Hoppe-Seyler's Z. Physiol. Chem. 355, 1515-1525 (1974)

Winnacker, E.L., Magnusson, G., Reichard, P.: Synthesis of polyoma DNA by isolated nuclei. Biochem. Biophys. Res. Comm. 44, 952-957 (1971)

Wintersberger, E.: DNA polymerases from yeast. Further purification and characterization of DNA-dependent DNA polymerases A and B. Eur. J. Biochem. 50, 41-47 (1974)

Wintersberger, U.: Absence of a low molecular weight DNA-polymerase from nuclei of the yeast, *Saccharomyces cerevisiae*. Eur. J. Biochem. 50, 197-202 (1974)

Wintersberger, U., Binder, M., Fischer, P.: Cytogenetic demonstration of mitotic chromosomes in the yeast *Saccharomyces cerevisiae*. Molec. Gen. Genet. (1975, in press)

Wintersberger, U., Hirsch, J., Fink, A.M.: Studies on nuclear and mitochondrial DNA-replication in a temperature-sensitive mutant of *Saccharomyces cerevisiae*. Molec. Gen. Genet. 131, 291-299 (1974)

Wintersberger, U., Smith, P., Letnansky, K.: Yeast chromatin. Preparation from isolated nuclei, histone composition and transcription capacity. Eur. J. Biochem. 33, 123-130 (1973)

Wintersberger, U., Wintersberger, E.: Studies on DNA polymerases from yeast. I. Partial purification and properties of two DNA-polymerases from mitochondria-free cell extracts. Eur. J. Biochem. 13, 11-19 (1970a)

Wintersberger, U., Wintersberger, E.: Studies on DNA polymerases from yeast. II. Partial purification and characterization of mitochondrial DNA polymerase from wild type and respiration deficient yeast cells. Eur. J. Biochem. 13, 20-27 (1970b)

6. Metabolism of Yeast Cells after UV-and X-Ray Irradiation

J. KIEFER, H. KOCH, B. LASKE, and H. WALLER

A. Introduction

Irradiation of a cell not only leads to the loss of reproductive integrity as measured by colony forming ability but also causes disturbances in many metabolic events and cell progression. Genetically determined repair and recovery processes may act on many of these parameters. It is, of course, not possible to measure all possible alterations, but the study of some key metabolic processes might help to understand the chain of events ultimately leading to cell death and the involvement of repair and recovery.

We undertook such investigations in diploid yeast of different sensitivities, making use of the relative ease with which this organism can be handled and of the availability of genetically defined sensitive mutants. It is our hope to add at least some information to the puzzling picture of cellular radiation damage.

B. Material and Methods

I. Cells and Culture Conditions

Two strains of diploid yeast *Saccharomyces cerevisiae* were used, the wild type 211 (originally obtained from W. Laskowski) and the ultraviolet (UV)-sensitive mutant 2094 (isolated by and obtained from D. Averbeck). The latter is mutated in the RAD2 gene (Game and Cox, 1971) and hence excision negative (Resnick and Setlow, 1972). Cells were grown in Wickerham's synthetic medium (Wickerham, 1951) at 30°C under vigorous aeration. For all investigations - except that of RNA synthesis - stationary phase cells from liquid cultures were used as starting material. For RNA studies, cells were taken from nutrient agar plates after 3 days incubation. After irradiation the cells were always transferred to fresh nutrient medium at a density of about 10^7 cells/ml and incubated at 30°C under aerobic conditions - if not otherwise stated - for the desired period of time. The cells were counted microscopically. Survival was assessed by standard plating techniques.

Irradiation: The x-ray source was a 100 kV tube (C.H.F. Müller, Hamburg) operated without external filtering and delivering a dose-rate of about 45 kR/min. Doses were determined by ferrous sulphate dosimetry.

A low-pressure mercury lamp Osram HNS 12 ofr provided the UV irradiation. Fluences were measured by chemical actinometry (Hatchard and Parker, 1956) or by a Hewlett-Packard radiometer. All irradiations were applied to aqueous suspensions under vigorous stirring.

II. Measurements of Biochemical Parameters

Energy Metabolism: Respiration and anaerobic fermentation were measured by standard manometric Warburg techniques, aerobic alcoholic fermentation by enzymatic determination of ethanol with ADH. Details have been given previously (Kiefer, 1974).

Protein Content and Synthesis: Cellular protein content was determined by the method of Lowry et al. (1951). Protein synthesis was followed by the incorporation of ^{3}H-lysine and ^{3}H-phenylalanine.

Details of the labelling and extraction procedure will be published elsewhere (Laske and Kiefer, in preparation).

RNA-Synthesis: Cells were incubated for one h in the presence of radioactive labelled uracil, disrupted, and the rRNA was isolated as described previously (Koch and Kiefer, 1972). 18-S and 28-S rRNA were separated by gel electrophoresis. Details of the method will also be published elsewhere (Koch and Kiefer, in preparation).

C. Results

Since we were interested in cellular metabolic activity after irradiation, all data were calculated on a per cell basis, thus eliminating the influence of the different proliferation rates in controls and irradiated samples. To this end it is absolutely necessary to count the actual cell number microscopically; the measurement of optical density is virtually meaningless because of the radiation-induced changes of cell shape and volume. We shall first report "early" metabolic changes, i.e., differences between irradiated and unirradiated cells as soon as they can be detected by our techniques. The times are different for the various parameters as summarized in Table 1. Because the UV experiments give more clear-cut results, we shall draw mainly upon these. A second reason is that the availability of a sensitive mutant that is at least partly also characterized at the molecular level may help to assess the importance of repair processes. Where possible, x-ray data are given for comparison. In another section we shall follow the time course of metabolic parameters after exposure.

Table 1. Metabolic parameters and the earliest time when irradiated cells differed from controls

Parameter	Time of "early" response (hours after exposure)
Incorporation of amino acids	0.75
rRNA synthesis	1[a]
Respiration	1
Fermentation	2

[a]Earliest time of measurement

D. "Early" Changes: UV-Irradiated Wild-Type Cells

The dose dependencies of various metabolic processes in strain 211 after UV exposure measured at the times given in Table 1 are shown in Figure 1. The sensitivities vary over a wide range. Respiration is obviously most resistant, RNA synthesis most sensitive. With the exception of respiration all the dose dependencies show "shoulder-curves". The final slopes for amino acid incorporation (both lysine (lys) and phenylalanine (phe)), alcoholic fermentation, and survival are equal within error limits. By far the most sensitive process is the synthesis of rRNA. In our stationary phase yeast cells 18-S and 28-S rRNA show the same sensitivity in contrast to exponential phase *E. coli* (Hackett and Sauerbier, 1974) or *Saccharomyces carlsbergiensis* (van den Eos et al., 1971). We have confirmed that this is not due to strain differences by showing that a differential sensitivity is obtained if exponential phase cells of strain 211 are used (Koch, 1975). We do not know at present the cause of this discrepancy; it may be related to phase dependent differences in precursor processing.

It is noteworthy that rRNA synthesis is already severely depressed at doses where survival is hardly affected. This might mean that rRNA synthesis plays a key role in cellular damage development after UV exposure.

Respiration constitutes the other extreme: it is only slightly depressed even when survival is as low such as 10%. This may be due to the fact that stationary phase cells are adapted to respiration, so that de novo enzyme synthesis is hardly necessary. The time course of cellular respiration (vide infra), however, demonstrates that this explanation is not able to account for all experimental data.

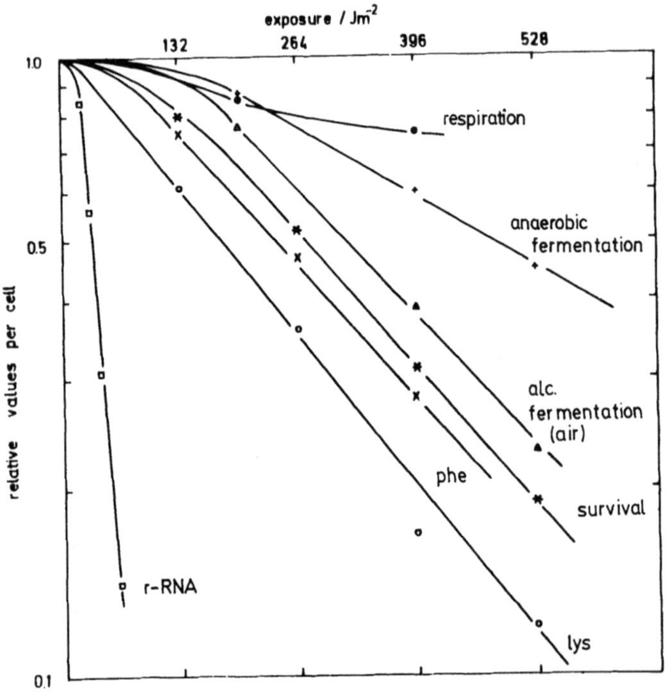

Fig. 1. UV-dose dependence of some metabolic parameters at "early" stages (see Table 1) in wild type

E. "Early" Changes: UV-Sensitive Cells

Some of the previously mentioned parameters have also been measured for the incision negative strain 2094 after UV exposure. Because of the higher sensitivity, lower doses had to be applied. The dose dependencies of rRNA synthesis and fermentation are compared in Figures 2 and 3, respectively. It is seen that for both processes pure exponential curves are obtained in the sensitive strain. The slopes may be used to estimate the target size of the responsible genome region. In the case of rRNA it corresponds to the precursor molecule of 18-S and 28-S rRNA (plus spacer regions), which are transcribed together if compared to the data of Udem and Warner (1972)(Koch, 1975). The dose dependence of rRNA synthesis in the repair proficient wild type displays a smaller slope and a shoulder, which may be indicative of repair that has taken place during the one-hour incubation period. The exponential part extrapolates to a dose D_Q = 20 J/m^2 at the 100% level. This corresponds to about 10,000 thymine dimers in diploid yeast (Unrau et al., 1972), which must at least have been removed during one hour. This figure constitutes only a lower estimate since we measured the <u>integral</u> synthesis over the whole period. Pulse labelling experiments will certainly reveal the pictures more clearly.

The dose response curve for cellular fermentation at 2 h changes from a pure exponential in strain 2094 to a shoulder type in strain 211. The slope in the sensitive strain is compatible with the structural gene of alcohol dehydrogenase, the key enzyme in alcoholic fermentation (Kiefer, submitted for publication). Taking again the D_Q value (132 J/m^2) as an estimate for the repair rates, one calculates about 30,000 dimers removed per hour. The discrepancy between the two figures, which is beyond experimental uncertainties, cannot at present be explained, although it must be admitted that the estimation is very crude.

F. "Early" Response: X-Rays

When looking at the metabolic parameters listed in Table 1 in x-irradiated wild type cells, the picture is more obscure than with UV. Energy metabolism is always increased, even soon after irradiation (Kiefer, 1974) and will not be discussed here. The behavior of the other processes is shown in Figure 4: rRNA synthesis is depressed in a monotonic manner with dose, but in contrast to UV it appears <u>less</u> sensitive than survival. The incorporation of amino acids falls below control values for low doses, paralleling survival more or less, but it passes a minimum around 60 kR and rises again up to 100%. It may be that the data do not really reflect protein synthesis but also changes in amino acid transport, which may be altered by high x-ray doses due to membrane damage. The high amount of incorporation may also be caused by a complete breakdown of cellular regulation, since with a dose of 90 kR (100% incorporation of lys or phe), there is only about 5% survival so that we are essentially dealing with a large majority of nonsurviving cells.

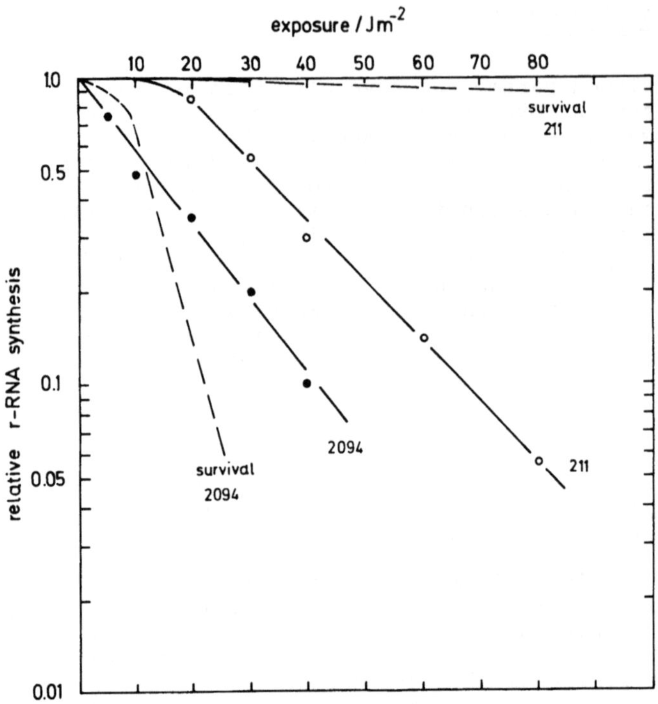

Fig. 2. UV-dose dependence of "early" rRNA synthesis in wild type and excision negative mutant 2094

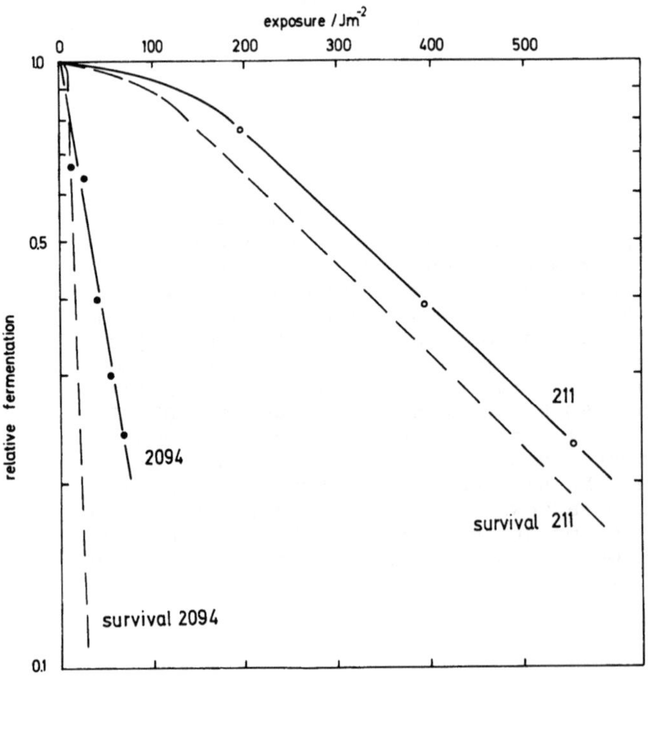

Fig. 3. Comparison of fermentation 2 h after UV exposure in wild type and sensitive mutants

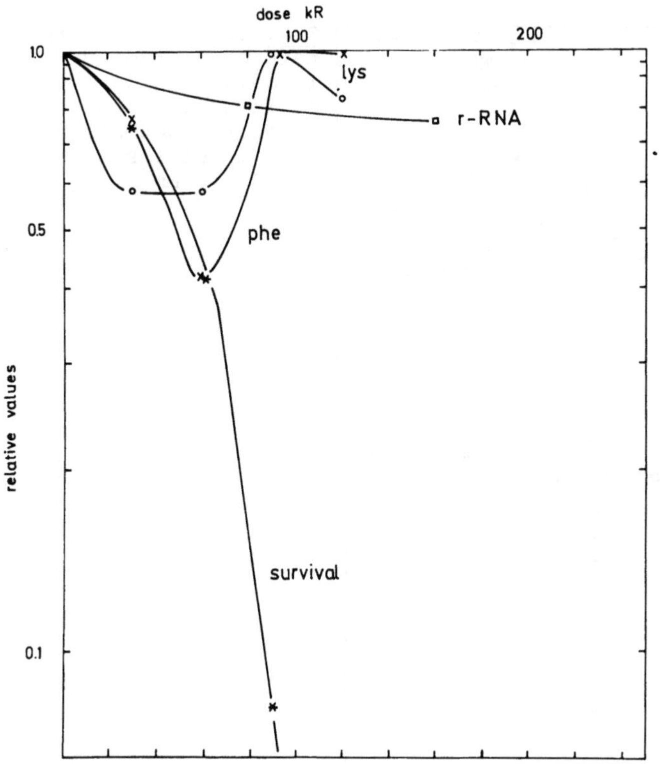

Fig. 4. "Early" response after x-irradiation in wild type cells

G. Changes during Prolonged Incubation

To follow synthetic activities in irradiated cells for longer incubation periods is a difficult task, since changes in the precursor pool may considerably obscure the real picture. This is the case, e.g., for protein synthesis as determined by the incorporation of lysine or phenylalanine (Laske, 1974), but plays only a minor role in rRNA synthesis (Koch, 1975). To obtain a rough estimate with regard to proteins, we measured the protein content per cell as a function of time after irradiation (Figs. 5 and 6). It may be seen that after UV as well as after x-irradiation, this parameter increases above control values after an initial slight depression. Lysine incorporation is also increased (Fig. 7) but this must not necessarily be due to increased protein synthesis as mentioned above. Pool size analysis, however, shows that cellular protein synthesis is in fact higher (Laske and Kiefer, to be published). Respiration proceeds also at a higher rate (Kiefer, 1974). rRNA synthesis and fermentation have been compared as a function of time in wild type and sensitive strains; respiration data are included for comparison (Figs. 8 and 9): it is seen that rRNA synthesis increases even above control values in repair proficient cells but drops drastically to very low levels in the sensitive strain 2094. Fermentation remains also below control level in these cells, but as in the wild type, respiration, on the other hand, is always increased. All these results taken together suggest that the coordination of synthetic activity breaks down as a consequence of the radiation insult. The picture that emerges indicates that the cellular radiation effect cannot be traced to single metabolic events.

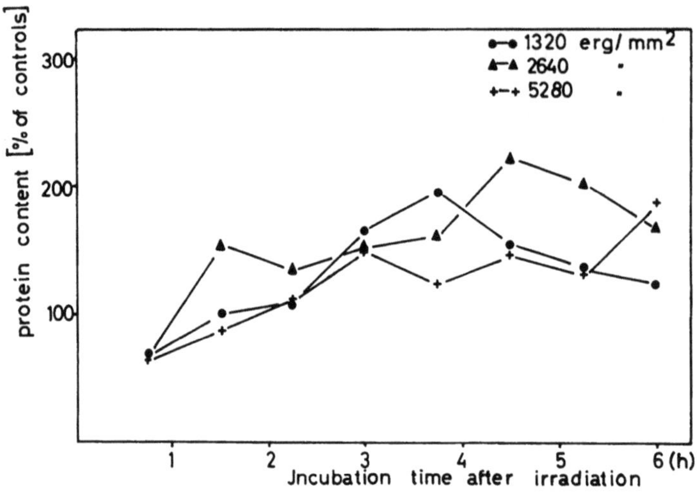

Fig. 5. Relative protein content of UV-irradiated wild type cells during post-exposure incubation (1 J/m^2 corresponds to 10 erg/mm^2)

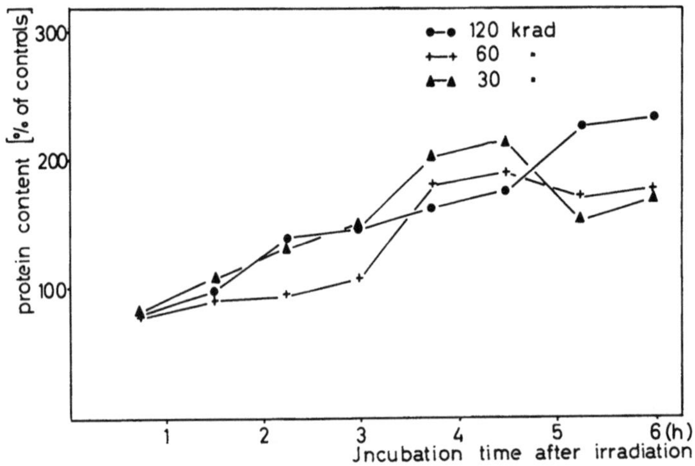

Fig. 6. Relative protein content of x-irradiated wild type cells

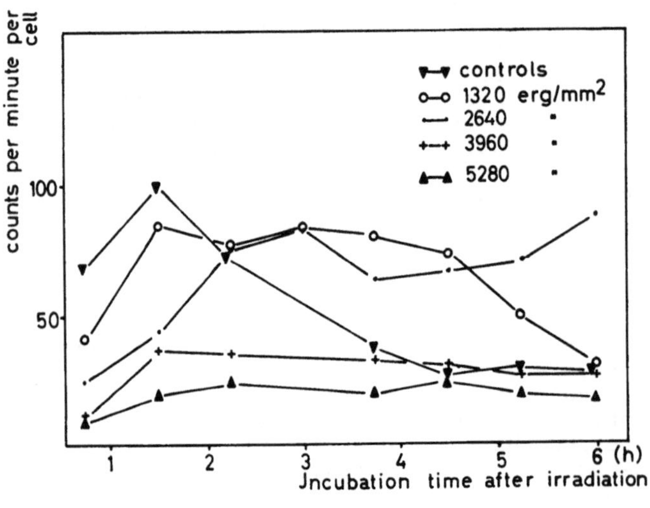

Fig. 7. Incorporation rate of ^{3}H-lysine as function of incubation time. Labelling time: 15 min; specific activity: 150 µCi/mmole

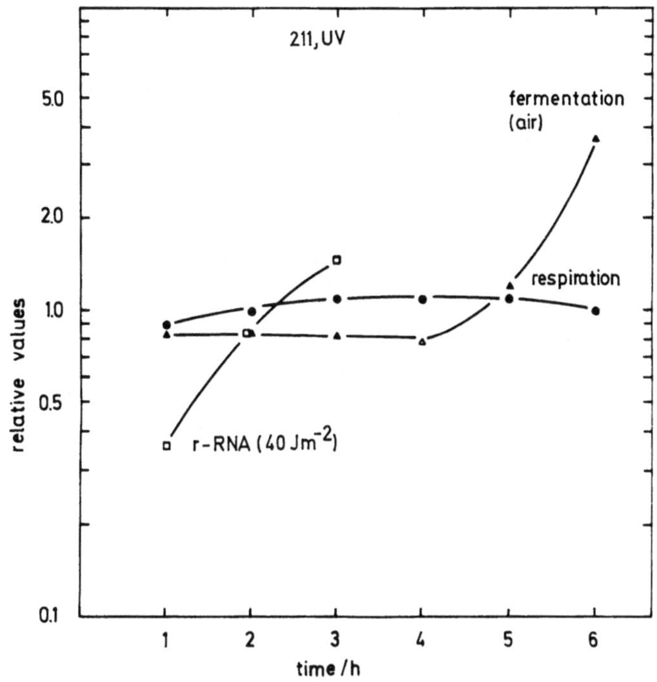

Fig. 8. Time course of respiration, fermentation, and rRNA synthesis per cell (UV, wild type). Dose for respiration and fermentation was 198 J/m^2

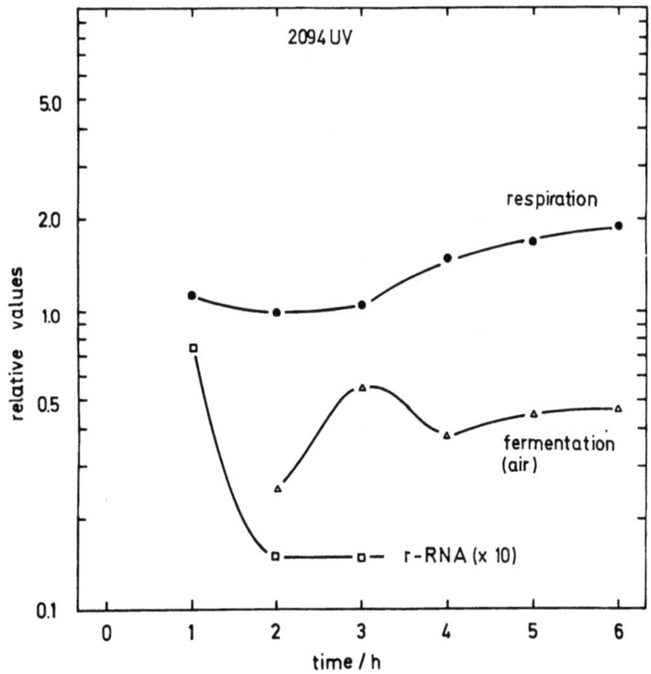

Fig. 9. As in Figure 8, but for sensitive strains. Doses: rRNA: 40 J/m^2; respiration and fermentation: 11 J/m^2

H. Discussion

The investigations partly reported in this paper were started several years ago to assess the behavior of key metabolic processes in irradiated cells. Yeast was chosen as model object because of its eukaryotic structure, its relative experimental simplicity, and the availability of a wide variety of mutants. An important shortcoming is the lack of thymidine kinase (Grivell and Jackson, 1968), which makes the selective incorporation of labelled thymidine and hence a simple investigation of DNA synthesis, impossible. Recent developments (see M. Brendel et al., 1975) promise an amelioration of this situation.

Four major points appear to evolve from our investigations:

1. The metabolic processes measured at an early stage after irradiation are differentially sensitive to irradiation. Their sensitivities may in some instances be related to the size of the structural genes coding for the relevant products that determine the cellular activity.

2. If isosurvival doses are used as a means of comparison, the effects of UV and x-rays are strikingly different. Metabolic parameters seem always to be less sensitive with x-rays.

3. During incubation after irradiation, most metabolic activities are increased in irradiated wild type cells over control values if calculated on a per cell basis.

4. The sensitive strain investigated shows similar effects at the early stage (but with lower doses) as compared to the wild type. On incubation, however, there is no increase in cellular activity of those processes that presumably require de novo synthesis.

It is not possible from the present observations to relate cell death in our cells to one particular metabolic parameter. rRNA synthesis, which is very sensitive to UV light if measured at an early stage, recovers its activity during postexposure incubation in the wild type. This is not the case, however, in the sensitive strain 2094, suggesting that repair of damage is a necessary prerequisite for the reattainment of normal cellular functionality. Fermentation, however, shows the same pattern. Respiration is always increased above control values. Taking all this together we feel that irradiation causes a general disturbance of cellular metabolic regulation. It may return to normality if part of the initial damage is repaired but may prove fatal for repair-deficient cells. The pronounced differences between the action of UV and x-rays cannot be explained. It cannot be excluded that in the latter case damage to cell constituents other than DNA may additionally be important. The "derailment" of cellular metabolism could be due to derepression of operons that are normally not active. This has been suggested previously (Kiefer, 1973; Swenson, 1975) but was not substantiated by experimental evidence. The recent finding of the synthesis of a specific protein ("protein x") in irradiated cells, not found in control cells (Segdwick, 1975), makes the hypothesis more likely. It is not to be expected, however, that radiation causes derepression directly because this ought to show up when assaying the induction of repressed operons in irradiated cells. In this instance evidence for derepression could not be unequivocally obtained (Gocke and Kiefer, 1975). A possible way could be the induction by the product of radiation-induced degradation (Gudas and Pardee, 1975) or the reduced synthesis of corepressors. Obviously, more research is needed to explore these possibilities. Radiation offers in this context a tool for the study of intracellular regulation.

I. Summary

To assess the significance of metabolic processes in the development of cellular radiation damage and their relation to repair phenomena, we investigated the following parameters in irradiated yeast strains of different sensitivities: protein content per cell, rate of protein synthesis, rRNA synthesis, respiration and alcoholic fermentation. Presumably the first effect of UV exposure is a severe depression of rRNA synthesis per cell within the first hour after irradiation. It is resumed at later times and exceeds that of the control in wild type, but not in sensitive mutant strains. All other parameters showed an increase after irradiation following a transient depression at an early stage of postexposure incubation if calculated on a per cell basis. In some cases this is also true for sensitive mutants, suggesting that the existence of repair processes is not causative for increased cellular metabolism after a radiation insult. We concluded from our experiments that radiation causes a general disturbance in cellular control processes that may be overcome in repair proficient, but may lead to inactivation in repair deficient cells. UV and x-ray results are compared.

References

Bos, R.C., van den, Retél, J., Planta, R.J.: The size and location of the ribosomal RNA segments in ribosomal precursor RNA of yeast. Biochim. Biophys. Acta (Amst.) 232, 494-508 (1971)

Brendel, M., Fäth, W.W., Toper, R.: DNA-specific labelling by deoxyribonucleoside 5'-monophosphates in *Saccharomyces cerevisiae*. In: Radiation and Cellular Control Processes (Kiefer, J., ed.). Berlin-Heidelberg-New York: Springer, 1976, pp. 35-40

Game, J.C., Cox, B.X.: Allelism tests of mutants affecting sensitivity to radiation in yeast and a proposed nomenclature. Mutation Res. 12, 328-331 (1971)

Gocke, E., Kiefer, J.: Inducible enzyme synthesis in irradiated yeast. In: Radiation and Cellular Control Processes (Kiefer, J., ed.). Berlin-Heidelberg-New York: Springer, 1976, pp. 59-67

Grivell, A.R., Jackson, J.F.: Thymidine kinase: evidence for its absence in *Neurospora crossa* and some other microorganisms, and the relevance of this to the specific labelling of deoxyribonucleic acid. J. Gen. Microbiol. 54, 307-317 (1968)

Gudas, L.J., Pardee, A.B.: Model for regulation of *Escherichia coli* DNA repair functions. Proc. Nat. Acad. Sci. USA 72, 2330-2334 (1975)

Hackett, P.B., Sauerbier, W.: Radiological mapping of the ribosomal RNA transcription unit in *E. coli*. Nature (Lond.) 251, 639-641 (1974)

Hatchard, C.G., Parker, C.A.: A new sensitive chemical actinometer. II. Potassium ferrioxalate as a standard chemical actinometer. Proc. Roy. Soc. A 235, 518-536 (1956)

Kiefer, J.: Metabolic changes in *Saccharomyces cerevisiae* after UV- and X-irradiation. Proc. 3rd Spec. Symp. on Yeast (Suomalainen, H., Waller, C., eds.). I, 46 (1973)

Kiefer, J.: Energy metabolism in yeast cells after UV- and X-irradiation. Int. J. Radiat. Biol. 26, 167-179 (1974)

Koch, H.: Synthese der ribosomalen Ribonukleinsäure nach Bestrahlung in stationären Hefezellen unterschiedlicher Empfindlichkeit. Ph.D. Thesis, University of Giessen, 1975

Koch, H., Kiefer, J.: RNA synthesis in irradiated diploid yeast at 50% survival level. Int. J. Radiat. Biol. 21, 223-234 (1972)

Laske, B.: Untersuchungen zur Proteinsynthese in bestrahlten Hefen. Diploma Thesis, University of Giessen, 1974

Lowry, O.H., Rosebrough, N.J., Farr, A.L., Randall, R.J.: Protein measurement with the folin phenol reagent. J. Biol. Chem. 193, 265-275 (1951)

Resnick, M.A., Setlow, J.K.: Repair of pyrimidine dimer damage induced in yeast by ultraviolet light. J. Bacteriol. 109, 979-989 (1972)

Segdwick, S.G.: Ultraviolet inducible protein associated with error prone repair in *E. coli* B. Nature (Lond.) 255, 349-350 (1975)

Swenson, P.A.: Physiological response of *Escherichia coli* to far-ultraviolet radiation. Photochem. Photobiol. Revs., in press

Udem, S.A., Warner, J.R.: Ribosomal RNA synthesis in *Saccharomyces cerevisiae*. J. Molec. Biol. 65, 227-242 (1972)

Unrau, P., Wheatcroft, R., Cox, B.S.: Methods for the assay of UV light induced pyrimidine dimers in *Saccharomyces cerevisiae*. Biochim. Biophys. Acta (Amst.) 269, 311-321 (1972)

Wickerham, L.J.: Taxonomy of yeasts. USDA Tech. Bull. 1029, p. 1-56 (1951)

7. Synthesis of the Inducible Enzyme Arginase in Irradiated Yeast Cells

E. Gocke and J. Kiefer

A. Introduction

For a conclusive description of the action of radiation on single cells, many links between energy absorption and the inability to form a colony are still missing. Studies of physiologic responses of irradiated cells should be helpful in the determination of the events that ultimately may lead to cell death. As part of our investigation of metabolic processes in irradiated yeast cell (see preceding paper), it seemed worthwhile to test if - and possibly how - irradiation interferes with the regulation of enzyme synthesis.

In this work we studied therefore the radiation effects on an enzyme, the regulation of which had been extensively investigated by Wiame and co-workers (Wiame, 1973). The activity of this enzyme arginase can be manipulated by supplementing the medium either with NH_4^+ ions or arginine as the sole nitrogen source. Arginase hydrolyzes arginine to ornithine and urea.

B. Material and Methods

The wild type strains (diploid and haploid) were obtained from W. Laskowski (S 211, S 211-1a). The *rad2* mutant (S 2094) was obtained from D. Averbeck. This strain is excision negative (Resnick and Setlow, 1972) and therefore extremely ultraviolet (UV)-sensitive. The medium is basically that of Wickerham (1951). Arginase is induced when the sole nitrogen source is arginine (2 g/l) and repressed when it is $(NH_4)_2SO_4$ (5 g/l).

As UV light source, a Hg-low pressure lamp (Osram HNS 12) was used. The fluence rate was determined to be 1.45 J/m^2/sec by chemical actinometry (Hatchard and Parker, 1956) and a fluxmeter (Fa. Hewlett-Packard 8330A).

For 100 kVp x-irradiation (Fa. Philips, Type FA 100/3, operated at 85 kV, 30 mA), the dose rate was 48 krad/min. The cells were grown in liquid medium at 30°C under rapid shaking for approximately 15 h. They were harvested in exponential phase of growth at a density of about 10^7 cells/ml and washed once with cold water. The cells were irradiated in aqueous suspension under rapid stirring and then resuspended in liquid medium (30°C). Samples were taken at given times and diluted in iced water, centrifuged, washed once, and stored at -20°C. The cells were permeabilized by freeze drying. This procedure renders the cell walls permeable for low-molecular-weight substances. Arginase activity assay was performed as described by Messenguy et al. (1971), except that urea was determined by color development after adding a mixture of diacetylmonoxime/thiosemicarbazide (Boehringer Biochemica Manual) by spectrophotometric measurement (Zeiss

PMQ 3) at 546 nm. The amount of urea formed is a measure of the arginase activity.

Cell concentration was determined by hemacytometer counts.

C. Results and Discussion

I. Dose Dependence

When exponentially growing cells are transferred from a repressing into an inducing medium, one observes a rapid increase of the enzyme activity in the culture. Irradiation reduces this enzyme production. Figure 1 shows the results after UV exposure of the wild type (S 211) and the *rad2* mutant (S 2094), which is considerably more UV-sensitive with respect to colony forming ability (CFA) (Averbeck, 1970). But the higher radiosensitivity of strain 2094 becomes apparent only after 1 1/2 h. At this time enzyme synthesis is stopped almost completely. Within the first hour there is, for the same physical doses, no significant difference in enzyme production between the two strains. For x-rays (Fig. 2), the applied doses have to be considerably higher - with respect to cell killing - to produce a measurable effect on the initial enzyme synthesis. While the maximal UV fluence of 430 J/m^2 (Fig. 1) reduces the CFA of the wild type to about 5%, the lowest x-ray dose of 120 krad (Fig. 2) leads to a survival of less than 1%.

To obtain better information about the dose dependence of the induction process, the relative enzyme synthesis during the first 20 min is plotted on semilog scale as a function of dose (Fig. 3). For UV the data obtained with the haploid wild type (S 211-1a) are also included. The curves clearly indicate an exponential type of dose dependence. The UV data show that the increase in activity during the first 20 min after irradiation does not depend on the radiosensitivity of the strain.

The results can be explained on the following basis: because radiation induces lesions into the DNA strands, there is a certain chance for the arginase gene to be hit, depending on size and structure of the gene. Brunschede and Bremer (1969) found that in *E. coli*, UV-induced dimers are critical lesions for transcription, i.e., the RNA polymerase cannot pass the dimer.

From in vitro experiments it has been shown (Hagen et al., 1970) that x-ray-induced single-strand breaks cause the RNA polymerase to fall off the template, and a shorter messenger molecule is produced.

This suggests that an arginase gene containing a radiation-induced lesion is not transcribed into an usable messenger, thus causing a reduction of total arginase production.

As one averages over a great number of cells ($> 10^7$) there will be a certain fraction of undamaged genes. This fraction is given by simple hit theory (e.g., Dertinger and Jung, 1969) to be $W_{(D)} = \exp(-KD)$, where $1/K$ is the dose (D_{37}) for which on the average one hit per gene occurs.

Assuming that radiation does not interfere with other steps involved in protein synthesis, e.g., initiation of transcription and translation, $W_{(D)}$ should also be the relative arginase synthesis for the dose D (the arginase production for the dose D divided by the arginase

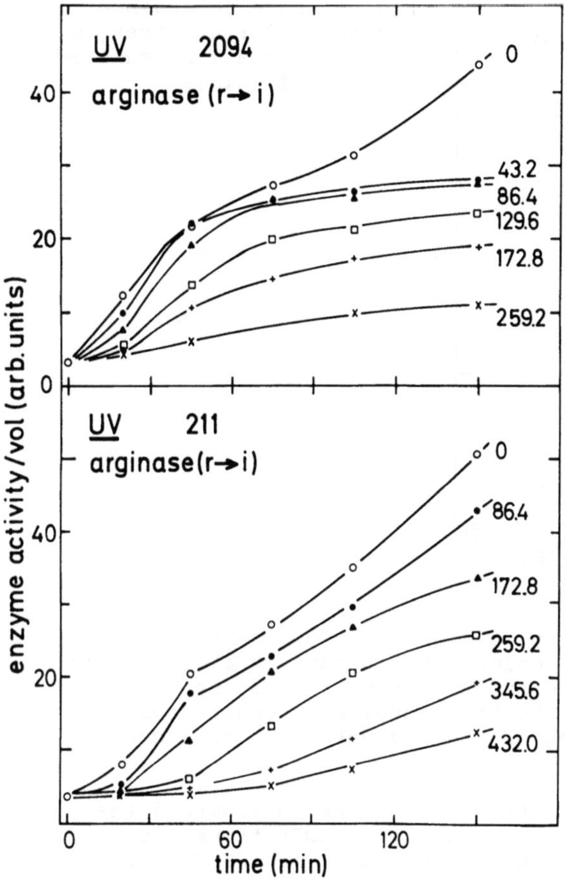

Fig. 1. Influence of UV light on increase of arginase activity in strain 211 (bottom) and strain S 2094 (top). Fluences are given in J/m^2 (see text)

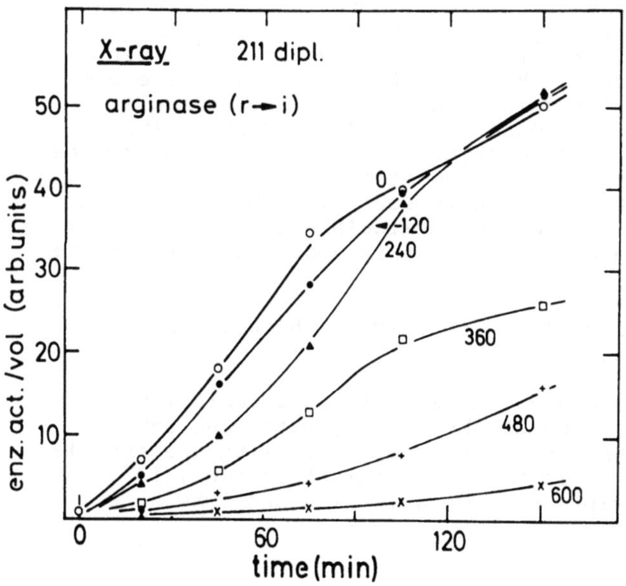

Fig. 2. Influences of x-rays on increase of arginase activity in strain 211. Doses in krad (see text)

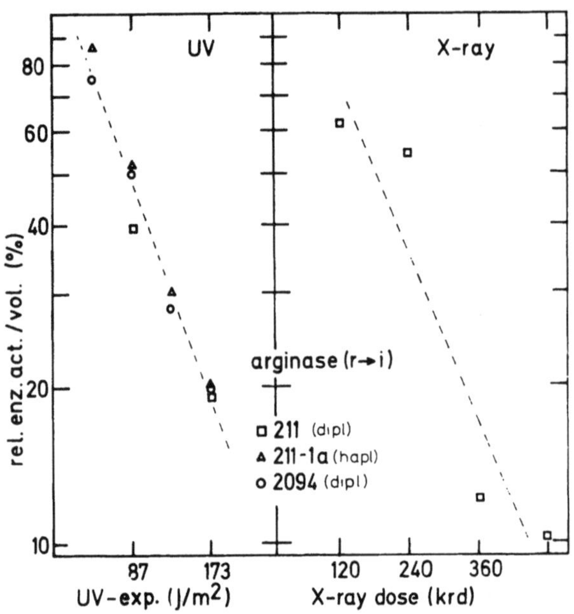

Fig. 3. Dose dependence of synthesis during first 20 min. Values are normalized to control values

production of the control) as given in Figure 3. This consideration does not include a repair of the lesions. But because the curves are the same for the wild type and the excision repair negative *rad2* mutant, it is apparent that during the time involved, repair does not reduce the number of damaged sites on the arginase gene to a considerable extent.

On the basis one can now calculate from the slopes in Figure 3 the number of hits. For UV exposure a value of 1.1×10^{-2} transcription-terminating lesions per arginase gene and per J/m^2 is found; for x-rays the value is 6×10^{-6} lesions per gene and rad.

For high UV fluences the number of induced dimers, the main UV lesions, can be determined chemically. Unrau et al. (1972) found a value of 240 dimers induced in the total haploid yeast genome per J/m^2. Using this value and knowing the size of the arginase gene one can test the validity of the foregoing assumptions. The molecular weight of the arginase enzyme is found to be 114,000 daltons (Messenguy et al., 1969). Taking an average amino acid weight of approximately 110 daltons, the arginase gene would consist of a stretch of about 3000 nucleotide base pairs. Given a statistical distribution of the dimers along the DNA molecules, one would calculate a value of 2.5×10^{-2} dimers per arginase gene and J/m^2. That is about twice the value obtained in this experiment.

The size of the gene might actually be smaller if the enzyme is a multimer of identical subunits.

From data of Hagen (1973) one would calculate for x-irradiation a number of 1.1×10^{-6} single-strand breaks per arginase gene and rad and about 6 to 9 times more base damages. Comparison with the experiment indicates that contrary to in vitro experiments (Hagen, 1973), also base damages are transcription-terminating lesions. The same result was obtained by Koch (1975).

These considerations show that the described technique permits the determination of radiation-induced lesions for moderate doses by a very simple method. Somewhat modified experiments could be helpful for many different studies of insults to the genetic material.

II. Effects on Regulation

For moderate doses (CFA ≥ 15% for wild type) the enzyme synthesis was followed for up to 8 h after exposure. Since irradiation blocks cell division (Fig. 4, also Faber and Kiefer, 1976), one has to consider another effect: after a short period the control culture contains considerably more cells than the irradiated ones. Also the number of gene copies should differ by approximately the same extent. Therefore one should expect a similar difference in the enzyme activity per unit volume, because fewer genes should imply fewer mRNA copies and fewer enzymes.

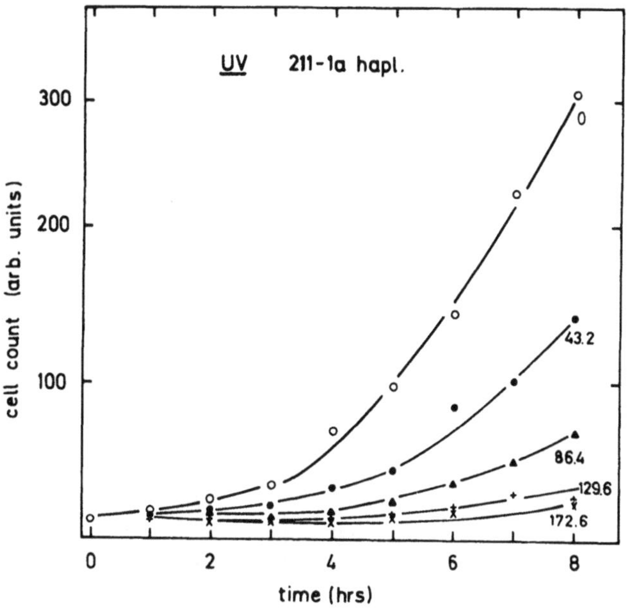

Fig. 4. Influence of UV on growth of strain 211-1a. Fluences in J/m^2

On the other hand the irradiation does not completely inhibit synthesis, as shown previously. So the enzyme activity per cell should be higher in the irradiated cultures since the cells do not divide. In the following, some representative experiments are analyzed on a per volume and a per cell basis. Figure 5 shows at the bottom the arginase activity per unit volume, at the tope the activity per cell (haploid wild type). In this experiment the media before and after irradiation lacked arginine, so that arginase synthesis was repressed. A similar experiment was performed transferring the cells after irradiation into medium containing arginine as the sole nitrogen source (arginase induced) (Fig. 6). After about 1 1/2 h, the enzyme activity in the control cells reached a value about ten times higher than that reached during the repressed state of synthesis. But apart from this the picture looks similar: per unit volume, the activity is smallest in the culture that received the highest fluence. If related to cell

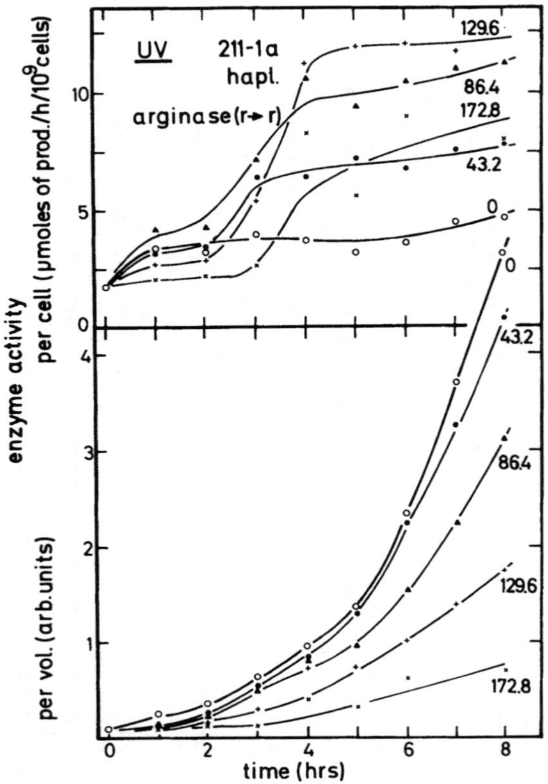

Fig. 5. Influence of UV light on arginase synthesis in strain 211-1a. Fluences in J/m^2. The arginase synthesis is repressed before and after irradiation (medium contains $(NH_4)_2SO_4$)

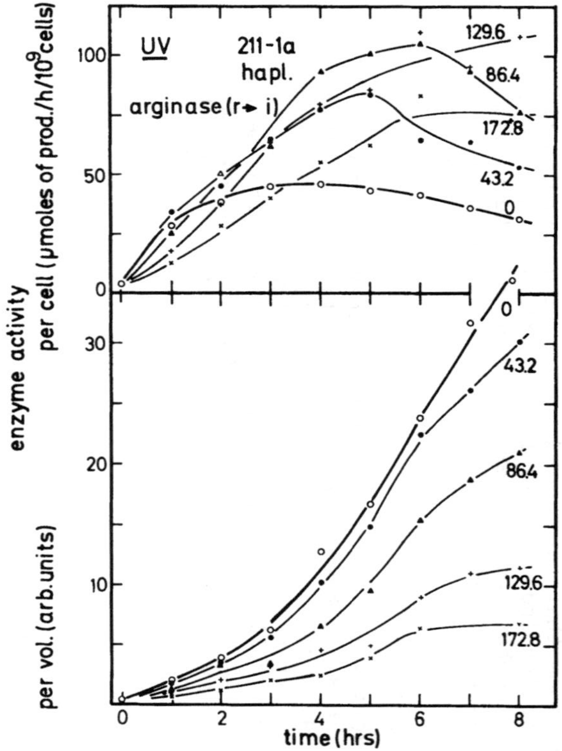

Fig. 6. Same as for Figure 4 except medium after irradiation contains arginine (arginase induced)

number, the activity is smallest in the control cells and highest in cultures that received intermediate fluences. The initial increase is slowed down with increasing fluence.

X-ray doses up to 100 krad (20% survival for diploid wild type) have almost no effect on enzyme synthesis during the first 5-6 h (Fig. 7, bottom). This is astonishing, since the difference in cell number reaches about a factor of 6 while enzyme production remains the same. Only after 6 h does enzyme synthesis in the irradiated cells show a rather sudden decline.

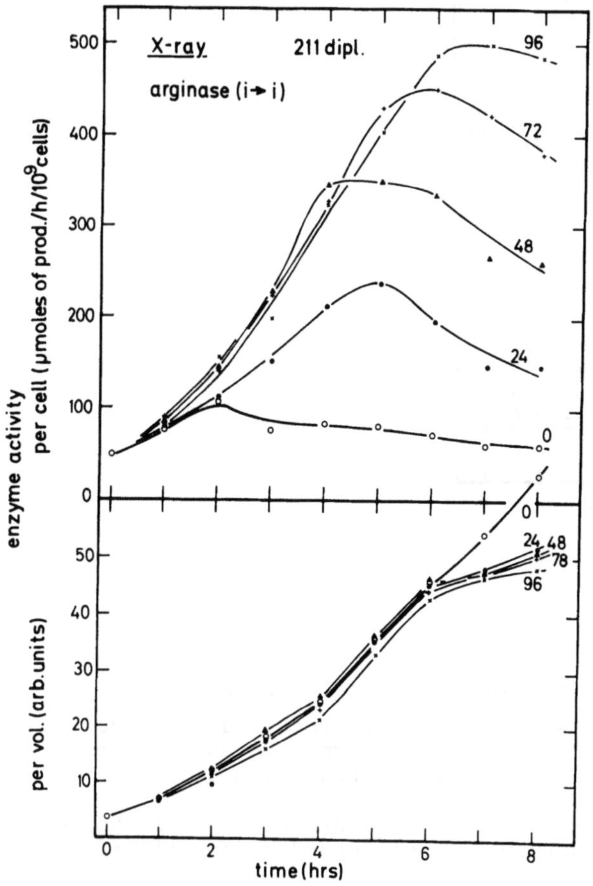

Fig. 7. Influence of x-rays on arginase synthesis in strain 211. Doses in krad. Arginase before irradiation is repressed, afterwards induced

Figure 8 shows the result obtained by exposing the *rad2* mutant to UV light. Also here the applied doses do not noticeably reduce initial enzyme synthesis. After about 1 1/2 h, however, the irradiated cultures almost completely stop arginase production (see also Fig. 1). Supposedly the whole cell metabolism breaks down at this time.

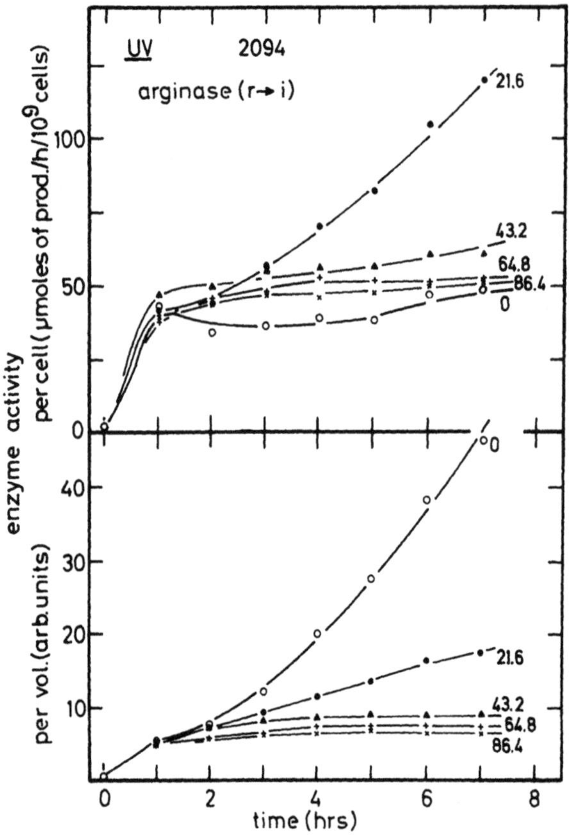

Fig. 8. Influence of UV lights on arginase synthesis in strain S 2094. Fluences in J/m^2. Arginase before irradiation is repressed, afterwards induced

D. Conclusions

In conclusion we should like to emphasize the following points:

a) Irradiation reduces arginase synthesis in the total culture. But due to simultaneous division inhibition, the enzyme content per cell may nevertheless become severalfold higher in irradiated cells.

b) The dose-dependent inhibition of the initial synthesis seems to be a function of gene size and structure and not dependent on radiation sensitivity of the strain.

c) For comparable levels of survival, x-irradiation is much less effective in reducing the initial synthesis.

d) Irradiation does not derepress the repressed arginase gene to a greater extent.

E. Summary

The enzyme arginase is regulated in yeast by its substrate arginine. The activity was measured in UV- and x-irradiated cells under repressing and inducing conditions. The results were assessed on a per cell and a per volume basis. The data show that shortly after irradiation the synthesis is dependent on the size of the arginase gene, while only at later times do differences in radiation sensitivity of the strains become effective.

References

Averbeck, D.: Isolierung und Charakterisierung von drei UV-sensiblen *Saccharomyces*-Mutanten. Ph.D. Thesis, Freie Universität Berlin, 1970

Brunschede, H., Bremer, H.: Protein synthesis in *E. coli* after irradiation with ultraviolet light. J. Molec. Biol. 41, 25-38 (1969)

Dertinger, H., Jung, H.: Molekulare Strahlenbiologie. Berlin-Heidelberg-New York: Springer, 1969

Faber, G., Kiefer, J.: Budding and division delay in irradiated yeast. In: Radiation and Cellular Control Processes (Kiefer, J., ed.). Berlin-Heidelberg-New York: Springer, 1976, pp. 264-270

Hagen, U.: Strahlenwirkung auf Struktur and Funktion der Desoxyribonukleinsäure. Biophysik 9, 279-289 (1973)

Hagen, U., Ulrich, M., Peterson, E.E., Werner, E., Kröger, H.: Enzymatic RNA synthesis on irradiated DNA. Biochim. Biophys. Acta (Amst.) 199, 115-125 (1970)

Hatchard, C.G., Parker, C.A.: A new sensitive chemical actinometer II. Proc. Roy. Soc. A 235, 518-536 (1956)

Koch, H.: Synthese der ribosomalen Ribonukleinsäure nach Bestrahlung in stationären Hefezellen unterschiedlicher Empfindlichkeit. Ph. D. Thesis, Strahlenzentrum, Gießen, 1975

Kiefer, J., Koch, H., Laske, B., Waller, H.: Metabolism of yeast cells after x-ray and UV-irradiation. In: Radiation and Cellular Control Processes (Kiefer, J., ed.). Berlin-Heidelberg-New York: Springer, 1976, p. 48-58

Messenguy, F., Wiame, J.M.: The control of ornithine transcarbomylase activity by arginase in *Saccharomyces cerevisiae*. FEBS letters 3, 47-49 (1969)

Messenguy, F., Penninchx, M., Wiame, J.M.: Interaction between arginase and ornithine carbomyltransferase. Eur. J. Biochem. 22, 277-286 (1971)

Resnick, M.A., Setlow, J.K.: Repair of pyrimidine dimer damage induced in yeast by ultraviolet light. J. Bacteriol. 109, 979-989 (1972)

Unrau, P., Wheatcroft, R., Cox, B.: Methods for the assay of UV light-induced pyrimidine dimers in *Saccharomyces cerevisiae*. Biochim. Biophys. Acta (Amst.) 269, 311-321 (1972)

Wiame, J.M.: The regulation of the biosynthesis and degradation of arginine in *Saccharomyces cerevisiae*. In: Proc. 3rd Intern. Spec. Symp. on Yeast, Otaniemi Congress Committee, 1973, pp. 307-330

Wickerham, L.J.: Taxonomy of yeast. US Dept. Agr. Techn. Bull. No. 1029 (1951)

8. Synthesis of RNA after Irradiation in Resistant and Sensitive Strains of the Slime-Mould *Dictyostelium discoideum*

P. E. Bryant

A. Introduction

There is considerable evidence to support the notion that RNA synthesis is involved in the recovery mechanisms of both bacteria and eukaryotic cells after ionizing radiation. For example Elkind et al. (1967) showed that very low concentrations of actinomycin D, an inhibitor of RNA synthesis, reduced split-dose recovery in Chinese hamster cells in culture. Cramp and Bryant (1975) showed that rifampicin, a specific inhibitor of RNA polymerase in bacteria, increased the sensitivity of *Escherichia coli* if given shortly before or after irradiation. Also there are a number of reports in the literature of increases in the rate of RNA synthesis after x-irradiation, for example in bacteria (Klamerth, 1964; Hudnik-Plevnik, 1965), in mammalian tissues (Hidvégi et al., 1966; Simic et al., 1967), and in an experimental tumour (Matsudaira et al., 1967). Koch and Kiefer (1972) showed that in yeast, increased RNA synthesis occurred after x- or ultraviolet (UV)-radiation but not after α-irradiation, and suggested that the increased synthesis could be part of the intracellular recovery process. Another line of research has demonstrated that supralethally irradiated amoebae can be rescued by the injection of polyribosomes (that is, ribosomes attached to messenger RNA) from unirradiated donor amoebae (Daniels and Breyer, 1970).

The work reported here supports the postulate that RNA synthesis is involved in the recovery of cells from radiation damage and shows a correlation between the appearance of a shoulder on the survival curve and the increased synthesis of RNA in slime-mould cells.

B. Materials and Methods

I. Cells and Culture Methods

Two cell lines of *Dictyostelium discoideum* were used in the work, the resistant wild-type line NC4, and the very sensitive line γs-13 (Deering et al., 1970). *Dictyostelium* is a cellular slime mould belonging to the Acrasiales, distinct from the Myxomycetes or true slime moulds, which form multinucleate plasmodia. Cells of *Dictyostelium* remain discrete at all times during the life cycle (see Fig. 1). After the food supply of a growing population becomes exhausted the amoebae aggregate to form a migratory slug. In the absence of nutrients about 10^5 cells aggregate under the influence of cyclic-AMP. The slug or cell aggregate then differentiates into two major tissue types, stalk cells and spore cells. The stalk cells elongate rapidly and form a cellulose-encased rod forming the stalk with spore cells at the top. Spores germinate when placed in favorable conditions (high humidity, presence of oxygen and certain chemicals, which can be supplied by commercial peptones).

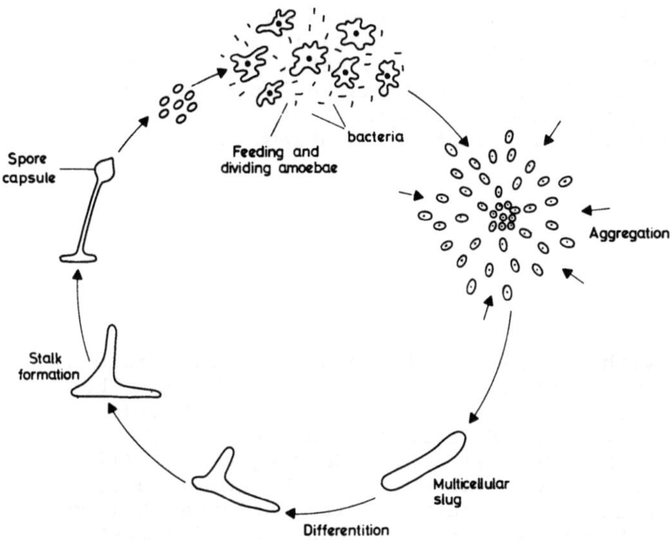

Fig. 1. Diagrammatic representation of the life cycle of *Dictyostelium discoideum*

For these experiments spores were germinated first on a bacterial lawn consisting of *E. coli* B/r, for about 30 h. Amoebae were then washed free of bacteria by differential centrifugation at 100 g in phosphate buffered saline (0.05 *M* Na_2HPO_4 : KH_2PO_4, 0.01 *M* NaCl, 0.01 *M* KCl, pH 6.5) and then grown in an aerated suspension of *E. coli* B/r (10^{10}/ml) to a concentration of about 5 x 10^6 amoebae per ml in buffered saline at 22°C. Under these conditions the cells grew with a doubling time of 3-5 h.

II. Method of Irradiation

Before irradiation bacteria were removed by differential centrifugation and concentrated to 2 x 10^7/ml. Cells were irradiated either with 8 MeV electrons or neutrons from the MRC cyclotron under aerobic conditions in a glass vessel. Cultures were kept chilled (~ 4°C) during irradiation, which in the case of neutrons lasted about 2 h.

III. Survival Assay

To assay for survivors, aliquots of diluted suspension were plated on nutrient agar together with about 10^9 bacteria. The bacteria formed an even lawn on which surviving clones grew and could be discerned as clear plaques after 2-5 days.

IV. Radioactive Labelling and RNA Extraction

Radioactive labelling was carried out in aerated suspensions of cells containing 2 x 10^7 cells/ml. ^{3}H-uridine or ^{3}H-adenine was added to a final concentration of 12 μCi/ml (400 μCi/m*M*). After labelling, cells were washed in ice-cold buffer-saline and lysed with buffered-saline citrate containing 1% sodium lauryl sulphate and 20 mg/ml bentonite (0.5 *M* NaCl, 0.02 *M* sodium citrate, 0.1 *M* Tris-HCl, pH 7.4). An equal volume of buffer-saturated phenol containing 0.1% 8-hydroxyquinoline was added and vortexed. The mixture was heated to 60°C for 5 min, cooled, and centrifuged at 6000 g for 10 min. The aqueous supernatant was removed and mixed with 2 volumes of ice-cold 100% ethanol. This was stored overnight at -20°C and then centrifuged at 1000 g for 10 min to pellet the RNA. The pellet was redissolved in an acetate buffer (0.05 *M* NaCl, 0.0001 *M* $MgCl_2$, 0.01 *M* Na acetate, pH 5.5) and reprecipitated with ethanol at -20°C. The RNA was pelleted and redissolved in acetate buffer; absorbance was measured at 260 nm. Radioactivity was assayed by liquid scintillation counting.

C. Results and Discussion

I. Survival Measurements

Survival curves obtained with the two strains are shown in Figure 2. The WT NC4 has a large shoulder with a D_{10} of about 300 krads. The sensitive strain γs-13 has a D_{10} of about 5 krads and no shoulder. The survival curve obtained for the WT strain with neutrons showed a much reduced shoulder on the survival curve. Deering et al. (1970) showed that the shoulder on the survival curve for WT cells could also be removed by post-irradiation treatment with caffeine at 1 mg/ml for 24 h although after 5 h treatment, the shoulder had almost disappeared.

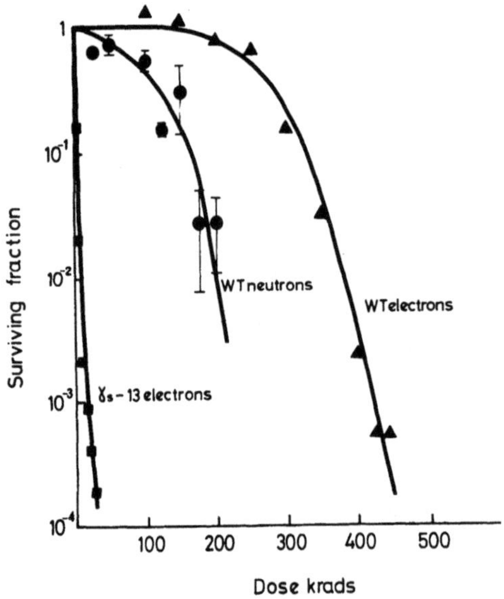

Fig. 2. Survival as function of dose for (▲) WT and (■) γs-13 and (●) WT cells irradiated with cyclotron neutrons. Vertical bars represent standard errors of mean surviving fractions

II. Measurements of RNA Synthesis

1. Using ³H-Uridine as a Precursor

I have measured RNA synthesis in these two strains after electron irradiation and in the WT strain after neutron irradiation and also after treatment with caffeine. It was found that when ³H-uridine was used to follow RNA synthesis, an apparent depression in rate of uptake was observed after irradiation (Fig. 3). This depression in rate of RNA synthesis appeared to be unrelated to cellular sensitivity,

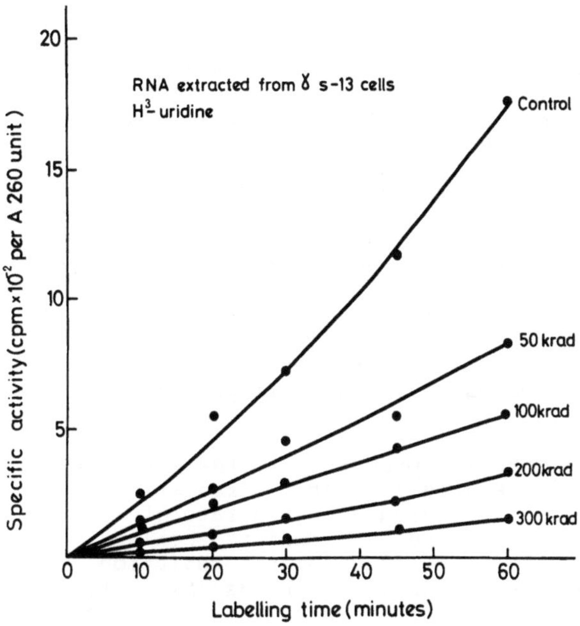

Fig. 3. Specific activity as a function of labelling time for γs-13 cells irradiated with electrons and labelled with ^{3}H-uridine

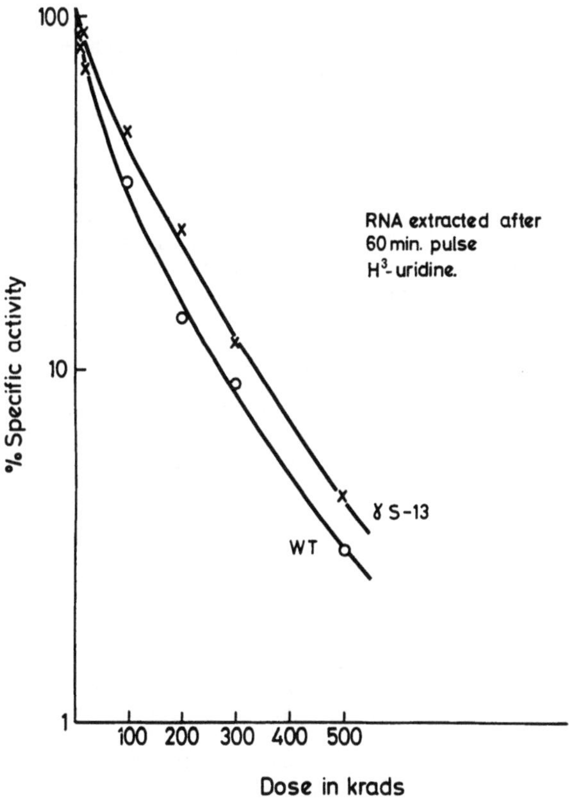

Fig. 4. % specific activity as function of dose for (O) WT cells (X) γs-13 cells labelled with ^{3}H-uridine for 60 min

both strains showing approximately the same depression of RNA synthesis as a function of dose (Fig. 4).

2. *Using ³H-Adenine as a Precursor*

Experiments with bacteria have shown that by altering conditions of growth before and after UV irradiation, the extent of accumulation of sublethal damage can be affected (Rudé and Alper, 1972). It has also been shown that these changes of culture conditions affect the synthesis of messenger RNA (Friesen, 1966; Gray and Midgeley, 1972). It was therefore of interest to separate messenger RNA from other species of RNA and label it to a high specific activity. This can be achieved by using ^{3}H-adenine since *Dictyostelium* messenger, like that of other eukaryotes, contains a region of polyadenylic acid at the 3' end (Fig. 5). In *Dictyostelium* this region is about 100 nucleotides long (Firtel et al., 1972). The function of the poly-A region is not known; however, it not only provides a method for labelling messenger to a high specific activity but also a means for separating it from other RNA species by attachment to polyuridylic acid immobilized on Sepharose-4B (Pharmacia Ltd).

The results using ^{3}H-adenine as a labelled precursor to RNA were very different from those obtained using ^{3}H-uridine. After irradiation (D_{10}, dose 300 kR) in the WT strain, an *increase* in the rate of RNA synthesis was observed above that for control unirradiated cells (Fig. 6). There was no similar increase observed for the radiation sensitive mutant γs-13 when given its D_{10} dose of 5 krads (Fig. 7), thus suggesting a possible correlation between the ability of cells to accumulate damage in sublethal form and ability to synthesize RNA more rapidly than normal after irradiation. However there was some increase above control rate in γs-13 when 300 krads were given. Treatment of WT cells with caffeine after irradiation reduced RNA synthesis to near the control level, caffeine alone reducing RNA synthesis below the control level.

III. Dose Effect Curves for RNA Synthesis

Figure 8 shows results of varying the dose of electrons to cells of both strains and the effect of various neutron doses on the synthesis of RNA. In the shoulder region it can be seen that there is substantially more RNA synthesis in the WT strain than in the sensitive γs-13. There was also less RNA synthesis after neutron irradiation. There appears to be a peak in the relative rate of RNA synthesis after 150 kR in the WT. These results again support the postulate that RNA synthesis is involved in the ability of cells to accumulate damage in sublethal form. Also the results with neutrons support the results of Koch and Kiefer (1972) who showed RNA synthesis not to be increased after high LET radiation in yeast whereas after x- or UV-radiation an increase in synthesis was observed.

IV. Analysis of RNA

I have attempted to analyze the total RNA samples to see whether one or other of the species becomes preferentially labelled after irradiation in the WT strain. Two methods were used: (1) sucrose gradient centrifugation and (2) poly-U Sepharose chromatography. The sucrose gradient method resolved RNA into 3 main peaks corresponding approximately to 28 S, 18 S and 4 S. In the case where ^{3}H-adenine had been used as a precursor the radioactivity profile did not correspond to the absorbance profile, indicating regions of very high specific activity that presumably corresponded with the messenger RNA. This was not the case when ^{3}H-uridine or ^{3}H-uracil was used as a precursor to

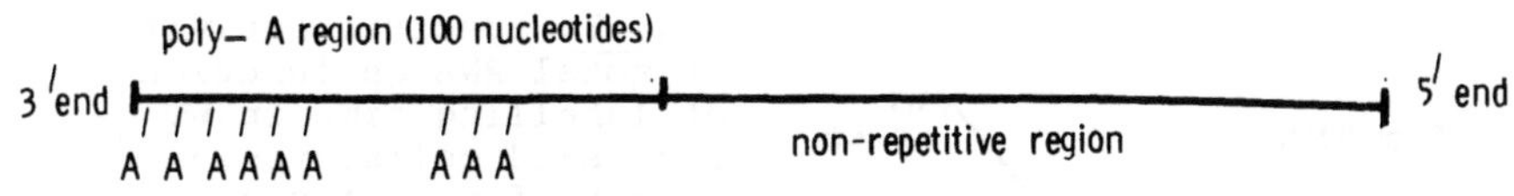

Fig. 5. Diagrammatic representation of *Dictyostelium* messenger RNA

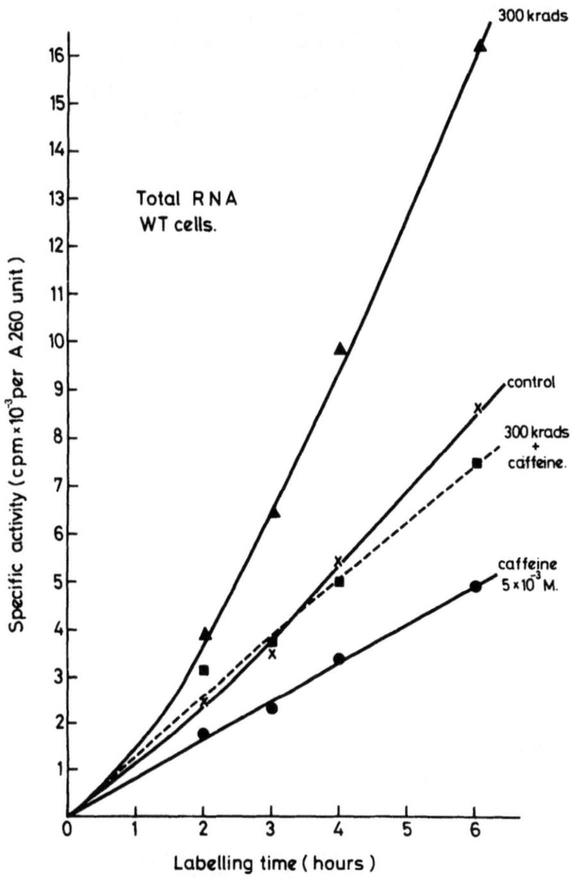

Fig. 6. Specific activity of total RNA as function of labelling time for (▲) WT cells given 300 krads; (X) unirradiated controls; (●) cells treated with caffeine alone; (■) cells given caffeine after irradiation. The concentration pf caffeine used was 5×10^{-3} *M*

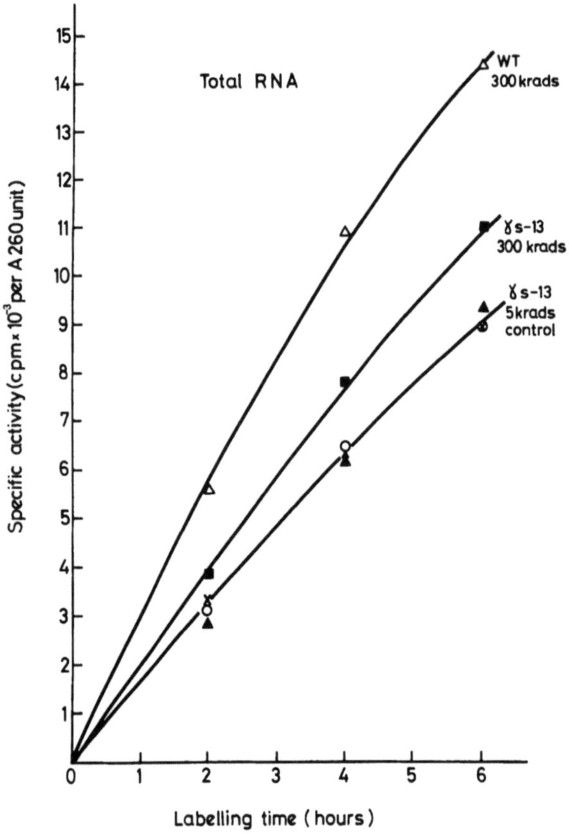

Fig. 7. Specific activity of total RNA as function of labelling time in WT and γs-13 cells. γs-13 results for controls have been normalized to correspond with control results for WT cells. Points for irradiated γs-13 cells have been adjusted by the same ratio. (O) WT controls (unirradiated); (X) γs-13 controls; (▲) γs-13 cells given 5 krads; (■) γs-13 cells given 300 krads; (Δ) WT cells given 300 krads

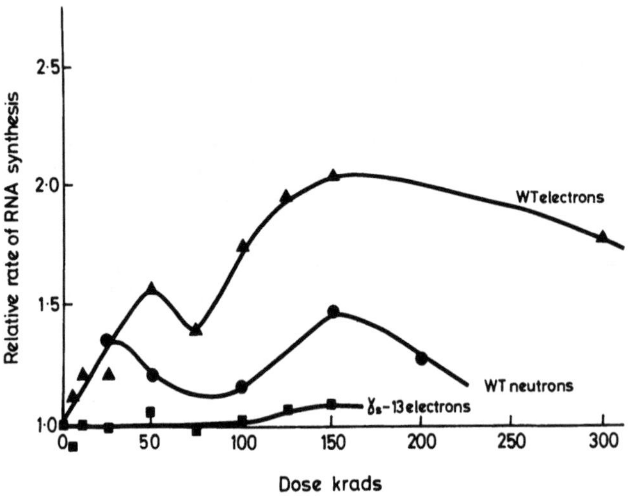

Fig. 8. Relative rate of total RNA synthesis as function of dose for WT cells; (▲) irradiated with electrons; (●) irradiated with neutrons; and (■) γs-13 cells irradiated with electrons

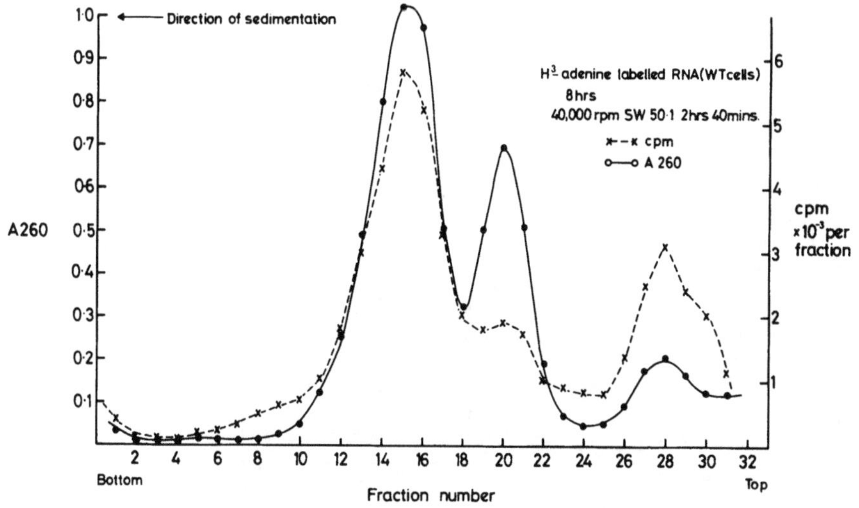

Fig. 9. Sucrose gradient sedimentation of RNA labelled with ^{3}H-adenine; (●) absorbance at 260 nm; (X) radioactive counts per fraction. Gradients were 5-20% sucrose in neutral acetate buffer

RNA (Fig. 10). If specific activity after ^{3}H-adenine labelling was calculated for each fraction, two main areas of high specific activity became evident (presumptive mRNA) (see Fig. 11). Mean specific activities calculated for each of these areas as a function of dose revealed a pattern similar to that for total RNA labelling, i.e., an increase after irradiation and inhibition of the increase by caffeine. This suggested that labelling of mRNA was increased after irradiation.

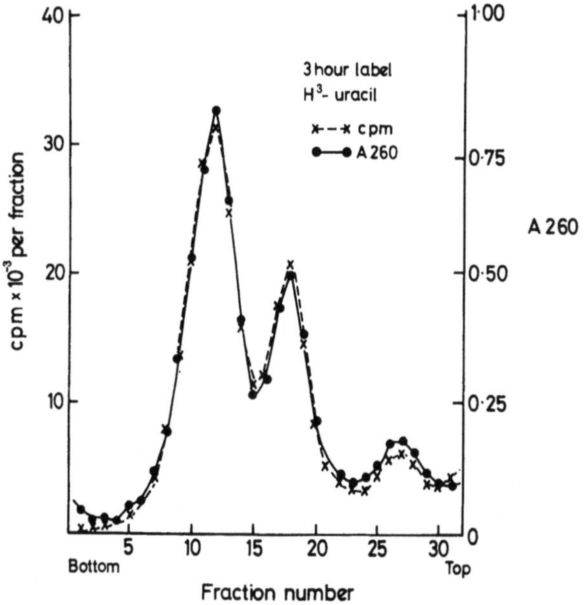

Fig. 10. Sucrose gradient sedimentation of RNA labelled with ^{3}H-uracil; (●) absorbance at 260 nm; (X) radioactive counts (5-20% sucrose in neutral acetate buffer)

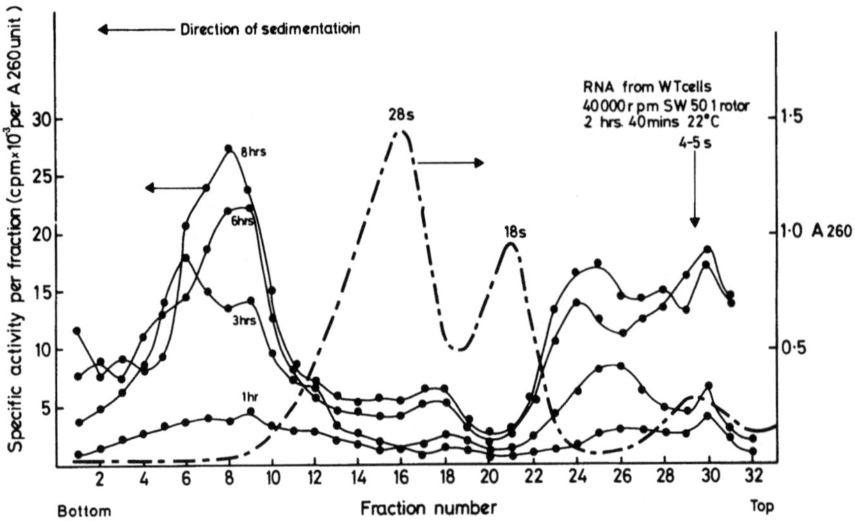

Fig. 11. Sucrose gradient sedimentation of RNA labelled with ^{3}H-adenine. (●) specific activity per fraction; dotted line represents absorbance at 260 nm. Different curves are for different times of labelling (5-20% sucrose in neutral acetate buffer)

Method 2 using poly-U Sepharose 4B chromatography is indicated diagramatically in Figure 12. Samples separated by poly-U chromatography were desalted using Sephadex G50 (Pharmacia Ltd); measurements were made of absorbance at 260 nm and radioactive counts using a liquid scintillation system. Preliminary results using this method (Fig. 13) show that the combined ribosomal and transfer RNA fraction (eluted in 25% formamide) followed labelling in the total RNA sample. The labelling in this first fraction was much slower than in the second mRNA fraction, and the rate appeared to increase gradually with time. In the second mRNA fraction, eluted with 90% formamide, the results were more difficult to quantitate due to the very small amounts of RNA separated. There was however clearly an increase after irradiation over the control values (Fig. 14). The results after caffeine treatment were less clear; the rate apparently increasing after 3 h labelling time to nearly the same value at 4 h as for the irradiated sample.

D. Conclusions

1. Quite different kinetics of labelling were obtained using ^{3}H-uridine and ^{3}H-adenine. It is possible that irradiation may be affecting the transport of uridine across the cell or nuclear membrane, thereby apparently reducing the apparent incorporation of uridine. Whatever the reason for the reduced uptake of uridine after irradiation, this reduction was found to be unrelated to the sensitivity of the cells to radiation or to their ability to accumulate damage in sublethal form. This difference in uptake between the two labels points to the need for extreme caution when deducing facts about the mechanism of radiation action based on incorporation studies.

2. The results of experiments with ^{3}H-adenine suggest a strong correlation between ability to accumulate damage in sublethal form and rapid RNA synthesis after irradiation. Where a shoulder was not apparent on

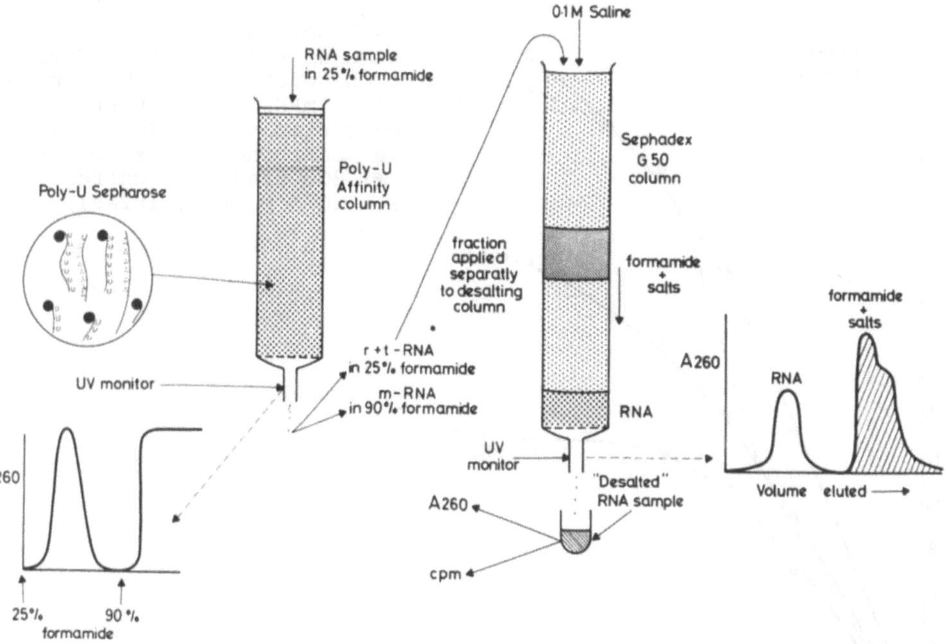

Fig. 12. Diagrammatic representation of mRNA separation technique. Samples were applied to poly-U column in either 25% or 90% formamide made with a buffer containing sodium lauroyl sulphate

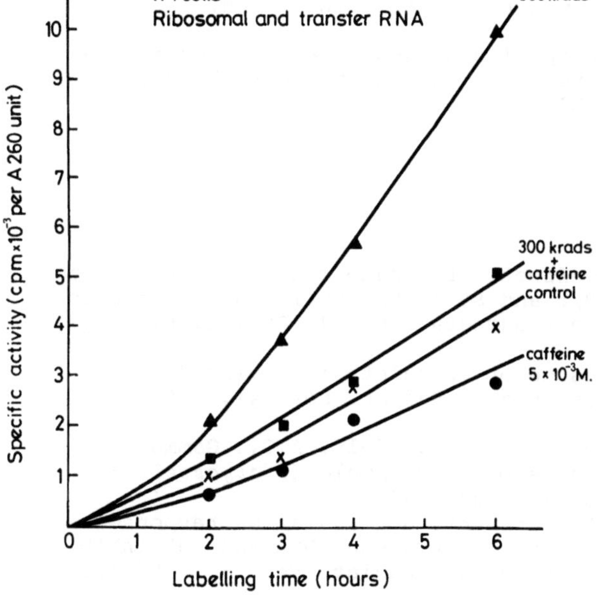

Fig. 13. Specific activity as function of labelling time with ^{3}H-adenine. Fractions eluted in 25% formamide (rRNA and tRNA). (X) unirradiated controls; (▲) 300 krads; (●) caffeine 5 x 10^{-3} *M*; (■) 300 krads + caffeine 5 x 10^{-3} *M* given after irradiation

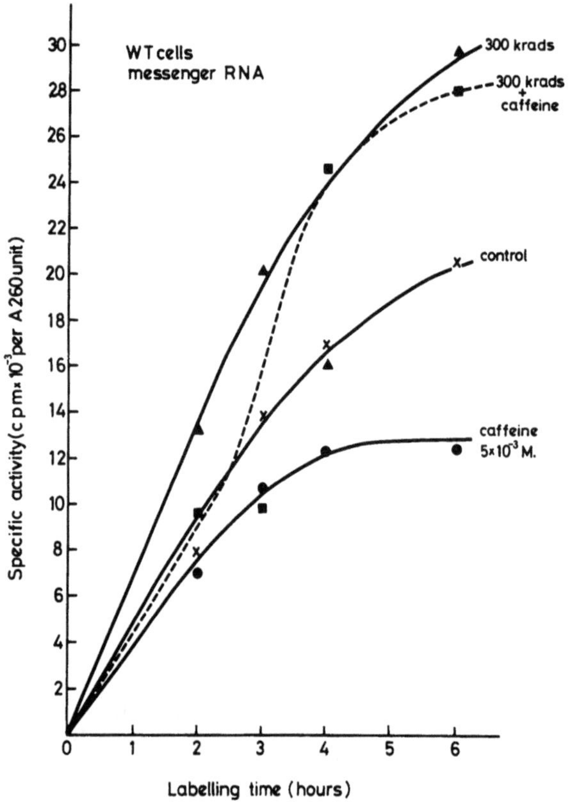

Fig. 14. Specific activity as function of labelling time with ^{3}H-adenine. Fractions eluted with 90% formamide (mRNA). (X) unirradiated controls; (▲) 300 krads; (●) caffeine at 5 x 10^{-3} *M*; (■) 300 krads and caffeine 5 x 10^{-3} *M* after irradiation

the survival curve or much reduced - (a) after neutron irradiation, (b) after caffeine treatment, (c) in the sensitive strain γs-13 - RNA synthesis after irradiation was not greatly increased.

3. Results of separation of RNA species suggest that synthesis of all species of RNA was increased after irradiation and that there is no preferential labelling of one RNA species.

References

Cramp, W.A., Bryant, P.E.: The effects of rifampicin on electron- and neutron-irradiated *E. coli* B/r and Bs-1: Survival, DNA degradation and DNA synthesis by membrane fragments. Int. J. Radiat. Biol. 27, 143-156 (1975)

Daniels, E.W., Breyer, E.P.: Rescue of supralethally x-irradiated amoebae with nonirradiated cytoplasm. Radiat. Res. 41, 326-341 (1970)

Deering, R.A., Smith, M.D., Thompson, B.K., Adolf, A.C.: Gamma-ray-resistant and -sensitive strains of slime mould (*Dictyostelium discoideum*). Radiat. Res. 43, 711-728 (1970)

Elkind, M.M., Moses, W.B., Sutton-Gilbert, H.: Radiation response of mammalian cells grown in culture. IV. Protein, DNA and RNA inhibition during the repair of X-ray damage. Radiat. Res. 31, 156-173 (1967)

Firtel, R.A., Jacobson, A., Lodish, H.F.: Isolation and hybridization kinetics of messenger RNA from *Dictyostelium discoideum*. Nature New Biol. 239, 225-228 (1972)

Friesen, J.D.: Control of messenger RNA synthesis and decay in *Escherichia coli*. J. Molec. Biol. 20, 559-573 (1966)
Gray, W.J.H., Midgeley, J.E.M.: The control of ribonucleic acid synthesis in bacteria. The synthesis and stability of ribonucleic acids in relaxed and stringent amino-acid auxotrophs of *Escherichia coli*. Biochem. J. 128, 1007-1020 (1972)
Hidvégi, E.J., Holland, J., Bölöni, E., Vârterész, V.: Enrichment of polysomes in the liver upon x-irradiation in vivo. Acta Biochim. Biophys. Acad. Sci. Hung. 1, 107-112 (1966)
Hudnik-Plevnik, T.M.: Effect of x-irradiation on pulse-labelling of ribonucleic acid in *Salmonella typhimurium*. Biochim. Biophys. Acta (Amst.) 103, 515-516 (1965)
Klamerth, O.: The synthesis of RNA in growing *E. coli* B after irradiation with X-rays. Int. J. Radiat. Biol. 8, 291-296 (1964)
Koch, H., Kiefer, J.: RNA-synthesis in irradiated diploid yeast at 50 per cent survival level. Int. J. Radiat. Biol. 21, 223-234 (1972)
Matsudaira, H., Sekiguchi, T., Nakamura, C., Hishizawa, T.: Effects of x-irradiation on the synthesis of rapidly labelled RNA in an experimental tumour of the rat. Int. J. Radiat. Biol. 13, 21-34 (1967)
Rudé, J., Alper, T.: Changes in UV survival curves of *Escherichia coli* B/r concomitant with changes in growth conditions. Photochem. Photobiol. 15, 51-60 (1972)
Simic, M.M., Hudnik-Plevnik, T., Kanazir, D.T.: Effect of x-irradiation on the synthesis and composition of pulse-labelled RNA of the liver of the rat foetus. Nature (Lond.) 214, 489-491 (1967)

Chapter II
Repair and Recovery

1. Repair in Yeast

E. MOUSTACCHI

The yeast *Saccharomyces cerevisiae* is one of the classic cellular systems studied by radiation biologists. The first step toward determining the nature of the repair systems and their mechanisms of action in such a unicellular eukaryotic organism consisted of isolating radiation-sensitive mutants. Complementation and recombination tests showed that these mutants, selected on the basis of their sensitivity for reproductive death, contain mutations corresponding to at least 30 distinct genetic loci (for review see Game, 1975).

Their phenotypic and genetic characterization included delineation of their pleiotropic effects with respect to biologic parameters including: spontaneous and radiation-induced nuclear and mitochondrial mutations, genetic recombination, sporulation and spore viability, cross sensitivity to chemicals, liquid-holding recovery in combination with photoreactivation.

The term *rad* is given to such radiosensitive mutants; the genetic loci that primarily confer ultraviolet (UV) sensitivity are numbered from 1 to 50; *rad50* and upward designates genes that principally affect sensitivity to ionizing radiation (Game, 1975).

Additional radiosensitive mutants selected by criteria other than killing, such as low inducibility of mutation (*rev*) or of gene conversion (*rec*), are also available (Lemontt, 1971; Resnick and Setlow, 1972; Rodarte-Ramon and Mortimer, 1972; Resnick, 1975). Some of them turned out to be allelic with already known *rad* genes.

Among the 30 loci mentioned, 17 govern the UV response, 8 affect principally the x-ray response, and 5 control both UV and x-ray sensitivity.

The existence of so many genes raises important questions regarding the multiplicity of factors that control cellular recovery and DNA repair pathways in wild type eukaryotic cells (Haynes, 1975). It cannot be excluded, however, that genetically controlled mechanisms other than DNA repair may affect radiosensitivity.

The unraveling of the interactions among these genes with the aim of establishing the number and interrelations of the various repair mechanisms has been the second task of yeast radiobiologists. A preliminary approach consisted of comparing the UV or x-ray survival curves of single *rad* mutants to those of multiple mutant strains. Three situations have been encountered: (1) epistasy; (2) additivity; (3) synergism. Ten genes have been studied thus far, and a tentative model of the possible relations has been proposed (Brendel and Haynes, 1973; Game, 1975; Haynes, 1975). It would appear that there may exist at least four distinct pathways for repair in yeast. Only one, the excision pathway, has been uniquivocally identified at the molecular level. Two others, the repair of single strand and of double strand breaks, have begun to receive some biochemical support. The recombinational repair pathway that has been proposed on the basis of biologic arguments is still undemonstrated.

The excision-repair pathway has received much attention in the last three years (Unrau et al., 1971; Resnick and Setlow, 1972; Waters and Moustacchi, 1975a). The lack of specific labelling of DNA in yeast has hampered such studies for some time. The quantum yield for pyrimidine dimers in yeast DNA is between 24 and 31 dimers per ergs/mm^2 (Cox, personal communication; Brendel, personal communication). For a fixed dose of UV radiation, the yield is similar in haploid and diploid yeast (Waters and Moustacchi, 1975b). Respiratory-deficient mutants demonstrate a lower capacity for pyrimidine dimer excision (Waters and Moustacchi, 1975c).

Above a certain incident threshold dose excision of the photoproducts from the nuclear DNA of cells of both ploidies decreases with increasing doses (Waters and Moustacchi, 1975b). Such a saturation of the excision process has been reported for several organisms. Post-irradiation incubation in buffer followed by further incubation in nutrient medium (conditions that mimic the "liquid holding recovery" type of experiments) increases excision in nuclear DNA as compared to that seen following incubation in either nutrient medium or saline alone (Waters and Moustacchi, 1975b).

The photoreactivability of UV-induced pyrimidine dimers in the mitochondrial genome has been demonstrated (Waters and Moustacchi, 1975a); a concomittant diminution in cytoplasmic "petite" mutants is observed. Such pyrimidine dimers are not removed in a controlled manner; extensive degradation, accompanied by an increase in "petite" mutants, of mitochondrial DNA is observed when irradiated stationary phase cells are kept in the dark. This degradation is upon storage in the dark. The fate of pyrimidine dimers in mitochondrial DNA from exponential phase cells (in the latter condition a diminution of "petites" occurs during maintenance in the dark) remains to be determined.

Detection of pyrimidine dimers by specific UV endonuclease, a test much more sensitive than the chromatographic technique, will probably clarify some conflicting aspects in the future.

Preliminary data demonstrate the induction and repair of single- (Cox and Haynes, paper submitted for publication) and double-strand breaks after x-ray exposure in yeast DNA (Ho and Mortimer, 1975; Resnick, 1975). It appears that 10 krads induce 114 breaks per haploid genome, each break requiring 103 eV of deposited energy. Within one hour of incubation in nongrowth medium, 90% of these breaks are repaired. The mutant *rad50* appears to be defective in this type of repair. A decrease in molecular weight is observed in neutral sucrose gradients after a dose of 30 krads (5.4×10^8 daltons for unirradiated cells to 3.8×10^8 daltons after 30 krads). After 3 1/2 hours of incubation in nutrient medium, 70% recovery is observed for wild type cells. The *rad50* mutants lack this rejoining capacity (0.7 double strand break/Do/haploid genome). The induction of double strand breaks appears to be related to dominant lethality (Ho and Mortimer, 1975).

References

Brendel, M., Haynes, R.H.: Interactions among genes controlling sensitivity to radiation and alkylation in yeast. Molec. Gen. Genetics 125, 197-216 (1973)

Game, J.C.: Radiation-sensitive mutants of yeast. In: Molecular Mechanisms for Repair of DNA (Hanawalt, P.C., Setlow, R.B., eds.). New York-London: Plenum Press, 1975, Part B, pp. 541-544

Haynes, R.H.: DNA repair and the genetic control of radiosensitivity in yeast. In: Molecular Mechanisms for Repair of DNA (Hanawalt, P.C., Setlow, R.B., eds.). New York-London: Plenum Press, 1975, Part B, pp. 529-540

Ho, K.S.Y., Mortimer, R.K.: X-ray induced dominant lethality and chromosome breakage and repair in a radiosensitive strain of yeast. In: Molecular Mechanisms for Repair of DNA (Hanawalt, P.C., Setlow, R.B., eds.). New York-London: Plenum Press, 1975, Part B, pp. 545-548

Lemontt, J.F.: Mutants of yeast defective in mutation induced by ultraviolet light. Genetics 68, 21-33 (1971)

Resnick, M.A.: The repair of double-strand breaks in chromosomal DNA of yeast. In: Molecular Mechanisms for Repair of DNA (Hanawalt, P.C., Setlow, R.B., eds.). New York-London: Plenum Press, 1975, Part B, pp. 549-556

Resnick, M.A., Setlow, J.K.: Repair of pyrimidine dimer damage induced in yeast by ultraviolet light. J. Bacteriol. 109, 979-987 (1972)

Rodarte-Ramon, U.S., Mortimer, R.K.: Radiation induced recombination in *Saccharomyces*: The genetic control of recombination in mitosis and meiosis. Radiat. Res. 49, 148-151 (1972)

Unrau, P., Wheatcroft, R., Cox, B.S.: The excision of pyrimidine dimers from DNA of ultraviolet irradiated yeast. Molec. Gen. Genetics 113, 359-362 (1971)

Waters, R., Moustacchi, E.: The fate of UV-induced pyrimidine dimers in the nuclear and mitochondrial DNAs of *Saccharomyces cerevisiae* on various post-irradiation treatments and its influence on survival and cytoplasmic "petite" induction. In: Molecular Mechanisms for Repair of DNA (Hanawalt, P.C., Setlow, R.B., eds.). New York-London: Plenum Press, 1975a, Part B, pp. 557-565

Waters, R., Moustacchi, E.: Excision of pyrimidine dimers from the nuclear DNA of a haploid respiration-deficient (ρ^-) strain of *Saccharomyces cerevisiae*. Photochem. Photobiol. 21, 441-444 (1975b)

Waters, R., Moustacchi, E.: Dose dependence of the excision of ultraviolet-induced pyrimidine dimers from nuclear deoxyribonucleic acids of haploid and diploid *Saccharomyces cerevisiae*. J. Bacteriol. 121, 901-906 (1975c)

2. Properties and Function of *Escherichia coli* DNA Polymerase I

K. D. JENTSCH and T. M. JOVIN

A. Introduction

DNA polymerases are ubiquitous in nature, because they are responsible for the replication and repair of cellular DNA. DNA polymerase I (Pol I) from the bacterium *Escherichia coli*, discovered by Kornberg and coworkers, is the best characterized and most intensely studied of the three polymerases that have been isolated from this organism. The following discussion is not intended as a complete review of the vast literature on Pol I but rather as a short survey of its chemical and physical characteristics and its mechanism of catalysis in vitro. The possible roles of Pol I in vivo will also be discussed and compared to the functions of the other two polymerases. Several excellent reviews and monographs on this subject have recently been published and the reader is referred to these for further references and details (Kornberg, 1974; Kornberg and Kornberg, 1974; Emmerson, 1974; Gefter, 1975; Grossman et al., 1975).

B. Some Physical and Chemical Properties of *E. coli* Polymerase I (Jovin et al., 1969; Kornberg, 1974)

A wild type *E. coli* cell contains about 400 molecules of Pol I (Kornberg, 1974; Gefter, 1975). The enzyme consists of a single intact polypeptide chain of about 1000 amino acids with a molecular weight of 109,000 daltons. It contains one disulfide bond and a free SH group that is not in the active site of the enzyme (Jovin et al., 1969). Pol I is thought to have an approximate diameter of 65 Å, which would make it possible for the enzyme to occupy a site on the DNA consisting of about 20 base pairs (Kornberg, 1969). The enzyme has been visualized directly in the bound state by electron microscopy (Griffith et al., 1971).

Limited proteolysis of the native structure produces two fragments, each with different catalytic activities (Brutlag et al., 1969; Klenow and Henningsen, 1970). The large fragment (Mol. wt. = 76,000) contains the polymerase and 3'-5' exonuclease activities (Setlow et al., 1972), while a smaller fragment (Mol. wt. = 36,000) contains a 5'-3' exonuclease activity (Setlow and Kornberg, 1972). The nuclease activities are thought to be especially important for the excision of damaged DNA or "mismatched" residues such as might arise by misincorporation during synthesis (Brutlag and Kornberg, 1972).

C. Functions and Mechanisms of Action of *E. coli* Polymerase I

Pol I is a multifunctional enzyme with the following five demonstrable catalytic activities:

1. 5'-3' Polymerization of deoxyribonucleotides
2. 3'-5' Pyrophosphorolysis of DNA
3. Pyrophosphate exchange
4. 3'-5' Exonucleolytic hydrolysis of DNA
5. 5'-3' Exonucleolytic hydrolysis of DNA

The second and third actions can be regarded as manifestations of the reversal of synthesis.

I. Polymerization

The turnover number for polymerization of Pol I in vitro is about 660 nucleotides per min under optimal reaction conditions (pH 7.4, 37°C, and activated poly d(AT) as template). The polynucleotide template determines the deoxynucleotide to be incorporated into the growing chain by directing the selection of the complementary deoxynucleoside triphosphate that is to be coupled by the enzyme to the 3'-OH terminus of the "primer".

Zn^{++} is stoichiometrically bound to the enzyme and is essential for the polymerization step (Slater et al., 1971; Springgate et al., 1973), presumably by interacting with the terminal 3'-OH of the primer. This interaction could help to catalyze the nucleophilic attack of the terminal 3'-OH group on the α-phosphate moiety of the incoming triphosphate (Mildvan, 1974). The polymerase activity requires in addition the divalent cation Mg^{++} at least partially in the form of a complex with the deoxynucleoside triphosphate substrates. Binding studies of the alternative divalent cation cofactor Mn^{++} and nucleotides to Pol I reveal multiple Mn^{++} binding sites on the enzyme. The tight and intermediate sites probably interact with the α-phosphorous of the deoxynucleoside triphosphate (Slater et al., 1972). Pyrophosphate is released after the formation of a phosphodiester bond, producing a new primer-terminus. The action of the enzyme can then be repeated resulting in the net extension of the primer strand in the 5'-3' direction.

Certain characteristics of the DNA structure are required for template recognition and function (Kornberg, 1974; Kornberg and Kornberg, 1974). The enzyme cannot bind tightly to intact duplex structures (circular double-stranded or linear DNA) except at the ends. Thus, double-stranded DNA must first be activated in order to serve as a template for polymerase I. This can be accomplished by the action of certain endonucleases, which cleave a phosphodiester bond resulting in a free 3'-OH terminus and a 5'-phosphate group. These "nicks" can then function as binding sites for the polymerase, presumably due to the free 3'-OH groups. Gapped duplexes, which can result from exonucleolytic degration of nicks (10-100 nucleotides), are even better templates. Such gaps can be produced by *E. coli* exonuclease III or the 5'-3' exonucleolytic activity of Pol I. Single-stranded DNA is only active if small regions of complementary oligonucleotides are present to act as primers. Pol I can also bind to free 3'-phosphate termini, but such a complex is inactive in polymerization.

At present there is no detailed knowledge of the mechanism by which Pol I recognizes specific binding sites on DNA. Also little is known about the physical processes involving substrate binding. McClure and Jovin (1975) showed in steady state kinetic experiments, using the synthetic alternating copolymer poly d(A-T) as template primer, that during synthesis the binding of dATP and dTTP were independent of each other. The kinetic results were consistent with an apparent "ping-pong" mechanism, in the sense that the template dictates the alternation between dATP and dTTP. No quaternary complex of the enzyme with template and the two triphosphates was observed. Further investigations of substrate saturation and product-inhibition led to the following scheme for the polymerization reaction:

During the first step the enzyme binds at the free 3'-OH end of the template-primer. Only then is it possible for the second substrate (dNTP) to bind specifically. Following phosphodiester bond formation pyrophosphate is released and the chain has been extended by one unit. At this point of the scheme there are two possibilities in the mechanism: (1) translocation of the enzyme along the template, or (2) dissociation of the enzyme from the complex. Thus one important question is whether the enzyme functions in a processive manner, that is, if it remains attached to the DNA, or dissociates from the template following a single (or a few) catalytic step(s).

This point was investigated by carrying out template challenge or competition experiments (McClure and Jovin, 1975). The synthesis of poly d(A-T) was initiated with enzyme, dATP, and dTTP. After 2.5 min, nicked calf thymus DNA and the other two substrates, ^{3}H-dCTP and dGTP, were added to the reaction mixture. To accurately measure the kinetic processes, the experiments were performed at 4°C to lower the turnover number to 3 nucleotides per min. The time dependence of further DNA synthesis was measured by the incorporation of radioactive nucleotides. Incorporation of ^{3}H-dCTP into the newly added template, calf thymus DNA, was found within the time of a single enzyme turnover, which is only possible if the enzyme had dissociated from the poly d(A-T) template-primer complex after a single polymerization step. The free polymerase was then able to distribute between the two templates. The fact that the rate of synthesis of poly d(A-T) was diminished upon addition of the competitive template was interpreted as strong evidence for nonprocessivity under the conditions used.

Similar interpretations have been presented by Chang (1975) for Pol I and the two DNA polymerases from calf thymus. Challenge experiments, similar to those mentioned above, showed immediate incorporation of nucleotides into the competing primer-template. The parallel incorporation of nucleotides into all molecules of template present in molar excess over enzyme was demonstrated by following the kinetics of incorporation with sucrose density gradient analysis. In complementary experiments, the early incorporation of inhibitor molecules into product was demonstrated. The nonprocessivity of the various polymerases was demonstrated further by product analysis of reactions in which the ratio of template to enzyme was 900 and only 1-2 nucleotide additions had taken place.

II. Exonucleolytic Hydrolysis

The 3'-5' exonucleolytic activity is present not only in Pol I but also has been found in the *E. coli* polymerases II and III (Gefter, 1974, 1975) and in the induced polymerase of the bacteriophage T4 (Goulian et al., 1968; Hershfield and Norsal, 1972). This exonuclease acts on single-stranded DNA and degrades in the 3'-5' direction, proceeding from the 3'-OH end and freeing 5'-nucleotides. Studies using synthetic polynucleotides with matched and mismatched ends suggest that this exonuclease acts as a proofreading enzyme, removing nucleotides that have been incorrectly incorporated at the 3'-OH end of the primer (Brutlag and Kornberg, 1972).

Bessman et al. (1974) and Goodman et al. (1974) studied the extent to which the corresponding nuclease from T4 bacteriophage can differentiate between adenine and the analog 2-aminopurine. In vivo, 2-aminopurine can cause a transition from A-T base pairs to G-C base pairs. Bessman et al. (1974) compared the induced nuclease activity of wild-type phages with the nuclease activity found with temperature sensitive mutants that showed either an increase in spontaneous mutations (mutators) or a decrease in spontaneous mutations (antimutators). It was found that the ratio of 3'-5' exonuclease activity to the polymerizing activity varied inversely with the frequency of spontaneous mutations. The antimutator contained the largest ratio of nuclease to polymerase.

The 5'-3' exonuclease activity of Pol I (Klett et al., 1968; Deutscher and Kornberg, 1969) is manifest only at base-paired regions. The products consist of oligonucleotides and mononucleotides (Kelly et al., 1969).

It is important to note that the 5'-3' exonuclease can also digest RNA fragments at the 5'-end of a DNA strand, such as are presumed to exist as initiators of Okazaki fragments during replication (Sugino et al., 1972). Roychoudhury (1973) carried out in vitro experiments on hybrid structures of poly(rA)-poly(dT). He showed that the enzyme degrades the ribonucleotide strand during DNA synthesis. The large fragment of Pol I (without the 5'-3' exonucleolytic activity) was able to synthesize DNA but unable to degrade the oligoribonucleotides. This finding and others to be described below suggest the involvement of the exonuclease activity in DNA replication.

A further function for the 5'-3' nuclease is the removal of damaged DNA fragments. It has been demonstrated in vitro that Pol I can remove thymine dimers from ultraviolet (UV)-irradiated DNA (Kelly et al., 1969), thus implying a role in excision repair in vivo. The intact Pol I enzyme together with added deoxyribonucleoside triphosphates can carry out the excision process concurrently with polymerization and with great efficiency (Friedberg and Lehman, 1974).

D. Functions of *E. coli* Polymerase II and III and the Possible Role of Pol I in vivo

Polymerases II and III were found in bacterial mutants that showed only traces of polymerase I, but that were still capable of synthesizing DNA (Knippers, 1970; Kornberg and Gefter, 1971; Gefter et al., 1972; Wickner et al., 1972). In normal *E. coli* cells there are about 40 molecules of polymerase II (Pol II). This enzyme, with a molecular weight of about 90,000 daltons, has about 10% of the specific activity of Pol I. Pol II functions poorly on nicked DNA, but duplexes with short gaps (10-100 nucleotides) are optimal templates. The physiologic role is unknown, but there is some evidence that the enzyme is involved in the repair of UV-irradiated DNA (Masker and Hanawalt, 1973; Masker et al., 1973), but is probably not required for DNA replication. Pol II may serve, however, in an auxiliary role.

Polymerase III (Pol III)(Kornberg and Gefter, 1972; Wickner et al., 1973), the product of the dnaE gene (now called polC), is essential for DNA replication. The molecular weight of this enzyme is 180,000 daltons and there are about 10 molecules per cell. The specific activity is approximately 15 times higher than that of Pol I. Pol III is active on DNA templates with short gaps and functions synergistically with additional replication proteins and factors. A higher polymer of Pol III (Pol III*) together with copol III* (another protein) is active on single-stranded DNA as template (Wickner and Kornberg, 1974). Its precise role in replication is still to be eludicated. It should be mentioned that Pol III has associated 3'-5' and 5'-3' nuclease activities (Livingston and Richardson, 1975).

The physiologic roles of Pol I have been investigated primarily by isolation and characterization of mutants deficient in the enzyme. The structural gene for Pol I, polA, is located at 75 min on the genetic map. Pol A mutants exist that are essentially devoid of the polymerizing activity but are viable. These strains retain, however, the 5'-3' exonuclease activity.

Pol A mutants do demonstrate pleiotropic phenotypic defects, the most important of which are:

1. Increased UV-sensitivity (De Lucia and Cairns, 1969; Gross and Gross, 1969)
2. Accumulation of short Okazaki fragments (Okazaki et al., 1971)
3. Extensive degradation of DNA in UV repair (Cooper and Hanawalt, 1972)
4. Increased frequency of deletions (Coukell and Yanowsky, 1970)
5. Increased frequency of recombination (Gross and Gross, 1969)
6. Loss of capacity for plasmid replication, e.g., of Col E_1 (Kingsbury and Helinski, 1970)

Bonhoeffer and Olivera (1974) have reported a temperature-sensitive Pol I mutant (BT 4113) that is conditionally lethal. In vivo experiments showed that these mutations are defective at high temperature not only in the polymerase but also in the 5'-3' exonuclease activity. Konrad and Lehman (1974) likewise found a mutant polAex1, which is defective in the 5'-3' exonuclease but not in the polymerase activity, and is conditionally lethal. These mutants show an abnormal accumulation of short Okazaki fragments, suggesting that in vivo not only the polymerase but also the 5'-3' nuclease activity is required for replication and repair. In addition, Konrad et al. (1973) detected a conditional lethal mutant with a temperature-sensitive DNA ligase, showing that this activity is also essential for normal DNA replication and repair in *E. coli* cells and that it is an important entity acting in combination with the polymerase.

Hendler et al. (1975) isolated and investigated a complex from *E. coli* cells, which they called P_3. This complex, with a molecular weight of about 390,000 daltons, contains Pol I as well as the recBC DNAse. The recBC enzyme is multifunctional and contains exonuclease as well as endonuclease and ATPase activities (MacKay and Linn, 1974). It appears that other nucleases and possible polymerases II and III may also be present in the complex. Thus it would seem that certain enzyme activities concerned with replication, repair, and recombination functions may be intimately associated in the bacterial cell.

The other important role for DNA polymerase I is its function in repair synthesis, a major topic of this conference. Pol I is involved in "short patch repair", while "large patch repair" is mediated by enzymes of recombination systems (Grossman et al., 1975). In addition, one may note the recent report of yet another nuclease function in *E. coli*, exonuclease VII, which may be involved in recombination and repair (Chase and Richardson, 1974a, 1974b).

Acknowledgements. We thank Dr. R. Clegg and Dr. G. Rhodes for assistance in the preparation of this manuscript.

References

Bessman, M.J., Muzyczka, N., Goodman, M.F., Schnaar, R.L.: Studies of the biochemical basis of spontaneous mutation. II. The incorporation of a base and its analogue into DNA by wild-type, mutator and antimutator DNA polymerases. J. Molec. Biol. 88, 409-421 (1974)

Bonhoeffer, F., Olivera, B.M.: Replication of *Escherichia coli* requires DNA polymerase I. Nature 250, 513-514 (1974)

Brutlag, D., Atkinson, M.R., Setlow, P., Kornberg, A.: An active fragment of DNA polymerase produced by proteolytic cleavage. Biochem. Biophys. Res. Comm. 37, 982-989 (1969)

Brutlag, D., Kornberg, A.: Enzymatic synthesis of deoxyribonucleic acid. XXXVI: A proofreading function for the 3'-5' exonuclease activity in deoxyribonucleic acid polymerases. J. Biol. Chem. 247, 241-248 (1972)

Chang, L.M.S.: The distributive nature of enzymatic DNA synthesis. J. Molec. Biol. 93, 219-235 (1975)

Chase, J.W., Richardson, C.C.: Exonuclease VII of *Escherichia coli*, purification and properties. J. Biol. Chem. 249, 4545-4552 (1974a)

Chase, J.W., Richardson, C.C.: Exonuclease VII of *Escherichia coli*, mechanism of action. J. Biol. Chem. 249, 4553-4561 (1974b)

Cooper, P.K., Hanawalt, P.C.: Role of DNA polymerase I and the rec system in excision-repair in *Escherichia coli*. Proc. Nat. Acad. Sci. USA 69, 1156-1160 (1972)

Coukell, M.B., Yanofsky, C.: Increased frequency of deletions in DNA polymerase mutants of *Escherichia coli*. Nature 228, 633-635 (1970)

De Lucia, P., Cairns, J.: Isolation of an *Escherichia coli* strain with a mutation affecting DNA polymerase. Nature 224, 1164-1166 (1969)

Deutscher, M.P., Kornberg, A.: Enzymatic synthesis of deoxyribonucleic acid. XXIX: Hydrolysis of deoxyribonucleic acid from the 5' terminus by an exonuclease function of deoxyribonucleic acid polymerase. J. Biol. Chem. 244, 3029-3037 (1969)

Emmerson, P.T.: DNA replication in *Escherichia coli*. In: Biochemistry of Nucleic Acids (Burton, K., ed.). London: Butterworths, 1974, pp. 141-164

Friedberg, E.C., Lehman, I.R.: Excision of thymine dimer by proteolytic and amber fragments of *Escherichia coli* DNA polymerase I. Biochem. Biophys. Res. Comm. 58, 132-139 (1974)

Gefter, M.L.: DNA polymerases II and III of *Escherichia coli*. Prog. Nucl. Acid Res. Mol. Biol. 14, 101-115 (1974)

Gefter, M.L.: DNA replication. Ann. Rev. Biochem. 44, 45-78 (1975)

Gefter, M.L., Molineux, I.J., Kornberg, T., Khorana, H.G.: Deoxyribonucleic acid synthesis in cell-free extracts. III: Catalytic properties of deoxyribonucleic acid polymerase II. J. Biol. Chem. 247, 3321-3326 (1972)

Goodman, M.F., Gore, W.C., Muzyczka, N., Bessman, M.J.: Studies on the biochemical basis of spontaneous mutation. III: Rate model for DNA polymerase-effected nucleotide misincorporation. J. Molec. Biol. 88, 423-435 (1974)

Goulian, M., Lucas, Z.J., Kornberg, A.: Enzymatic synthesis of deoxyribonucleic acid. XXV: Purification and properties of deoxyribonucleic acid polymerase induced by infection with phage T4. J. Biol. Chem. 243, 627-638 (1968)

Griffith, J., Hubermann, J.A., Kornberg, A.: Electron microscopy of DNA polymerase bound to DNA. J. Molec. Biol. 55, 209-214 (1971)

Gross, J., Gross, M.: Genetic analysis of an *Escherichia coli* strain with a mutation affecting DNA polymerase. Nature (Lond.) 224, 1166-1168 (1969)

Grossman, L., Broun, A., Feldberg, R., Mahler, J.: Enzymatic repair of DNA. Ann. Rev. Biochem. 44, 19-43 (1975)

Hendler, R.W., Pereira, M., Scharff, R.: DNA synthesis involving a complexed form of DNA polymerase I in extracts of *Escherichia coli*. Proc. Nat. Acad. Sci. USA 72, 2099-2103 (1975)

Hershfield, M.S., Nossal, N.G.: Hydrolysis of template and newly synthesized deoxyribonucleic acid by the 3' to 5' exonuclease activity of the T4 deoxyribonucleic acid polymerase. J. Biol. Chem. 247, 3393-3404 (1972)

Jovin, T.M., Englund, P.T., Bertsch, L.L.: Enzymatic synthesis of deoxyribonucleic acid. XXVI: Physical and chemical studies of a homogeneous deoxyribonucleic acid polymerase. J. Biol. Chem. 244, 2996-3008 (1969)

Kelly, R.B., Atkinson, M.R., Huberman, J.A., Kornberg, A.: Excision of thymine dimers and other mismatched sequences by DNA polymerase of *Escherichia coli*. Nature (Lond.) 224, 495-501 (1969)

Kingsbury, D.T., Helinski, D.R.: DNA polymerase as a requirement for the maintenance of the bacterial plasmid colicinogenic factor E. Biochem. Biophys. Res. Comm. 41, 1538-1544 (1970)

Klenow, H., Henningsen, I.: Selective elimination of the exonuclease activity of the deoxyribonucleic acid polymerase from *Escherichia coli* B by limited proteolysis. Proc. Nat. Acad. Sci. USA 65, 168-175 (1970)

Klett, R.P., Cerami, A., Reich, E.: Exonuclease VI, a new nuclease activity associated with *Escherichia coli* DNA polymerase. Proc. Nat. Acad. Sci. USA 60, 943-950 (1968)

Knippers, R.: DNA polymerase II. Nature (Lond.) 228, 1050-1053 (1970)

Konrad, E.B., Lehman, I.R.: A conditional lethal mutant of *Escherichia coli* K12 defective in the 5'-3' exonuclease associated with DNA polymerase I. Proc. Nat. Acad. Sci. USA 71, 2048-2051 (1974)

Konrad, E.B., Modrich, P., Lehman, I.R.: Genetic and enzymatic characterization of a conditional lethal mutant of *Escherichia coli* K12 with a temperature-sensitive DNA ligase. J. Molec. Biol. 77, 519-529 (1973)

Kornberg, A.: Active center of DNA polymerase. Science 163, 1410-1418 (1969)

Kornberg, A.: DNA Synthesis. San Francisco: Freeman, 1974

Kornberg, T., Gefter, M.L.: Purification and DNA synthesis in cell-free extracts: Properties of DNA polymerase II. Proc. Nat. Acad. Sci. USA 68, 761-764 (1971)

Kornberg, T., Gefter, M.L.: Deoxyribonucleic acid synthesis in cell-free extracts. IV: Purification and catalytic properties of deoxyribonucleic acid polymerase III. J. Biol. Chem. 247, 5369-5375 (1972)

Kornberg, T., Kornberg, A.: Bacterial DNA polymerase. In: The Enzymes (Boyer, P.D., ed.). New York: Academic Press, 1974, 2nd ed., Vol. X, pp. 119-144

Livingston, O.M., Richardson, C.C.: J. Biol. Chem. (in press)(1975)
MacKay, V., Linn, S.: The mechanism of degradation of duplex deoxyribonucleic acid by the rec BC enzyme of *Escherichia coli* K-12. J. Biol. Chem. 249, 4286-4294 (1974)
Masker, W.E., Hanawalt, P.C.: Ultraviolet-stimulated DNA synthesis in toluenized *Escherichia coli* deficient in DNA polymerase I. Proc. Nat. Acad. Sci. USA 70, 129-133 (1973)
Masker, W.E., Hanawalt, P.C., Shizuya, H.: Role of DNA polymerase II in repair replication in *Escherichia coli*. Nature New Biol. 244, 242-243 (1973)
McClure, W.R., Jovin, T.M.: The steady state kinetic parameters and non-processivity of *Escherichia coli* deoxyribonucleic acid polymerase I. J. Biol. Chem. 250, 4073-4080 (1975)
Mildvan, A.S.: Mechanism of enzyme action. Ann. Rev. Biochem. 43, 357-399 (1974)
Okazaki, R., Arisawa, M., Sugino, A.: Slow joining of newly replicated DNA chains in DNA polymerase I-deficient *Escherichia coli* mutants. Proc. Nat. Acad. Sci. USA 68, 2954-2957 (1971)
Roychoudhury, R.: Transcriptional role in deoxyribonucleic acid replication. Nature of primer function of newly synthesized ribonucleic acid in vitro. J. Biol. Chem. 248, 8465-8473 (1973)
Setlow, P., Brutlag, D., Kornberg, A.: Deoxyribonucleic acid polymerase: two distinct enzymes in one polypeptide. I. A proteolytic fragment containing the polymerase and 3'-5' exonuclease functions. J. Biol. Chem. 247, 224-231 (1972)
Setlow, P., Kornberg, A.: Deoxyribonucleic acid polymerase: two distinct enzymes in one polypeptide. II. A proteolytic fragment containing the 5'-3' exonuclease function. Restoration of intact enzyme function from the two proteolytic fragments. J. Biol. Chem. 247, 232-240 (1972)
Slater, J.P., Mildvan, A.S., Loeb, L.A.: Zinc in DNA polymerase. Biochem. Biophys. Res. Comm. 44, 37-43 (1971)
Slater, J.P., Tamir, J., Loeb, L.A., Mildvan, A.S.: The mechanism of *Escherichia coli* deoxyribonucleic acid polymerase I. Magnetic resonance and kinetic studies of the role of metals. J. Biol. Chem. 247, 6784-6794 (1972)
Springgate, C.F., Mildvan, A.S., Abramson, R., Engle, J.L., Loeb, L.A.: *Escherichia coli* deoxyribonucleic acid polymerase I, a zinc metalloenzyme, nuclear quadrupolar relaxation studies of the role of bound zinc. J. Biol. Chem. 248, 5987-5993 (1973)
Sugino, A., Hirose, S., Okazaki, R.: RNA-linked nascent DNA fragments in *Escherichia coli*. Proc. Nat. Acad. Sci. USA 69, 1863-1867 (1972)
Wickner, R.B., Ginsberg, B., Berkower, J., Hurwitz, J.: Deoxyribonucleic acid polymerase II of *Escherichia coli*. I. The purification and characterization of the enzyme. J. Biol. Chem. 247, 489-497 (1972)
Wickner, W., Kornberg, A.: A holoenzyme form of deoxyribonucleic acid polymerase III. Isolation and properties. J. Biol. Chem. 249, 6244-6249 (1974)
Wickner, S., Wright, M., Hurwitz, J.: Studies on in vitro DNA synthesis. Purification of the dna G gene product from *Escherichia coli*. Proc. Nat. Acad. Sci. USA 70, 1613-1618 (1973)

3. Action of Sublethal Concentrations of SO_2 on Repair by Budding Yeast after Ionizing and UV Radiation

J. L. Dorange

A. Introduction

Cells of a number of eukaryotic organisms display marked changes in radiosensitivity during the cell cycle. Budding cells of yeast are more resistant to inactivation by ultraviolet (UV) and ionizing radiation than nonbudding cells (Beam et al., 1954; Moustacchi, 1964; Perper, 1975). These cells are synthesizing DNA (Esposito, 1966; Wildenberg, 1970). In the yeast *Saccharomyces*, the radioresistance attained during early budding is largely independent of ploidy (Beam, 1955) and is dependent on growth (Moustacchi, 1964) and starvation conditions (Perper, 1975). Administration of sulfurous acid (SO_2) sensitizes budding yeast cells to γ-ray treatment. This SO_2 action was first observed by Dupuy and Tremeau (1964) and, in analogy to other sensitizing agents such as N-ethylmaleimide or iodoacetic acid groups (Bridges, 1969), was termed sensitization. If present during irradiation at sublethal concentrations, iodoacetamide produces approximately the same effects (Dupuy and Tremeau, 1966). But, while sensitization by iodoacetamide is a general process (Dean and Alexander, 1962), γ-sensitization by SO_2 is specific for the budding yeast cell (Dorange and Dupuy, 1974). All yeast strains are not sensitized by SO_2.

This SO_2-sensitization is linked to a high resistance of the wild strains such as S132. Genetically it is a dominant character (Dorange and Dupuy, 1974). We have investigated the physiologic conditions under which budding yeast is sensitized to the action of γ-rays when γ and SO_2 treatments are dissociated. We have also investigated whether treatment with sublethal amounts of SO_2 sensitizes budding yeast to UV light.

B. Materials and Methods

I. Strains of Yeast

Diploid strain S132 of *Saccharomyces cerevisiae* was isolated in our laboratory from wine. Haploid strain SII/a was derived from a culture of one S132 ascospore. Diploid strain BII and haploid strain 59R were gifts of Dr. P. Galzy (Chaire de Génétique E.N.S.A. Montpellier).

II. Media, Irradiations, and Incubation Procedures

Routine growth and incubation medium was Wickerham's malt (Wickernam, 1951). Strains were grown aerobically in liquid malt at 28°C. Cells were harvested after 24 h of incubation (late log phase) by centrifugation and washed. Stock suspensions starved at 0°C in MacIlvaine buffer, pH 3.8 contained 2×10^6 cells/ml. SII/a and 59R cultures were previously treated by sonication (Moustacchi, 1964). Irradiation and SO_2 treatment were carried out at 0°C. Colony formation was determined by dilutions in phosphate buffer (0.02 *M*, pH 7) and plating on Wickerham's malt agar. Colonies were counted after 4-6 days incubation at 28°C.

γ-rays were obtained from a ^{60}Co source (350 krad/h) and UV light from a germicide lamp. UV dosimetry was measured with a LATARGET dosimeter.

Sulfurous acid was obtained from 0.05 *M* potassium disulfite solutions. Conditions (pH, concentration, time) for SO_2 administration were chosen so as to prevent a lethal effect. This was verified in every manipulation.

C. Results

I. γ and UV Irradiation during SO_2 Sublethal Treatment

Optimal conditions for the expression of SO_2 sensitization are: (1) addition of SO_2 to stock suspension; (2) irradiation of samples 1 h after SO_2 incorporation; (3) starvation of cells in irradiated media. All of these manipulations were performed at 0°C.

Figure 1 shows the survival rates for diploid strains S132, BII and haploid SII/a, 59R during postirradiation incubation. These strains received 100 krad and 1.5×10^{-3} *M* (diploid, Fig. 1A) and 1×10^{-3} *M* SO_2 (haploid, Fig. 1B), respectively. Under these conditions, cells of strains S132 and SII/a rapidly lost their ability to form colonies, while cells of strains BII and 59R were unaffected, in fact, protected by SO_2 treatment.

Figure 2 presents survival curves for these strains incubated in various SO_2 concentrations, then in dilutions of phosphate buffer, pH 7, and plated on Wickerham's malt 2 h after termination of irradiation. Without SO_2, the four survival curves obtained with late log-phase cultures are typical for yeast. The shoulders at about 10% survival, representing the presence of a resistant subpopulation, are approximately equal with the subpopulation of budding cells (Beam et al., 1954). The budding cell populations of strains S132 and SII/a are more resistant than the corresponding subpopulations of strains BII and 59R.

While all nonbudding and budding cells of strains BII and 59R are protected by SO_2 treatment, budding cells of S132 and SII/a are sensitized.

Similar experiments with UV radiation did not show any evidence of protection or sensitization by SO_2.

II. γ and UV Irradiation before SO_2 Sublethal Treatment

SO_2 sensitization was also obtained when SO_2 was added to diploid strain S132 and haploid strain SII/a after γ-irradiation. Results were comparable to those in Figure 1. Resistant cells lost their ability to form colonies after approximately 2 h of incubation with SO_2 at 0°C (pH = 3.8). No SO_2 sensitizing effect was observed with strains BII and 59R under the same experimental conditions.

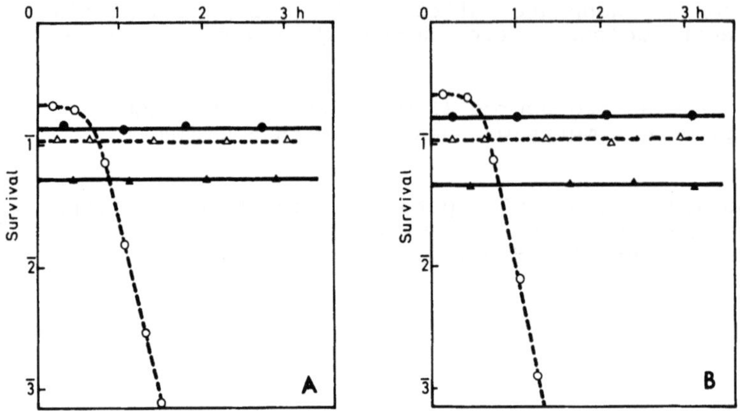

Fig. 1

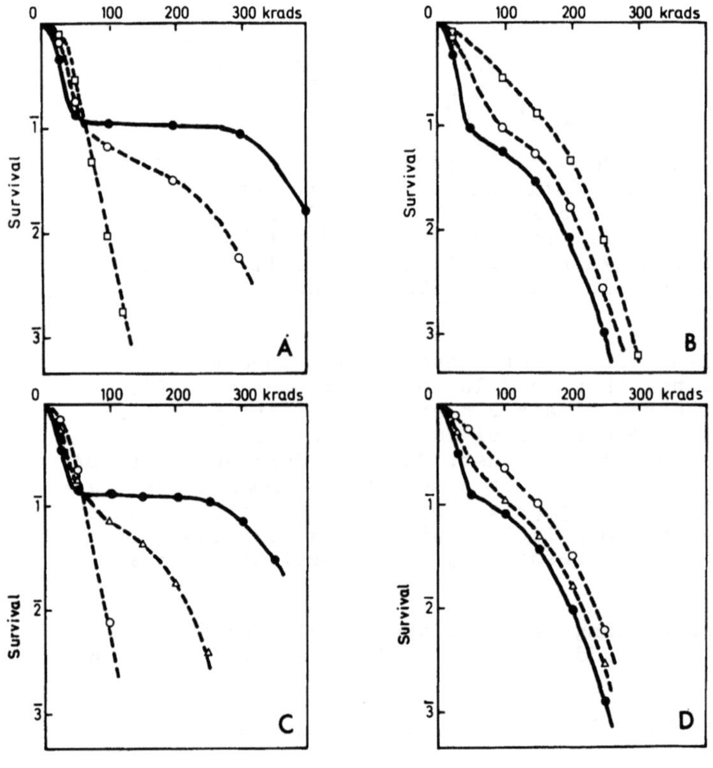

Fig. 2

Legends see opposite page

With UV rays (1,200 ergs/cm^2), there was no SO_2 sensitization for strains BII, 59R and strains S132 and SII/a.

III. γ and UV Irradiation after SO_2 Sublethal Treatment

Table 1 contains the results obtained when S132 cells were treated with SO_2 before irradiation.

Table 1. Effect of incubation before γ-irradiation on survival rate of *S. cerevisiae* S132 immediately and 2 h after the end of irradiation

Time of SO_2 incubation before irradiation		1 h		3 h	
Time of post-irradiation incubation		0 h	2 H	0 h	2 h
NS	NI	1.00	1.00	1.00	1.00
	I	0.10	0.10	0.10	0.10
S	NI	0.98	1.00	0.96	0.97
	I	0.08	0.09	0.08	0.01

Stock cell suspensions in buffer, pH 3.8 (10^6 cells/ml) were incubated 1 or 3 h with 1.5×10^{-3} *M* SO_2. Then, cells were washed twice, irradiated (100 krads) and plated immediately or after 2 h incubation in irradiated media. All treatments were performed at 0^oC. NS: no treatment with SO_2; S: SO_2 incubation 1.5×10^{-3} *M*; NI: no γ-irradiation; I: irradiation, 100 krad

Fig. 1 A and B. SO_2 sensitization to γ-radiation of *S. cerevisiae* diploid strains S132, BII and haploid strains SII/a, 59R. Survival rate of diploid strains S132, BII (left curves) and haploid strains SII/a, 59R (right curves) irradiated with SO_2 at sublethal concentrations versus time between end of irradiation and plating. Stock suspensions of cells in MacIlvaine buffer, pH 3.8, harvested in late exponential phase (10^6 cells/ml), incubated 1 h with SO_2, then irradiated and kept in irradiated medium until plating on Wickerham's malt. Samples were maintained at 0^oC during irradiation and during incubation before and after irradiation. N: number of colonies of treated cells; N_0: control cells. A. ●, S132 without SO_2; o, S132 with SO_2 1.5×10^{-3} *M*; ▲, BII without SO_2; Δ, BII with SO_2 1.5×10^{-3} *M*. B. ●, SII/a without SO_2; o, SII/a with SO_2 1×10^{-3} *M*; ▲, 59R without SO_2; Δ, 59R with SO_2 1×10^{-3} *M*. In these experiments no significant action of SO_2 treatment without ionizing radiation was observed

Fig. 2 A-D. Survival curves of *S. cerevisiae* diploid strains S132, BII and haploid strains SII/a, 59R following γ-irradiation in presence of sublethal concentrations of SO_2. Stock suspensions of cells in MacIlvaine buffer, pH 3.8, harvested in late exponential phase, incubated 1 h with SO_2, irradiated, and kept 2 h in irradiated medium before plating on Wickerham's malt. Samples were maintained at 0^oC during irradiation and incubation before and after irradiation. A. Survival curves of diploid S132: ●, without SO_2; o, with SO_2 1×10^{-3} *M*; □, with SO_2 1.5×10^{-3} *M*. B. Survival curves of diploid BII: ●, without SO_2; o, with SO_2 1×10^{-3} *M*; □, with SO_2 1.5×10^{-3} *M*. C. Survival curves of haploid SII/a: ●, without SO_2; Δ, with SO_2 0.75×10^{-3} *M*; o, with SO_2 1×10^{-3} *M*. D. Survival curves of haploid 59R: ●, without SO_2; Δ, with SO_2 0.75×10^{-3} *M*; o, with SO_2 1×10^{-3} *M*. In these experiments no significant action of SO_2 treatment without radiation was observed

In these experiments, the cells in stock suspensions were first incubated at 0°C in MacIlvaine buffer, pH 3.8, in presence of 1.5 x 10^{-3} *M* SO_2 for diploids and 1 x 10^{-3} *M* SO_2 for haploids for 1 h and 3 h. After SO_2 treatment, cell suspensions were harvested, washed twice in phosphate buffer, pH 7, and resuspended in same volume of MacIlvaine buffer (pH 3.8). Samples were γ-irradiated; the temperature during irradiation and in the irradiated medium during postirradiation incubation was 0°C.

SO_2 treatment before irradiation was without effect when cells were plated in malt agar immediately after termination of γ-irradiation (Table 1). There was sensitization when cells were plated 2 h after γ irradiation.

No SO_2 sensitizing effect was observed after UV irradiation (1,200 ergs/cm^2) of SO_2-pretreated cells.

IV. Effect of Incubation between Irradiation in SO_2 Treatment

Experimental conditions were the same as those described in section II, except that SO_2 was not added immediately after γ-irradiation. Following termination of irradiation, samples were incubated in irradiated medium at 0 and 28°C. Table 2 contains results obtained with strain S132. SO_2 was added after 80 to 160 min of postirradiation incubation. The SO_2 sensitization effect was attenuated when the cells were incubated at 0°C and was weak or nonexistent when cells were incubated at 28°C.

Table 2. Survival rate of *S. cerevisiae* diploid strain S132 after irradiation and SO_2 treatment

t	0 min	80 min		160 min	
θ	0°C	0°C	28°C	0°C	28°C
NI, NS	1.00	0.98	1.00	1.10	0.99
NI, S	0.99	0.98	1.08	0.97	1.02
I, NS	0.18	0.16	0.19	0.17	0.20
I, S	0.003	0.003	0.025	0.005	0.16

t: time between end of irradiation and beginning of SO_2 treatment; irradiation: 100 krad 0°C, pH = 3.8; incubation: at 0°C or 28°C; SO_2 treatment: 1.5 x 10^{-3} *M* at 0°C for 2 h; NI, NS: no irradiation, no SO_2; NI, S: no irradiation, SO_2 treatment; I, NS: irradiated, no SO_2; I, S: irradiated, SO_2 treatment

D. Discussion

Sensitization to γ-rays by SO_2 is not a general characteristic of yeast. It has been discovered for strain S132. As it is also expressed in the haploid derivative, this character is dominant. It has not been encountered in the other diploid strains tested. It seems to be connected with the existence of the peculiarly high radioresistance of budding cells.

The observed sensitization is not due to the action of SO_2 on the reactions likely to modify the indirect effects of ionizing radiation (radiolysis of water or subservient cell components), because it occurs even when both treatments are separated in time.

At the present state of our knowledge, the resistance of budding cells of S132 to ionizing radiation seems to be related to the existence of one or several repair processes, peculiarly efficient in this strain. Assuming that incubation for 160 min at 28°C is enough to allow the repair of radiation-induced damage in DNA, sulfurous acid probably interfers with this process. This repair process is not related to UV dark repair of budding yeast.

To explain the resistance of budding yeast cells to ionizing radiation, de Langguth and Beam (1973) presented a pattern that includes the repair enzymes. Like numerous other enzymes, e.g., DNA polymerase, the repair enzymes are synthesized periodically, and therefore their concentrations would fluctuate during the mitotic cycle. Budding cells possess an enzymic level high enough to allow the radio-induced damage to be repaired before being "fixed" by cell metabolism. Thus, in yeast, the repair process appears to be linked to a metabolic state in unstable balance; it is thus easily understandable that the presence of sulfurous acid, with the ability to act in numerous ways, could damage this process.

E. Summary

In contrast to other less resistant strains, a diploid strain of *Saccharomyces cerevisiae* resistant to ionizing radiation was sensitized by SO_2 at acid pH. Resistant subpopulation cells (budding cells), when starved in buffer solution, lost their ability to form colonies after irradiation when treated with SO_2 before, during, or after irradiation. This sensitization disappeared when cells were incubated in buffer solution for 160 min at 28°C between irradiation and SO_2 treatment. No SO_2 sensitization was observed following UV irradiation. From this dominant genetic character, it has been hypothesized that SO_2 interfers in a budding repair process specific for ionizing radiation.

References

Beam, C.A.: The influence of ploidy and division stage on the anoxic protection of *Saccharomyces cerevisiae* against x-ray inactivation. Proc. Nat. Acad. Sci. USA 48, 857-861 (1955)

Beam, C.A., Mortimer, R.K., Wolfe, R.G., Tobias, C.A.: The relation of radioresistance to budding in *Saccharomyces cerevisiae*. Arch. Biochem. Biophys. 49, 110-122 (1954)

Bridges, B.A.: Sensitization of organisms to radiation by sulfhydryl-binding agents. In: Advances in Radiation Biology (Augenstein, L.G., Mason, R., Zelle, M., eds.). Academic Press 3, 123-176 (1969)

Dean, C.J., Alexander, P.: Sensitization of radioresistant bacteria to x-rays by iodoacetamide. Nature (Lond.) 196, 1324-1325 (1962)

Dorange, J.L., Dupuy, P.: Effet sensibilisateur de l'acide sulfureux sur les levures résistantes aux radiations ionisantes. C.R. Soc. Biol. 168, 496-501 (1974)

Dupuy, P., Tremeau, O.: L'acide sulfureux sensibilisateur des levures aux radiations ionisantes. In: Microbial Inhibitors in Food (Molin, M., Erichsen, A., eds.). Stockholm-Göteborg-Uppsala: Almgvist and Wiksells Broktryckeri AB, 1964, pp. 118-129
Dupuy, P., Tremeau, O.: Radiosensibilisation des levures par l'iodo-acêtamide. Int. J. Appl. Radiat. Isotopes 17, 521-529 (1966)
Esposito, R.: Genetic recombination in synchronised cultures of *Saccharomyces*. Genetics 59, 191-210 (1966)
Langguth, E.N., de, Beam, C.A.: Repair mechanisms and cell cycle dependent variations in x-ray sensitivity of diploid yeast. Radiat. Res. 53, 226-234 (1973)
Moustacchi, E.: Thèses, Facultê des Sciences, Paris, 1964
Perper, T.: Cyclic x-ray resistance to lethal and nonlethal damage in *Saccharomyces cerevisiae*. Radiat. Res. 63, 97-118 (1975)
Wickerham, L.J.: Taxonomy of yeasts. U.S. Dept. Agr. Tech. Bull. No. 1029, 1951
Wildenberg, J.: The relation of mitotic recombination to DNA replication in yeast pedigrees. Genetics 66, 291-304 (1970)

4. Reparable and Irreparable Damage in Yeast Cells after Low LET-Irradiation

D. FRANKENBERG

A. Abstract

Diploid yeast cells in early G1 phase were irradiated with ^{60}Co-γ-rays or 30 MeV electrons (mean LET, $\bar{L}$ = 0.2 keV/μm) and with x-rays (effective photon energy = 40 keV; $\bar{L}$ = 3 keV/μm). The dose rates of the irradiation were so low ($\dot{D}$ = 27 rad/min) that maximum repair occurred during irradiation.

^{60}Co-γ-rays produce no irreparable damage up to absorbed doses D of about 120 krads. X-rays on the other hand cause irreparable damage in the same dose range, yielding a survival fraction of 0.25 at D = 120 krads.

Comparing the cumulative fractional absorbed dose distributions in L_{100} for ^{60}Co-γ-rays and for x-rays (E_{eff} = 40 keV), the irreparable damage due to x-irradiation cannot be caused by the high LET fraction of the absorbed dose (L ≥ 10 keV/μm). In fact, the x-ray-induced irreparable damage is caused by the fraction of the absorbed dose in the LET range from about 1 keV/μm to 10 keV/μm.

B. Introduction

Ionizing radiation causes reparable and irreparable damage in living cells. The amount of damage per absorbed dose in a living cell depends on several physical and biologic parameters. The most important physical parameters are absorbed dose, LET, and dose rate. Generally, the degree of irreparable damage per absorbed dose will increase with increasing values of these physical parameters. The most important biologic parameter is the metabolic situation of the cell. The efficiency of biologic counterreactions such as repair of sublethal damage depends heavily on the metabolic state of the cell.

During irradiation maximum repair of sublethal damage may occur provided the rate at which reparable damage is generated is smaller than the rate of the repair of such damage. Hence, at sufficiently low absorbed dose rates, dose-effect curves are obtained that are due exclusively to irreparable damage. On the other hand, for high absorbed dose rates, the dose-effect curves reflect the sum of reparable and irreparable damage. The reparable damage may then be eliminated by liquid-holding repair following the irradiation.

^{60}Co-γ-rays and x-rays down to 50 or 40 kVp are considered as low LET-radiation without discrimination. In fact, no or only small differences in the dose-effect curves are obtained at moderate or high dose rates, which are applied generally in cell inactivation studies. In Figure 1, such dose-effect curves are shown for diploid yeast cells in the early G1 phase. Irradiation was performed at high dose rates with 70 kVp x-rays and 30 MeV electrons (30 MeV electrons have the

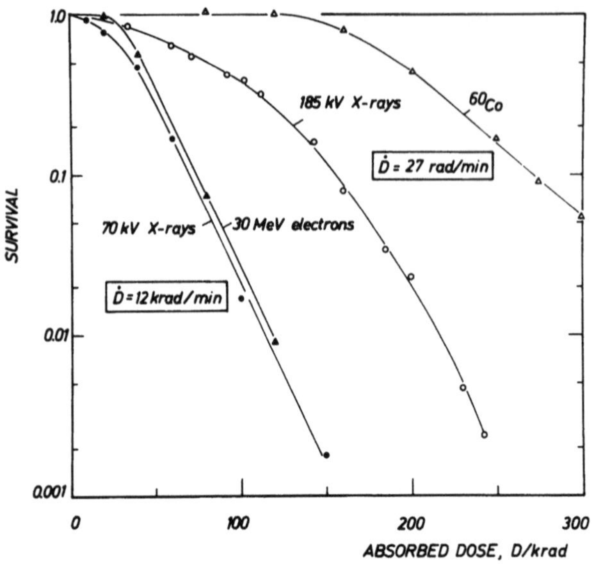

Fig. 1. Dose-effect curves for diploid yeast cells in early G1 phase for ^{60}Co-γ-rays and 185 kVp x-rays at different absorbed dose rates. Cells are irradiated in suspensions bubbled with oxygen

same distribution of absorbed dose in LET as ^{60}Co-γ-rays; see Figure 2). Apart from a small difference in the shoulder region, the dose-effect curves for 70 kVp x-rays and 30 MeV electrons look similar, i.e., the sum of reparable and irreparable damage per absorbed dose is about equal for both types of radiation at high dose rates.

The amount of reparable and irreparable damage in yeast cells for these two types of radiation can be determined by irradiation at extremely low absorbed dose rates. Such experiments were performed in our laboratory and are reported in this paper.

C. Results and Discussion

Synchronized diploid yeast cells in the early G1 phase were irradiated with ^{60}Co-γ-rays and 185 kVp x-rays (E_{eff} = 40 keV), respectively. Cell suspensions bubbled with oxygen were irradiated at a dose rate of 27 rads/min. This dose rate was so low that maximum repair during irradiation was guaranteed. This was deduced from the finding that the twofold dose rate resulted in the same survival curve.

The most interesting results of the low dose rate experiments are:

1. ^{60}Co-γ-rays generate no irreparable damage up to an absorbed dose of 120 krads. This means that the initial slope of the shoulder-type dose-effect curve is equal to zero.

2. In contrast, 185 kVp x-rays applied at the same dose rate produce irreparable damage leading to a surviving fraction of 0.25 at 120 krads. In this case the initial slope of the dose-effect curve seems to be different from zero. Further clarification of this point is in progress using x-rays of various effective photon energies.

The different effects of ^{60}Co-γ-rays and 185 kVp x-rays on cell survival at extremely low dose rates up to an absorbed dose of 120 krads must be due to the different LET spectra of these two types of radiation, since all other parameters are kept constant. In Figure 2 the

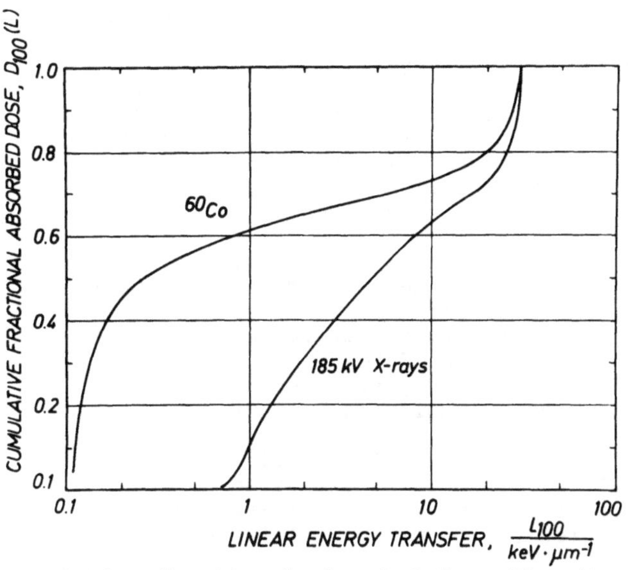

Fig. 2. Cumulative fractional absorbed dose distributions in LET for ^{60}Co-γ-rays and 185 kVp x-rays in water for energy cutoff of 100 eV

cumulative fractional absorbed dose distributions in LET are shown for ^{60}Co-γ-rays and 185 kVp x-rays in water for an energy cutoff of 100 eV (ICRU, 1970).

The LET distributions of the two types of radiation are quite different. For ^{60}Co-γ-rays, 62% of the whole imparted energy $\varepsilon = D \cdot m$ (D: absorbed dose; m: mass) is deposited in the LET range between 0.1 keV/μm and 1 keV/μm. For 185 kVp x-rays, however, only 10% of the whole imparted energy ε is deposited in this LET range. In the LET range from 1 keV/μm up to 10 keV/μm, only 11% of ε is deposited in the case of ^{60}Co-γ-rays, whereas for 185 kVp x-rays, 54% of ε is deposited in this LET range. In the LET range between 10 keV/μm and 30 keV/μm, 27% of ε is deposited for ^{60}Co-γ-rays and 36% of ε for 185 kVp x-rays. Table 1 summarizes the fractions of the total imparted energy ε, which are deposited in the 3 LET ranges L^i_{100} (i = 1, 2, 3).

Table 1. The fractions of the total imparted energy deposited in the LET ranges L^i_{100} (i = 1, 2, 3) for ^{60}Co-γ-rays and 185 kVp x-rays

	L^1_{100}	L^2_{100}	L^3_{100}
L_{100}-range	$0 < L_{100} \leq 1\ \frac{keV}{\mu m}$	$1\ \frac{keV}{\mu m} < L_{100} \leq 10\ \frac{keV}{\mu m}$	$10\ \frac{keV}{\mu m} < L_{100} \leq 30\ \frac{keV}{\mu m}$
^{60}Co	0.62	0.11	0.27
185 kVp x-rays	0.10	0.54	0.36

The LET range L^1_{100} cannot be responsible for the irreparable damage caused by 185 kVp x-rays, because the fraction of ε deposited in L^1_{100} by this radiation is only one-sixth of the fraction of ε deposited in L^1_{100} by ^{60}Co-γ-rays. Nor can the LET range L^3_{100} cause the irreparable damage. The fraction of ε deposited in the LET range L^3_{100} by 185 kVp x-rays is greater by a factor of 1.33 in relation to ^{60}Co-γ-rays. If this LET range were responsible for the irreparable damage produced by 185 kVp x-rays, absorbed doses of ^{60}Co-γ-rays 1.33 times higher than the absorbed doses for 185 kVp x-rays should result in the same surviving fractions. This is not the case, as can be seen in Figure 1. For example, 90 krads of 185 kVp x-rays result in a surviving fraction of 0.45; a 1.33 times higher absorbed dose of ^{60}Co-γ-rays, i.e., 120 krads, should result in the same surviving fraction. However, for ^{60}Co-γ-rays the surviving fraction at 120 krads is $1._2$ From these considerations it must be concluded that the LET range L_{100} (1 keV/μm < $L_{100} \leq$ 10 keV/μm) is responsible for the irreparable damage generated by 185 kVp x-rays up to absorbed doses D = 120 krads.

The energy depositions for ^{60}Co-γ-rays and 185 kVp x-rays are made exclusively by secondary electrons. From the relation between L_{100} and electron energy it can be derived that the irreparable impairment is produced by secondary electrons with energies in the range from about 1.5 keV to about 15 keV. Taking 3 keV/μm as the mean L_{100} for this LET range and 30 eV as the mean deposited energy in a spur, the mean distance between spurs in this LET range amounts to 10 nm.

In conclusion, the irreparable damage caused in diploid yeast cells by absorbed doses of up to 120 krads cannot be explained by the high LET fraction of the total imparted energy ε of ^{60}Co-γ-rays and 185 kVp x-rays. Rather it is radiation of the medium LET range that is responsible for this effect.

References

International Commission on Radiation Units and Measurements: Linear Energy Transfer, ICRU Report 16, p. 12, Washington, 1970

5. Survival of Yeast after Heavy Ion Irradiation

U. BERTSCHE and H. LIESEM

A. Abstract

Survival curves for diploid yeast after irradiation with particles of different charge and energy have been fitted to models developed by Katz et al. (1971) and by Kappos and Pohlit (1972).

B. Introduction

Many models have been developed recently to explain reactions of the living cell after irradiations of various qualities. Models on a more physical basis consider the deposition of energy along the path of the ionizing particle and calculate from there multitarget single-hit curves, as done by Katz et al. (1971). Biologic models, on the other hand, characterized by considering time-dependent recovery and repair effects, cannot predict survival in different LET ranges.

We used the model of Katz et al. to predict survival in the investigated LET range and fitted our experimental data to the cybernetic model developed by Kappos and Pohlit (1972). The data are given from irradiation of the diploid wild type yeast strain 211 (*Saccharomyces cerevisiae*).

C. Materials and Methods

The experiments were performed using the compact cyclotron of the Deutsches Krebsforschungszentrum Heidelberg. Protons and deuterons were accelerated to 20 and 10 MeV respectively and reduced to a certain energy by aluminium foils to give LET values in yeast ranging from 20 to 200 MeV cm^2/g. The LET values of yeast for ions and x-rays are given in Table 1. The setup is shown in Figure 1 and has been described in detail previously (Liesem and Bertsche, 1975). Dosimetry was performed with a transmission-type ionization chamber and checked by a ferrous sulfate dosimeter system.

α-particles from radioactive Am-241 were reduced by a thin gold foil to an energy of 3.5 MeV corresponding a LET value in yeast of 1170 MeV cm^2/g. Yeast cells were irradiated also with 70 kV x-rays. The LET value is nearly identical to that of 20 MeV protons.

The survival following immediate and delayed plating after 48 h was measured in all experiments. Due to the low ranges of ionizing particles all irradiations were performed with a monolayer of cells placed on a wet Millipore filter. From the measurement of Bragg curves it was found that the maximum energy loss occurred after transition through the cells. However, 3.5 MeV α-particles loose a significant

Table 1

Radiation	Energy	Mass stopping power for yeast $(1/\rho)\cdot(dE/dx)$ in $MeV\cdot cm^2\cdot g^{-1}$
X-rays	70 kV	29
Protons	20.2 MeV	26
	14.7 MeV	34
	8.7 MeV	51
Deuterons	8.0 MeV	94
	3.5 MeV	180
α-particles	3.5 MeV	1170

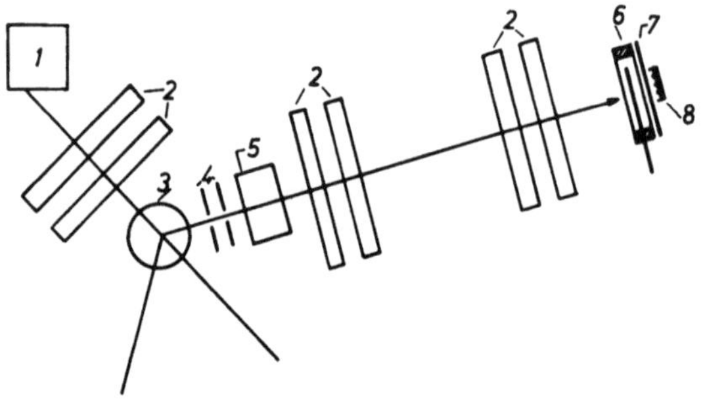

Fig. 1. Schematic drawing of experimental setup for irradiations with heavy ions: (1) fixed energy cyclotron, (2) quadrupole lenses, (3) analyzing magnet, (4) beam aperture, (5) beam-steerer, (6) transmission-type ionization chamber, (7) aluminium absorber, (8) remote-control sample changer

portion of their energy in the cells and cover therefore a LET range from about 1050 to 1250 MeV cm^2/g.

Further experimental and culture conditions have been described previously (Liesem and Bertsche, 1975).

D. Results and Discussion

Dose-survival measurements for diploid yeast, irradiated with x-rays in air, fast protons of 20, 15, and 9 MeV, deuterons of 8 and 3.5 MeV, and α-particles of about 3.5 MeV are shown in Figures 2-5. The plotted points result from different runs taken over a period of six months.

The experiments show nearly identical survival following proton, deuteron, and x-ray irradiations as expected by calculations according to the theory of Katz et al. (1971).

We have fitted the measured x-ray survival N/N_0 with the parameters $m = 6$ and $E_0 = 1.7 \times 10^6$ erg/cm^3 from the formula

$$N/N_0 = 1 - (1-\exp(-D/E_0))^m$$

given by Katz et al. (1971) where D represents the absorbed dose.

The probability of lethal damage by direct hits instead of prior accumulation of sublethal damage by δ-rays becomes higher in the upper LET range, where the survival is described by a four parametric formula:

$$N/N_0 = \exp(-\sigma_0\cdot P\cdot D/L)\cdot\left[1-\{1-\exp(-(1-P)\cdot D/E_0)\}^m\right]$$

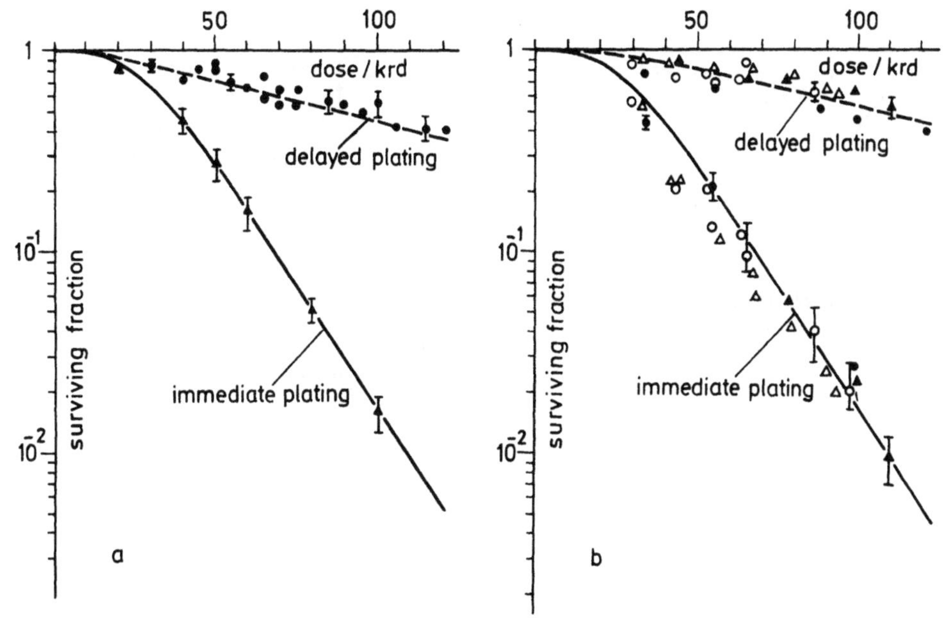

Fig. 2 a and b. Survival of diploid yeast strain 211. (a) Irradiation with 70 kV x-rays filtered with 0.7 mm aluminium. Dose-rate: 75 krad/ min, LET = 29 MeV cm^2/g. The solid line has been fitted to the model of Katz et al. (1971). (b) Irradiation with 20.2 MeV protons. LET = 26 MeV cm^2/g. The solid line for immediate plating indicates a least square fit of experimental points

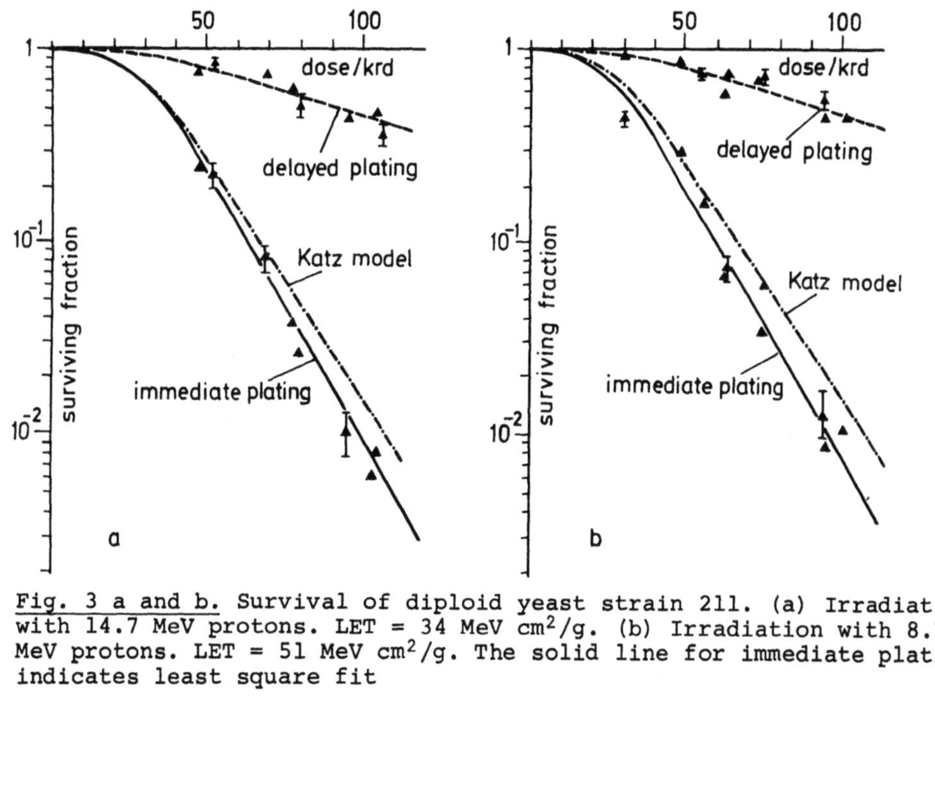

Fig. 3 a and b. Survival of diploid yeast strain 211. (a) Irradiation with 14.7 MeV protons. LET = 34 MeV cm^2/g. (b) Irradiation with 8.7 MeV protons. LET = 51 MeV cm^2/g. The solid line for immediate plating indicates least square fit

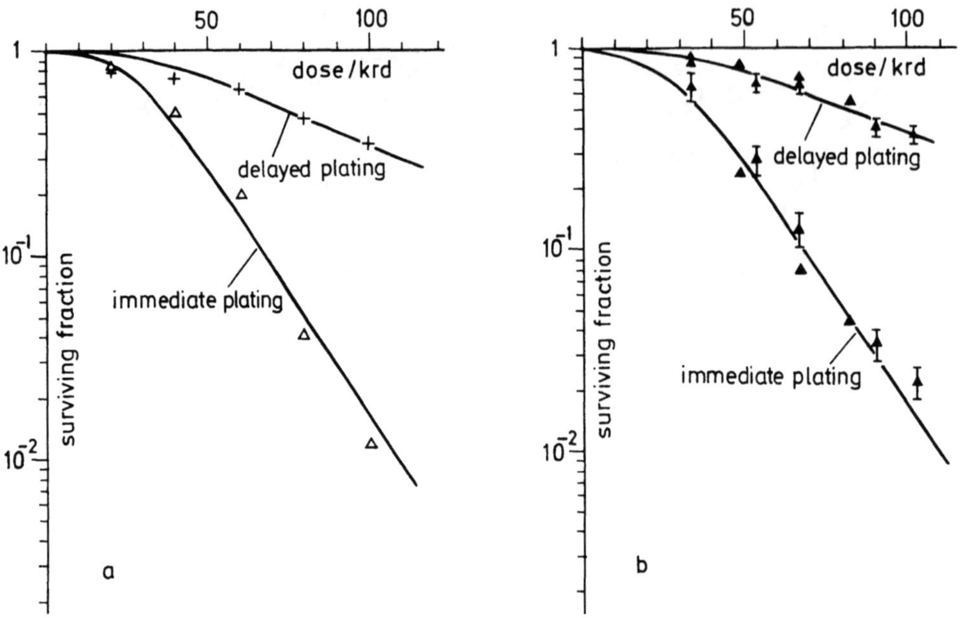

Fig. 4 a and b. Survival of diploid yeast strain 211. (a) Irradiation with deuterons of about 8 MeV. LET = (94±5) MeV cm^2/g. (b) Irradiation with deuterons of about 3.5 MeV. LET = (180±15) MeV cm^2/g. The solid line for immediate plating has been fitted to the model of Katz et al. (1971)

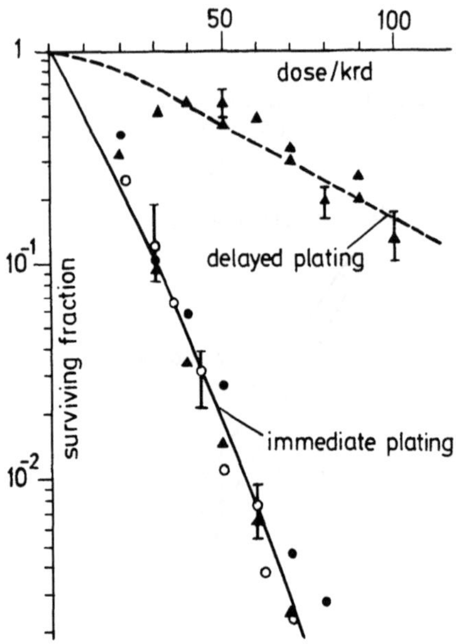

Fig. 5. Survival of diploid yeast strain 211 following 3.5 MeV α-irradiation. LET = (1170 ± 150) MeV cm^2/g. The solid line for immediate plating has been fitted to the model of Katz et al. (1971)

L is the value of linear energy transfer in yeast, and P the probability of inactivation by direct hits:

$$P = \left[1-\exp(-z^2/k\beta^2)\right]$$

where z represents the effective nuclear charge and $\beta = v/c$, the velocity coefficient of the penetrating ion (Katz et al., 1971). The measured survival following α-irradiation was fitted with $\sigma_o = 3.4 \times 10^{-9}$ cm^2, $k = 1100$, and the two parameters mentioned above. There is good agreement between theoretical prediction and experimental data for other particles and energies, especially for deuterons, whereas the survival following proton irradiation is somewhat lower than calculated. However, the differences are near the limit of experimental error.

The cybernetic model of cell survival developed by Kappos and Pohlit (1972) distinguishes between three cell states after irradiation, as shown in Figure 6. At low absorbed doses in the shoulder region, a state A is reached where the cells are in the most insensitive condition. No investigations of recovery have been carried out in our experiments.

At higher absorbed doses the straight exponential form of the dose-effect curve is reached where the cells are most sensitive. Now many of the cells are either in state B, where repair of lethal damage is possible, or in state C with irrepairable damage. Failure of repair leads from state B to C. The transitions are described by reaction rate coefficients η_{AB}, η_{AC}, and η_{BC}. Whereas η_{AC} can be assumed to be correlated with direct radiation effects, the other coefficients are generally dose-dependent. At high dose rates, however, there is no time for recovery, and repair of potentially lethal damage, and these coefficients become constant. They are measured by slope evaluation of the exponential part of the dose-effect curve taken at high dose rates above 1 krad/min. Reduction of the total cell number is due to $\eta_{AB} + \eta_{AC}$. If, however, the cells are allowed to repair their potentially lethal damage, a repair curve is obtained, the slope of which gives η_{BC}.

There are two ways to determine η_{AC}: (1) a measurement at extremely low absorbed dose rates, where η_{AB} approaches zero, is possible. This was carried out with x-rays and α-rays. (2) The empirical relation

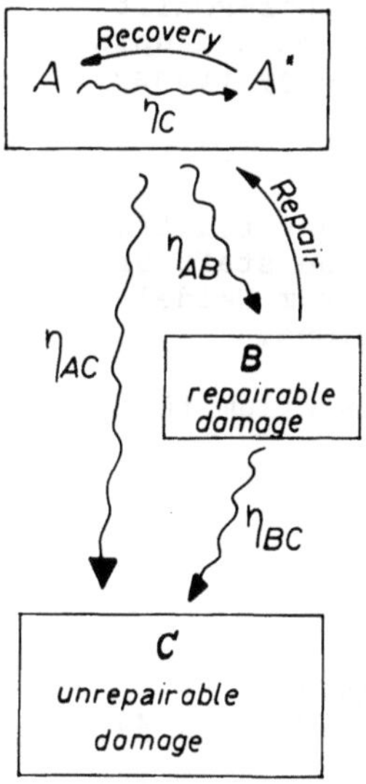

Fig. 6. Cybernetic model of cellular reaction developed by Kappos and Pohlit (1972)

proposed by Pohlit (personal communication, 1974)

$$\eta_{BC} = \eta_{AC} + \eta_{AB}^2$$

was used in experiments with heavy ions where irradiations at very low absorbed dose rates cannot be performed because of prolonged irradiation times. In both cases the determination of η_{AC} for x-ray and α-ray survival gives an agreement within 15% in η_{AC}. Within this limit of uncertainty the calculation of irrepairable damage using the relation mentioned above seems to be justified in the investigated region of linear energy transfer. Figure 7 shows LET-dependent values of the reaction rate coefficients in a double logarithmic plot. The values of η_{AC} are somewhat higher than those measured by Kappos and Pohlit (1972), which seems to result from different experimental techniques. The rate constants also give values of the magnitude of repair measured as a dose modifying factor DMF = $(\eta_{AB} + \eta_{AC})/\eta_{BC}$. These values show a slight LET dependence.

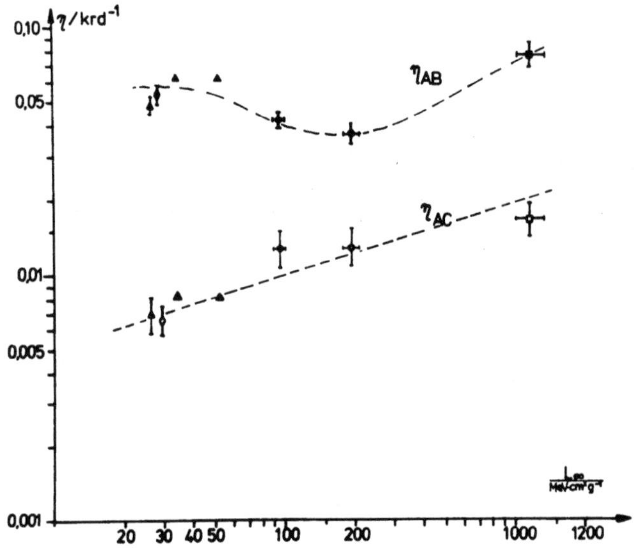

Fig. 7. Reaction rate coefficients η_{AB} and η_{AC} of Kappos-Pohlit model, dependent upon linear energy transfer

Acknowledgements. We wish to thank Prof. W. Pohlit for many fruitful discussions during the experiments. We are grateful to the staff of the Cyclotron Group of the Deutsches Krebsforschungszentrum Heidelberg for their help and hospitality during the irradiations.

This study was supported by a grant from the Bundesministerium für Forschung und Technologie.

References

Kappos, A., Pohlit, W.: Int. J. Radiat. Biol. 22, 51 (1972)
Katz, R., Ackerson, B., Homayoonfar, M., Sharma, W.S.: Rad. Res. 47, 402 (1971)
Liesem, H., Bertsche, U.: VII. Int. Conf. on Cyclotrons, Zürich (1975)

6. Incorporation of (Methyl ^{3}H) 5'-dTMP into DNA of Diploid Yeast Cells as Tracer of Liquid Holding Recovery after X-Ray Irradiation[1]

G. W. HÖLTZ and M. BRENDEL

A. Introduction

It is known that the cells of the yeast *Saccharomyces cerevisiae* exhibit a postirradiation increase in colony-forming ability after delayed plating. This phenomenon has been termed liquid holding recovery (LHR) (Patrick and Haynes, 1964). LHR is dependent upon the availability of metabolic energy (Jain and Pohlit, 1972). Respiratory deficient yeast strains must be supplemented with glucose in the LHR buffer to perform LHR.

Isolation of a diploid yeast strain, auxotrophic for 5'-dTMP, makes it possible to study biochemically the LHR phenomenon by measuring the incorporation of (methyl ^{3}H) 5'-dTMP into the DNA while holding in buffer.

B. Material and Methods

Strain. Throughout the work strain 211*B ρ^- of *S. cerevisiae* was used. This diploid yeast strain has the genotype:

a ade2 arg4-17 his5-2 lys1-1 ilv2 tlr tmp1-3 typl ρ^-
α + + + + *ilv2 tlr tmp1-3 typl* ρ^-.

Strain 211*B ρ^- was derived from strain MB1052 by screeing for 5'-dTMP low-requirement. Strain MB1052 was constructed by mating haploid strains 211-1aM *tmp1-3* α and MB1051-13C *tmp1-3 a* (Brendel and Fäth, 1974). Control of mating of the 5'-dTMP auxotrophs necessitated screening for diploid zygotes by means of relative x-ray sensitivity.

Media. The synthetic medium contains per ml: 6.7 mg Difco yeast nitrogen base w/o amino acids, 2 mg Difco casamino acids w/o vitamins, 20 mg D(+)-glucose and 7.2 µg 5'-dTMP.

Irradiation. Yeast was grown on synthetic agar for 4 days at 30°C. The cells were suspended in phosphate buffer (8 g KH_2PO_4 and 1.43 g Na_2HPO_4 in 1000 ml H_2O, pH 6), washed 3 times, and resuspended in the same buffer. The cells were irradiated at a concentration of 5 x 10^7 cells/ml with a high voltage x-ray tube. The irradiation was performed at room temperature and a dose-rate of 80 krad/min.

Labelling of Cells. After irradiation the cells were diluted 1:5 in the phosphate buffer, which contained 10 mmole/ml glucose, 7.2 µg/ml 5'-dTMP, and 4 µCi/ml (methyl ^{3}H) 5'-dTMP (Amersham, specific activity approx. 20 Ci/mmole). Uptake of label was monitored as follows: 0.1 ml

[1] This work is part of the Ph.D. thesis of the first author.

aliquots were withdrawn and suspended in 1 ml ice-cold 5% trichloracetic acid (TCA). Precipitation was allowed at 0°C for 30 min. To determine alkali stability, the aliquots were incubated in 0.3 N NaOH at 37°C overnight before neutralization by HCl and precipitation by ice-cold TCA.

Extraction and Purification of DNA. The cells were washed, frozen, and broken in an Eaton press (Eaton, 1962). The resulting homogenate was thawed at 60°C; pronase (B grade, Calbiochem; preheated at 60°C for 30 min in 0.5 *M* NaCl, 0.1 *M* EDTA) was added to a final concentration of 2 mg/ml. The solution was made 1% for Sarkosyl NL97 (Geigy) and incubated first at 60°C for 1 h and then at 37°C overnight.

Centrifugation was performed as follows: 7.2 g CsCl were dissolved in 3 ml of distilled water and 3 ml of the homogenate was added. The gradients were run at 32 x 10^3 rpm using a Ti50 rotor in a Spinco L3-50 centrifuge (Beckman Instruments) at 25°C for 60 h. Fractions were collected by puncturing the polyallomer tubes at the bottom and collecting the drops in portions of 20. The fractions were precipitated with ice-cold TCA 30 min at 0°C.

Radioactivity was measured in a liquid scintillating system containing 4 g PPO, 200 mg POPOP, 20 ml ethyleneglycol, 100 ml methanol, 60 g naphthalene per 1 l dioxane. All samples were counted in identical volumes (10 ml) of scintillator. The efficiency of the scintillator was controlled by measuring the external standard ratios.

C. Results

Cell survival after LHR is shown in Figure 1: cells of strain 211*B ρ^- were irradiated with 100 krad and incubated in phosphate buffer with 10 mmole/ml glucose and 7.2 µg/ml 5'-dTMP. After the third hour of incubation in the LHR buffer, the cells show a significant increase in colony-forming ability.

Incorporation of (methyl ^{3}H)5'-dTMP into the DNA after LHR treatment is shown in Figure 2. Repair of x-ray damage occurs during the first three hours of LHR treatment. Radioactivity measured with nonirradiated cells is not located in the DNA. There is a reversible binding of (methyl ^{3}H)5'-dTMP with the cell membrane or cell wall that cannot be removed by the TCA treatment. This binding slowly decreases during holding in phosphate buffer. Irradiated yeast cells show a lower binding of radioactivity to the cell wall or cell membrane. From Figures 1 and 2 we can conclude that repair of DNA damage alone is not sufficient for restoration of cellular viability.

Radioactivity was measured in TCA-treated whole cells. It had to be determined, therefore, whether the radioactivity was located in DNA and not in a residual 5'-dTMP pool in the cytoplasm. As described in Material and Methods, cells of strain 211*B ρ^- were washed, frozen, and broken in an Eaton press. The DNA was extracted and purified in an isopycnic CsCl gradient (Fig. 3). Of the total 5000 cpm in the homogenate, 2900 cpm banded in CsCl as yeast nuclear DNA. There is no mitochondrial DNA, which is in agreement with the results of Brendel and Fäth (1974), who found that introduction of the *tmp1* marker led to neutral cytoplasmatic "petites".

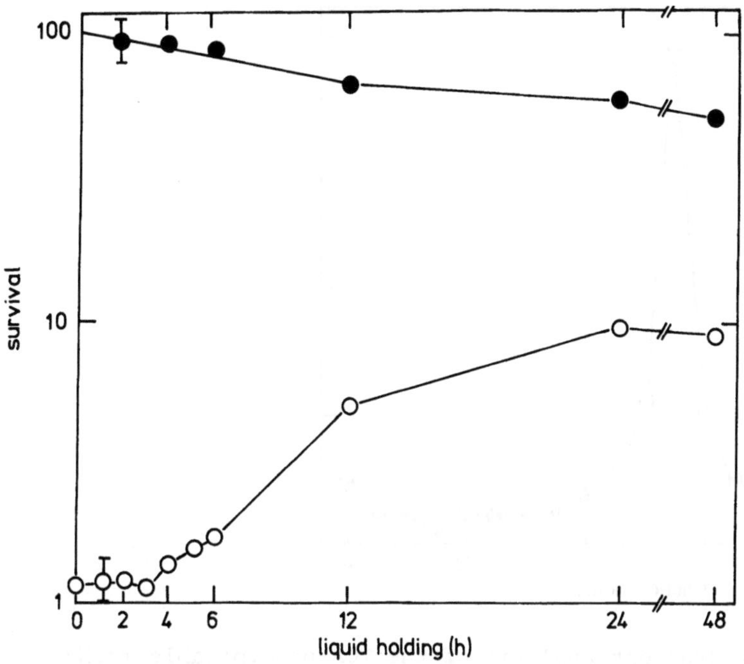

Fig. 1. LHR of strain 211*B ρ^- irradiated with 100 krad. Phosphate buffer was supplemented with 10 mmole/ml glucose and 7.2 µg/ml 5'-dTMP. —●—●— nonirradiated cells; —o—o— irradiated cells (standardized to control)

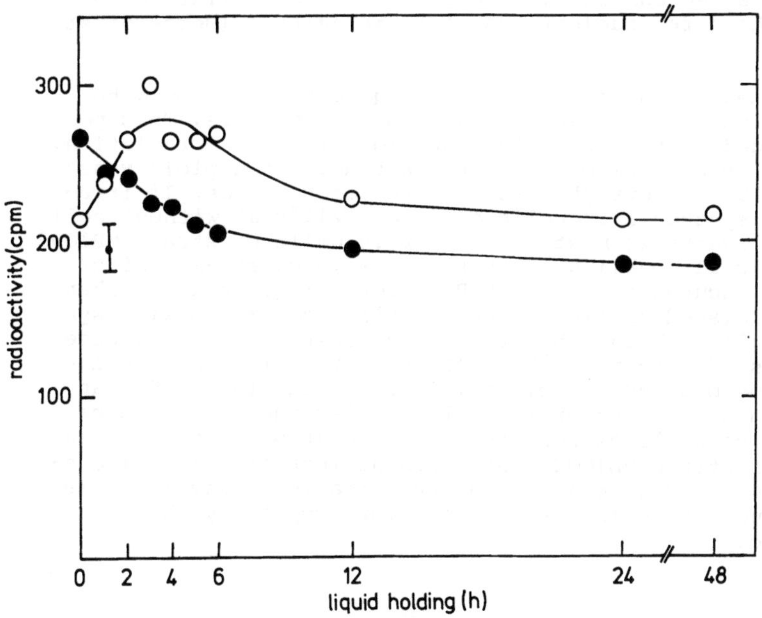

Fig. 2. Incorporation of (methyl ^3H)5'-dTMP into DNA of strain 211*B ρ^- irradiated with 100 krad. After irradiation cells were suspended in phosphate buffer with 10 mmole/ml glucose, 7.2 µg/ml 5'-dTMP, and 4 µCi/ml (methyl ^3H)5'-dTMP. —o—o— irradiated cells; —●—●— nonirradiated cells

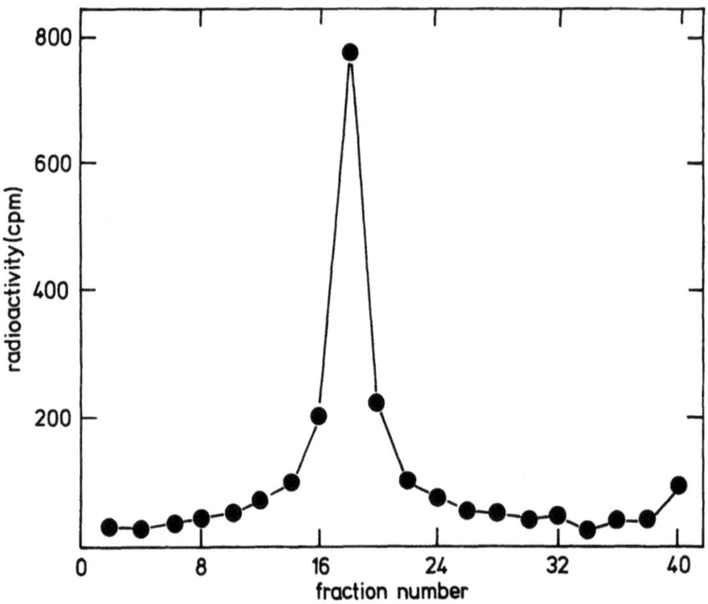

Fig. 3. CsCl density gradient analysis of the TCA-precipitable methyl-^{3}H label incorporated by strain 211*B ρ⁻ after LHR. Cells were irradiated with 100 krad and suspended in phosphate buffer with 10 mmole/ml glucose, 7.2 μg/ml 5'-dTMP and 20 μCi/ml (methyl ^{3}H) 5'-dTMP. After 3rd hour of LHR treatment, cells were withdrawn from phosphate buffer

Density of the extracted DNA is 1.699 g/cm^3. This is similar to results of other authors for nuclear DNA of *S. cerevisiae* (Bernardi et al., 1968).

Incorporation of (methyl ^{3}H) 5'-dTMP during liquid holding could be caused by DNA replication in a small fraction of the cells. This possibility was excluded by the following experiment: after irradiation, cells of strain 211*B ρ⁻ were mixed with nonirradiated diploid yeast cells of strain MB1092 auxotrophic for several amino acids. If irradiation causes DNA synthesis in cells of strain 211*B ρ⁻ without cell doubling, this DNA synthesis must also occur in cells of strain MB1092 and should lead to cell doubling in this unirradiated strain. After different times of incubation in the LHR buffer, aliquots were taken and after dilution plated on two different solid media: one with synthetic medium plus 5'-dTMP and the other with yeast extract, glucose plus 5'-dTMP. Cells of strain 211*B ρ⁻ can form colonies on both agar plates with the same plating efficiency. Cells of strain MB1092 can grow only on medium with yeast extract. The difference in the number of all colonies on both plates permits the calculation of the concentration of cells of strain MB1092. There was no increase in cell concentration, as shown in Figure 4. During LHR treatment, any increase in the number of budding cells was checked microscopically. No increase was observed.

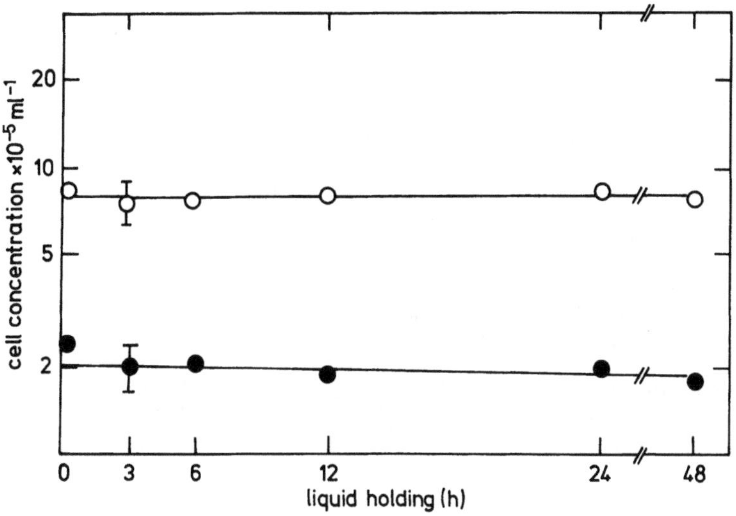

Fig. 4. Concentration of cells of strain MB1092 during LHR. Cells of strain 211*B ρ⁻ were mixed with two different cell concentrations of strain MB1092 during LHR. —o—o— 8 x 10^5 cells/ml; —●—●— 2 x 10^5 cells/ml

D. Discussion

The amount of radioactivity incorporated during LHR enables us to determine the average size of newly synthesized DNA strands per damage site. By comparing radioactivity incorporated during one round of normal DNA replication and that incorporated during maximal LHR of cells irradiated with 100 krad, we may conclude that about 4% of the yeast genome is newly synthesized. Frey and Hagen (1973) reported that 36 single-strand breaks are induced per 100 krad per 1000 nucleotides. These results, obtained with native T1 and calf thymus DNA, are used to estimate the number of nucleotides incorporated into DNA per break. If every lesion were repaired, only one nucleotide would be incorporated into DNA. This, however, cannot be expected because only 10% of all cells after LHR are able to form colonies (see Fig. 1). The average size of each newly synthesized DNA strand per break would be ten nucleotides if all DNA damage were repaired only by this 10%. This also cannot be expected, because it is unlikely that 90% of the irradiated cells that do not regain viability after LHR treatment are unable to repair any DNA damage. The number of nucleotides incorporated into DNA per DNA break, therefore, should be between 2 and 9. Painter and Young (1972) reported that 3 to 5 nucleotides are incorporated when a single-strand break is repaired. Their result, obtained with mammalian DNA, seems to hold for yeast cells also, if the results obtained in vitro are also valid in vivo.

Repair of DNA damage alone does not seem to be sufficient for restoration of the colony forming ability of the cells. It is possible that a different, non-DNA damage after irradiation with x-rays must be repaired. The spindle plaque is a likely candidate, because it is a sensitive target and because plaque duplication is mandatory in the budding process (Moens and Rapport, 1971; Byers and Goetsch, 1973).

E. Summary

The liquid holding phenomenon was examined in diploid yeast strain 211*B ρ^-, which is auxotrophic for 5'-dTMP. For maximal LHR this strain requires 10 mmole/ml glucose in LHR buffer.

Repair of x-ray damage in the DNA as measured by incorporation of labelled 5'-dTMP occurs during the first 3 h of LHR treatment. After the third hour of LHR treatment, the number of colony forming units increases significantly.

Incorporation of 5'-dTMP into the DNA during the first 3 h of LHR is not caused by normal DNA replication of some cells.

Comparing the incorporation of 5'-dTMP due to repair to that due to normal replication, it was concluded that 2 to 9 nucleotides per DNA break are required for repair under LHR conditions in diploid yeast.

References

Bernardi, G., Carnevali, F., Nicolaieff, A., Piperno, G., Tecce, G.: Separation and characterization of a satellite DNA from a yeast cytoplasmic "petite" mutant. J. Molec. Biol. 37, 493-505 (1968)

Brendel, M., Fäth, W.W.: Isolation and characterization of mutants of *Saccharomyces cerevisiae* auxotrophic and conditionally auxotrophic for 5'-dTMP. Z. Naturforsch. 29c, 733-738 (1974)

Byers, B., Goetsch, L.: Duplication of spindle plaques and integration of the yeast cell cycle. Cold Spring Harbor Symp. Quant. Biol. 28, 123-131 (1973)

Eaton, N.R.: New press for disruption of microorganism. J. Bacteriol. 83, 7359-7360 (1962)

Frey, R., Hagen, U.: Oxygen effect on γ-irradiated NDA. Radiat. Environ. Biophys. 11, 125-133 (1974)

Jain, V.K., Pohlit, W.: Influence of energy metabolism on the repair of x-ray damage in living cells. Biophysik 8, 254-263 (1972)

Moens, P.B., Rapport, E.: Spindles, spindle plaques and meiosis in the yeast *Saccharomyces cerevisiae* (Hansen). J. Cell Biol. 50, 344-361 (1971)

Painter, R.B., Young, B.R.: Repair replication in mammalian cells after x-irradiation. Mutation Res. 14, 225-235 (1972)

Patrick, M.H., Haynes, R.H.: Dark recovery phenomena in yeast. II. Conditions that modify the recovery process. Radiat. Res. 23, 564-579 (1964)

7. Influence of Intracellular Adenosine-Triphosphate Concentration on Survival of Yeast Cells Following X-Irradiation

R. D. REINHARD and W. POHLIT

A. Introduction

Experiments on the effect of ionizing radiation on yeast cells have lead to the development of a cybernetic model, which allows complete description of the radiation response of living cells (Kappos and Pohlit, 1972). If the interpretation of the radiation effect is based on the dose-effect curve and hence on the colony forming ability of the population, two different reactions have to be considered: (1) damage to the cell by irradiation and (2) the counterreaction of reactivation of damaged cells by repair processes.

Studies on the repair of irradiated yeast cells using inhibitors of energy metabolism show that liquid holding reactivation (Jain and Pohlit, 1972; Jain et al., 1973) and split dose recovery (Jain and Pohlit, 1973; Kiefer, 1971) are dependent on the energy providing systems of the cell. Especially the glucose antimetabolite 2-deoxy-D-glucose (2-DG) causes significant changes in radiation sensitivity and repair capacity of x-irradiated yeast cells under different experimental conditions, as has been shown recently (Jain et al., 1975). This paper is a report of the effect of 2-DG and glucose on the ATP concentration in the cell and possible changes of radiation sensitivity connected to variations of the ATP pool, if cells are plated on nutrient agar directly after irradiation.

B. Materials and Methods

I. Organisms

The yeast *Saccharomyces cerevisiae* diploid strain 211 (Laskowski, 1960) and its respiratory deficient mutant 211 pμ were used in these experiments.

II. Methods

Cell cultivation, irradiation apparatus, and measurement of the surviving fraction were carried out as described earlier (Jain, 1970; Jain and Pohlit, 1972).

Preincubation and Irradiation of Cells. Stationary cells from agar plates or from liquid culture were suspended in phosphate or phthalate buffer, containing 2-DG or glucose and shaken at 30°C. Cell concentration during preincubation was 10^8/ml, during irradiation 10^8/ml for the wild type cells, and 10^6/ml for the mutant. Thus 2-DG was present during irradiation of the wild type, while it was diluted by a factor of 100 just before irradiation of the mutant. After dilution of the irradiated samples for plating on agar the concentrations of the preincubation agents were biologically ineffective in all cases.

Cell volume was measured with an electronic cell counter (Telefunken); the mean volume was determined from the volume distribution spectrum for each population.

ATP was extracted by a modification of the procedure described by Polakis and Bartley (1966) and determined by the luciferin-luciferase assay system (Strehler, 1970). Luciferase and 2-DG were supplied by Serva, Heidelberg.

C. Results

I. ATP Concentration in Wild Type Yeast, Strain 211

Figure 1 shows the effect of different incubation procedures on the ATP concentration in yeast cells of the diploid wild type. Treatment of stationary cells with 1 mmol/l 2-DG in buffer decreases the ATP level rapidly to about 25% of the control (Fig. 1a). Use of a 10-fold concentration of 2-DG enhances the reduction. It is not possible to obtain a further significant decrease of ATP by prolonging the incubation times. Treatment of wild-type cells with glucose does not increase the ATP concentration beyond the value of the control.

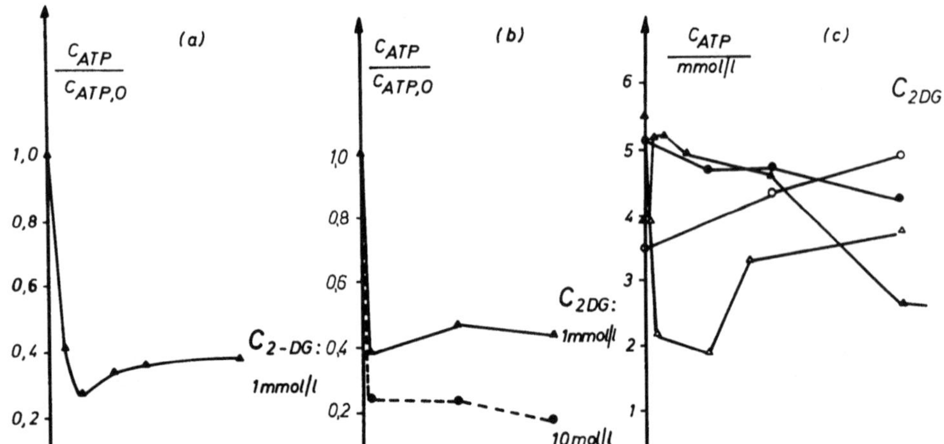

Fig. 1 a-c. (a) Relative concentration of ATP in stationary yeast cells, strain 211, c_{ATP}, as function of incubation time, t, in buffer containing 1 mmol/l 2-DG. (b) As (a), for two concentrations of 2-DG. (c) Absolute concentration of ATP in stationary yeast cells incubated in phosphate buffer (cells from agar (▲) and cells from liquid culture (△)), in buffer with 10 mmol/l glucose (cells from agar (●) and cells from agar after 18 days storage at 4°C (o))

Obviously 5 mmol/l is the highest concentration attainable inside the cell (Fig. 1c). However, cells that have become deprived of ATP can be enriched again by incubation with glucose. Moreover Figure 1c shows the ATP concentrations of stationary cells, which have been cultivated on agar and in a liquid medium, after having been suspended in phosphate buffer.

II. ATP Concentration in the Respiratory Deficient Mutant 211 pμ

Stationary cells of the respiratory deficient mutant have an ATP content of about 14% of that of the wild type.

Incubation with 2-DG also decreases intracellular ATP in this strain very rapidly (Fig. 2a). Incubation for longer times causes the ATP to increase again, possibly by the breakdown of ATP-consuming processes (Fig. 2b).

Figure 2c shows the accumulation of ATP in 211 pμ upon incubation in glucose; after about 30 min a value can be obtained that is about half that of the wild type. Incubation of this mutant with glucose causes a limited increase of the endogenous carbohydrate reservoir of trehalose and a small increase of glycogen.

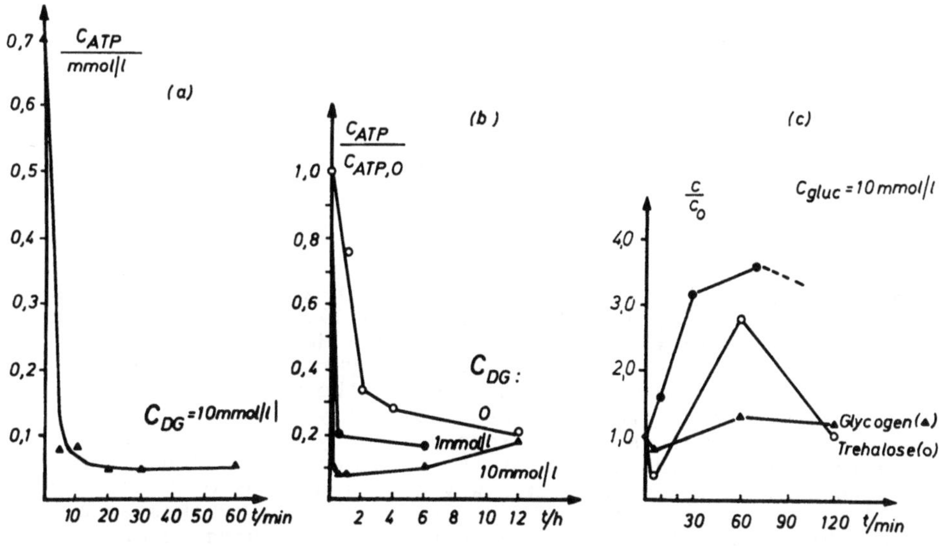

Fig. 2 a-c. ATP concentration in stationary respiratory deficient yeast cells, strain 211 pμ, c_{ATP}, as function of incubation time t. (a) Absolute ATP concentration in cells (211 pμ) suspended in phosphate buffer containing 10 mmol/l 2-DG. (b) As (a), for different concentrations of 2-DG. (c) Relative concentration of ATP, glycogen, and trehalose in stationary cells (211 pμ) as function of incubation time t in phosphate buffer with 10 mmol/l glucose

III. ATP Concentration and Dose-Effect Curves

Stationary cells of the wild type show a dose-effect curve with a shoulder (D_q = 29 krad) and with a slope in the exponential part of D_o = 20.5 krad. Incubation for 20 min with 2-DG (1 mmol/l) before irradiation causes D_q to decrease to 17.5 krad, whereas D_o remains constant (Fig. 3). The corresponding values for ATP concentrations in the cell before irradiation are 5.0 mmol/l and 0.8 mmol/l, respectively.

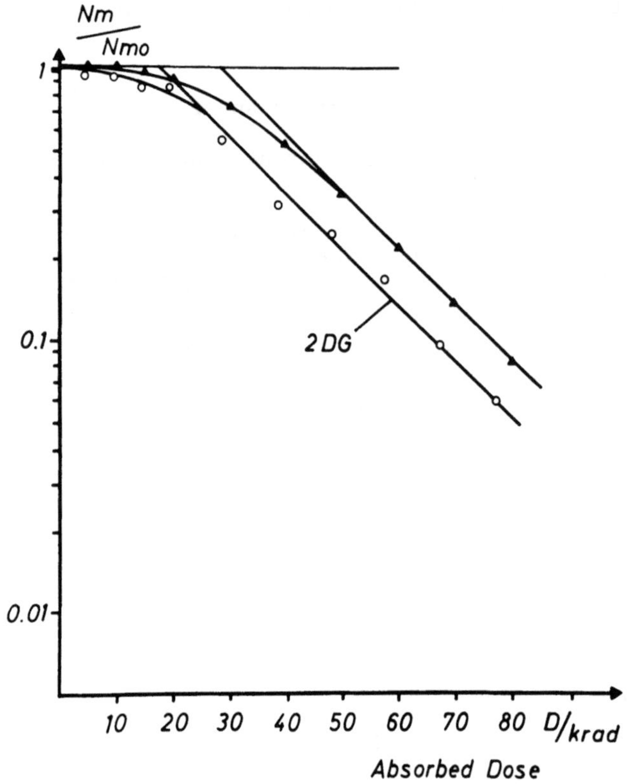

Fig. 3. Relative number of cells capable of growing to macrocolony Nm as function of absorbed dose D for stationary yeast cells, strain 211, kept in buffer (▲) and in buffer with 2-DG before irradiation (o)

Corresponding data for stationary cells of strain 211 pμ have been obtained: C_{ATP} = 0.7 mmol/l, D_q = 18.0 krad, D_o = 13.0 krad. 2-DG has a similar effect on the mutant as on strain 211: a decrease of ATP down to 0.06 mmol/l is accompanied by a corresponding reduction of D_q to 12.0 krad. On the other hand, preincubation of the mutant cells in glucose-containing buffer provides the possibility of increasing the ATP content up to the values of stationary wild-type cells (data summarized in Table 1). The extent of the shoulder in the dose-effect curve of a population treated in this way corresponds exactly to that of the wild type. Figure 4 shows the relation of all D_q values as a function of the intracellular ATP concentration. The curve is drawn through the points for both strains.

D. Discussion

The results presented here show the possibility of manipulation of the shoulder length D_q by varying the intracellular ATP concentration. The way in which C_{ATP} is changed is not essential: lowering of ATP concentration by 2-DG or by starvation causes shortening of the shoulder (with both strains), increasing ATP by glucose stretches the shoulder (with the mutant strain). The conclusion is: there is a direct influence of C_{ATP} on D_q, at least in the region from about 10 to about 30 krad.

The experiments described do not give any information about the molecular mechanisms by which the shoulder of a dose-effect curve is estab-

Table 1. ATP concentrations in *S. cerevisiae*, strains 211 and 211 pμ, after different incubation procedures

Strain	Preincubation (phosphate buffer, pH 6.0 cell concentration 10^8/ml	C_{ATP} / mmol/l
211	-	5.2[a]
	10 min 2-DG (1 mmol/l)	0.7
	20 min 2-DG (1 mmol/l)	0.85
	30 min 2-DG (10 mmol/l)	1.0
	24 h 2-DG (10 mmol/l) + glucose (10 mmol/l)	1.2
	10 days on agar plates at 4°C	4.2
211 pμ	-	0.7
	20 min 2-DG (10 mmol/l)	
	30 min 2-DG (10 mmol/l)	
	12 h 2-DG (10 mmol/l) + glucose (10 mmol/l)	0.05-0.07
	12 h 2-DG (1 mmol/l)	
	10 min glucose (10 mmol/l)	1.1
	30 min glucose (10 mmol/l)	2.2
	70 min glucose (10 mmol/l)	2.5
	13.5 h glucose (10 mmol/l) cell concentration: 10^6/ml	5.5

[a]ATP concentration for cells that have been analyzed immediately after having been grown on agar for 3 days at 30°C. During storage of agar plates in refrigerator, ATP is lost at a rate of about 2% per day

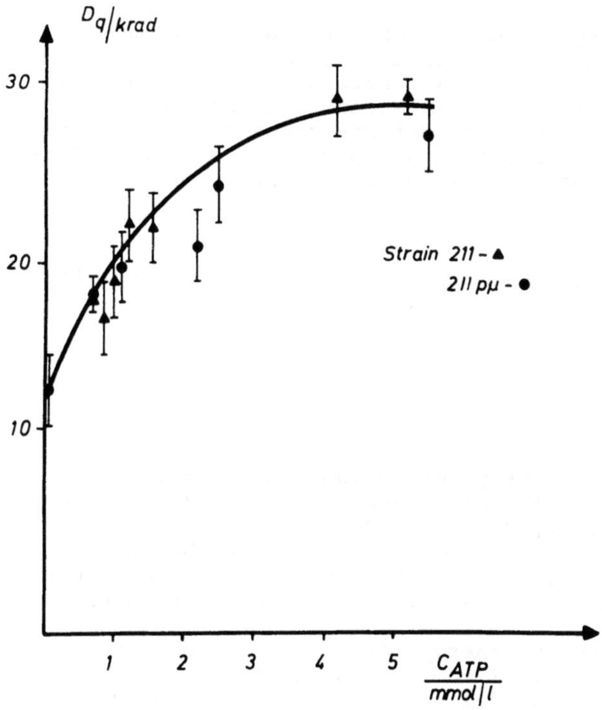

Fig. 4. Length of shoulder of diploid yeast cells, strains 211 and 211 pμ, D_q, as function of ATP concentration before irradiation, C_{ATP}

lished under the influence of ATP. But it can be derived: provided that ATP does not influence the number of primary lesions at the target, its effect could only be participation in a biochemical reaction running counter to the effect of radiation. The extent of this reaction is a steady function of C_{ATP}. A reaction like this is, by virtue of its effect, a repair reaction. The reaction involving ATP could be either a very fast reaction taking place only during irradiation and thus causing the different survival for populations with different ATP concentrations before irradiation, or it influences in addition to its effects during irradiation the repair behavior of the cells after irradiation, when the cells have been plated on nutrient agar and incubated at 30°C.

The results presented in this communication suggest that *S. cerevisiae* possesses an ATP-dependent repair system, which essentially influences the survival following small doses. A part of this hypothetical repair system could be the polynucleotide ligase, which has been shown to require ATP as a cofactor (Lindahl and Edelman, 1969).

This hypothesis is in line with the findings that under certain experimental conditions the ATP content of a cell decreases as a function of the absorbed dose, and that the ATP pool is refilled again if the cells are held under liquid holding reactivation conditions after irradiation. Further information is expected from studies on the influence of the ATP flux on the shape of the dose-effect curves (Reinhard, Pohlit and Jain, in preparation); experiments on this question are currently under way.

E. Summary

The effect of D-glucose, 2-deoxy-D-glucose, and starvation in buffer on the ATP concentration of yeast cells has been studied. In both the wild type and a respiratory-deficient mutant strain, 2-deoxy-D-glucose decreases the value for ATP, while it is enhanced by glucose only in the mutant strain. Populations with different ATP-concentrations have been irradiated. The results suggest, that ATP may be an essential factor in the system, that determines the extent of the shoulder in the dose-effect curves.

Acknowledgement. We are indebted to Dr. Jain for many stimulating discussions.

References

Jain, V.K.: Relationship between energy metabolism and growth. I. Glucose dependance of the exponential growth rate of *Saccharomyces cerevisiae*. Arch. Mikrobiol. 72, 252-259 (1970)

Jain, V.K., Pohlit, W.: Influence of energy metabolism on the repair of x-ray damage in living cells. I. Effect of respiratory inhibitors and glucose on the liquid holding reactivation in yeast. Biophysik 8, 254-263 (1972)

Jain, V.K., Pohlit, W.: Influence of energy metabolism on the repair of x-ray damage in living cells. II. Split-dose recovery, liquid holding reactivation and division delay reversal in stationary populations of yeast. Biophysik 9, 155-165 (1973)

Jain, V.K., Pohlit, W., Purohit, S.C.: Influence of energy metabolism on the repair of x-ray damage in living cells. III. Effect of 2-deoxy-D-glucose on the liquid holding reactivation in yeast. Biophysik 10, 137-142 (1973)
Jain, V.K., Pohlit, W., Purohit, S.C.: Influence of energy metabolism on the repair of x-ray damage in living cells. IV. Effect of 2-deoxy-D-glucose on the repair phenomena during fractionated irradiation of yeast. Radiat. Environ. Biophys. 13, 1-6 (1975)
Kappos, A., Pohlit, W.: A cybernetic model for radiation reactions in living cells. I. Sparsely ionizing radiations; stationary cells. Int. J. Radiat. Biol. 22, 51-65 (1972)
Kiefer, J.: The importance of cellular energy metabolism for the sparing effect of dose fractionation with electrons and UV-light. Int. J. Radiat. Biol. 20, 325-336 (1971)
Laskowski, W.: Inaktivierungsversuche mit homozygoten Hefestämmen verschiedenen Ploidiegrades. I. Aufbau homozygoter Stämme und Dosiseffektkurven für ionisierende Strahlen, UV und organische Paroxide. Z. Naturforsch. 15b, 495-506 (1960)
Lindahl, T., Edelman, G.M.: Polynucleotide ligase from myeloid and lymphoid tissues. Proc. Nat. Acad. Sci. USA 61, 680 (1968)
Polakis, E.S., Bartley, W.: Changes in the intracellular concentration of adenosin phosphates and nicotinamide nucleotides during the aerobic growth cycle of yeast on different carbon sources. Biochem. J. 99, 521-533 (1966)
Strehler, B.L.: Adenosin-5'-triphosphat und Creatinphosphat. Bestimmung mit Luciferase. In: Methoden der enzymatischen Analyse (Bergmeyer, H.U., ed.). Weinheim: Verlag Chemie, 1970

8. Split Dose Recovery and Liquid Holding Recovery in Diploid Yeast Exposed to Fractionated X-Irradiation

I. WIENHARD and J. KIEFER

A. Introduction

In cells exposed to irradiation two important recovery phenomena have been demonstrated. Split dose recovery (SDR, also defined as recovery of sublethal damage) can be detected by the split dose technique (Elkind and Sutton, 1960). Liquid holding recovery (LHR, also termed recovery from potential lethal damage) is observed if the cells are kept under nonnutrient conditions after irradiation.

In wild-type strains of diploid yeast both SDR and LHR occur after x- and ultraviolet (UV) irradiation (Bacchetti and Mauro, 1965; Kiefer, 1968, 1970; Patrick et al., 1964). LHR and SDR are strongly dependent on a functioning energy metabolism of the cells (Patrick and Haynes, 1964; Jain and Pohlit, 1972; Kiefer, 1971). This indicates that the processes are of enzymatic nature. SDR is further reduced in the presence of DNA-binding substances (Elkind et al., 1967; Kiefer, 1973), suggesting that DNA repair mechanisms may be involved in the recovery process.

By using radiosensitive mutants blocked in specific repair pathways it was shown for *E. coli* that LHR after UV irradiation is due to excision-repair of pyrimidine dimers in DNA (Ganesan and Smith, 1969; Harm, 1966).

In UV-irradiated yeast, LHR seems also to be due to excision-repair since excision negative mutants do not show LHR (Parry and Parry, 1969; Averbeck, 1970); for SDR, however, also other repair pathways seem to be important (Kiefer, 1975). After x-irradiation, the nature of repair mechanisms for SDR and LHR is still unknown.

Results of investigation of a wild-type strain and a radiosensitive *rad9-4* mutant of diploid yeast *Saccharomyces cerevisiae* with respect to SDR after x-irradiation are reported. Parallel to the investigation of SDR, the cells were examined for their ability to perform subsequent LHR in order to determine whether there exists an interrelationship between the two recovery processes.

B. Materials and Methods

I. Strains

The following diploid strains of *S. cerevisiae* were used: (1) the prototrophic wild-type strain 211, originally obtained from W. Laskowski; (2) the radiosensitive mutant S 2095 homozygous for the *rad 9-4* allele (Game and Mortimer, 1974), provided by D. Averbeck. Strain S 2095 requires valine and isoleucine as growth factors.

II. Media

Nutrient Medium. A modified synthetic Wickerham-medium (Wickerham, 1951) was used, containing asparagine as nitrogen source and a 5-fold higher concentration of ferrous chloride $FeCl_3$ (0.25 mg/l instead of 0.05 mg/l). For plates the medium was solidified with 2% Difco agar. In the experiments with strain S 2095 100 mg/l L-isoleucine and L-valine were added.

Buffer Solution. For irradiation and liquid holding, cells were suspended in potassium phthalate buffer (pH 6)(Documenta Geigy: Wissenschaftliche Tabellen).

Culture Conditions. To obtain stationary phase cultures the cells were grown at 30°C for 3 days on Wickerham agar plates (depth 1 cm). After this the cell population consists of about 98% of single cells with a rather homogenous cell size distribution.

Exponential phase cultures were obtained by growing cells in batch-cultures. Stationary phase cells were inoculated at a concentration of 10^5/ml into liquid nutrient medium and incubated at 30°C under vigorous shaking. After about 16 h of cultivation, when the cell concentration in the culture has reached a value of 10^7/ml, the cells were used for the experiment. The fraction of budding cells was about 60%.

For irradiation the cells were washed twice with buffer, sonicated for 5-10 sec to remove clumping and resuspended in buffer at a concentration of $2\text{-}3 \cdot 10^7$/ml.

III. Irradiation

X-irradiation was performed with a 100 kV x-ray machine (FA 100/3, Philips). The absorbed dose rate was determined by ferrous sulphate dosimetry to be 30 krad/min (strain 211) and 11.8 krad/min (strain S 2095). The cell suspensions were exposed in open-glass petri dishes (diameter 3 cm) and vigorously stirred during the exposure.

IV. Posttreatment of the Irradiated Cells

Immediate Plating (IP). The cells were spread immediately after irradiation in appropriate dilutions on nutrient agar plates and incubated at 30°C for 5-6 days.

Liquid Holding (LH). Ten (10) ml of the irradiated cell suspension were placed at a concentration of $5 \cdot 10^5 - 2 \cdot 10^6$/ml in 100 ml Erlenmeyer flasks and stored at 30°C for 4 days in the dark. Cell concentration of the samples was monitored at the beginning and end of the storage period by microscopic counting.

V. Split Dose Experiment

The experimental procedure is schematically given in Figure 1. A cell suspension ($3 \cdot 10^7$ cells/ml) in phthalate buffer was prepared. To determine the single dose survival curve of the cell population several doses were applied. One part of the irradiated sample was immediately plated, the other was submitted to liquid holding treatment before plating. For fractionated irradiation, cells were exposed to

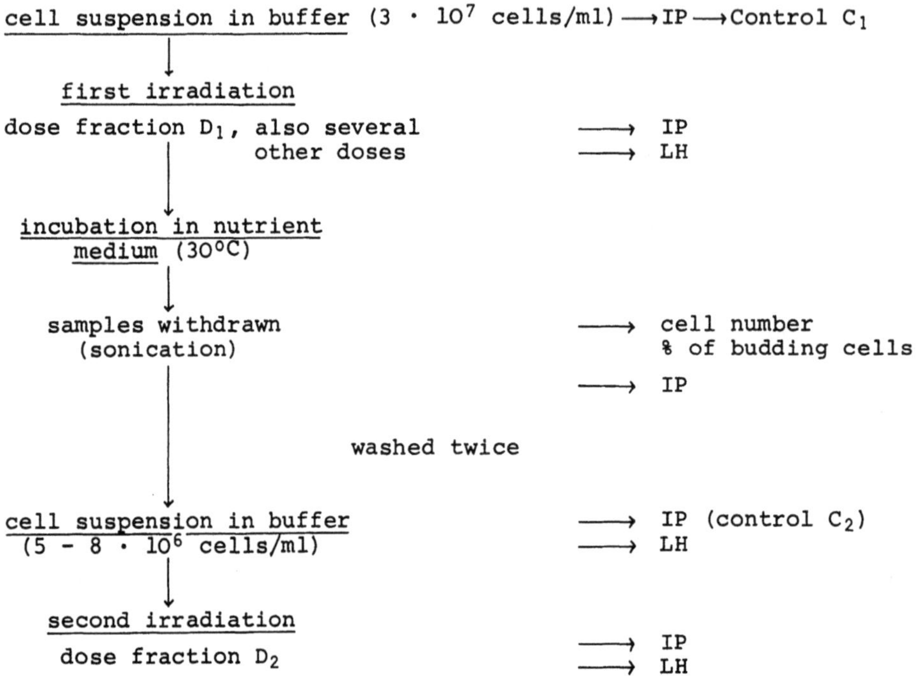

Fig. 1. Experimental procedure of split dose experiment (IP = immediate plating, LH = liquid holding)

a first dose D_1, centrifuged, and immediately incubated under optimal growth conditions in nutrient medium at a concentration of $4 - 5 \cdot 10^6$ cells/ml. After different incubation times, samples were withdrawn and exposed to sonication (5 sec) in order to separate adherent cells. One ml of the sample was used to determine cell number and percentage of budding cells of the preirradiated cell culture as well as the colony forming ability by plating. The remaining cell suspension was washed twice and resuspended in buffer at a concentration of $5 - 8 \cdot 10^6$/ml. One part of this suspension was immediately plated (to determine control C_2); another part was submitted to liquid holding. The rest was irradiated with the second dose fraction. For complete survival curves, several different doses were applied. After this the cells were plated immediately and also tested for subsequent LHR.

C. Results

I. Fractionated X-Irradiation of the Wild-Type Strain 211

1. Split Dose Recovery (SDR)

The experiments were performed with stationary phase cell populations. The cells were mainly in G_1 phase, as was shown by quantitative DNA measurements (Jentsch, 1973). In Figure 2, x-ray survival curves of the cell population are given for immediate plating (IP) and after subsequent LHR. The IP-survival curve is characterized by an extrapopulation number of 5 and a D_o of 18 krad in the terminal exponential

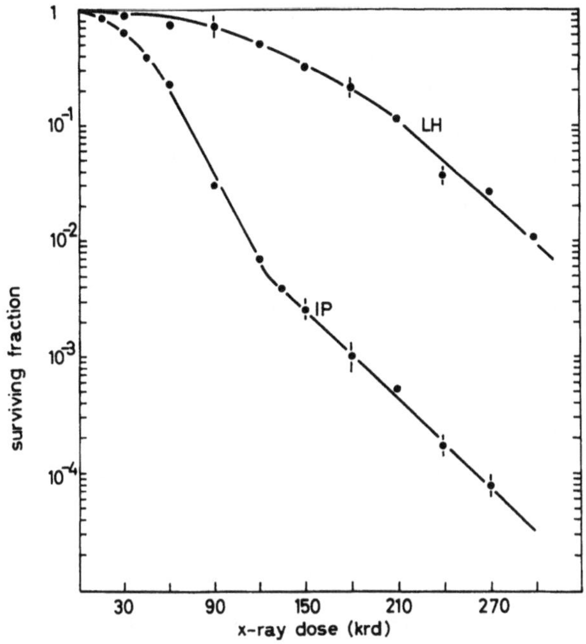

Fig. 2. X-ray survival curves of stationary phase cells of wild-type strain 211 at immediate plating (IP) and after liquid holding treatment (LH)

part. Its radioresistant tail at higher doses seems to be due to population heterogeneity with respect to radiosensitivity.

For dose fractionation experiments, the cells were irradiated with a first dose of 60 krad, which reduces cell survival to 20%. The results for split doses of 60 krad are shown in Figure 3. A pronounced division delay is induced in the population by the first dose. Cell number (curve N in Fig. 3) and colony forming ability (curve IP(D_1) in Fig. 3) of the preirradiated cells remain constant over an incubation period of 12 h, whereas unirradiated stationary phase cells start to divide after 3-4 h of incubation. In contrast cell budding is only slightly delayed. In preirradiated cell populations cell budding (curve B in Fig. 3) starts after 2 h, in unirradiated populations after 1-1.5 h of incubation. With increasing incubation time in the culture, cells accumulate with buds the same size as the mother cell (curve B' in Fig. 3).

Since the buds cannot be separated by sonication, this situation may be considered as a morphologic indication of an x-ray-induced G_2-phase block. Cell survival after application of two dose fractions of 60 krad as a function of the incubation time between them is given by curve IP (D_1+D_2) in Figure 3. It decreases with short incubation intervals, passes a minimum at about 3.5 h, and increases with longer incubation times.

Complete survival curves measured after different incubation intervals are shown in Figure 4.

As can be seen, stationary phase cells do not show SDR immediately, but become even more radiosensitive at the beginning of incubation. After 3.5 h an exponential survival curve was obtained with a higher initial slope than that obtained without dose fractionation. With increasing incubation intervals compound survival curves were measured, showing an exponential part at low doses followed by a shoulder curve toward higher doses. After 23 h of incubation, when the cell concen-

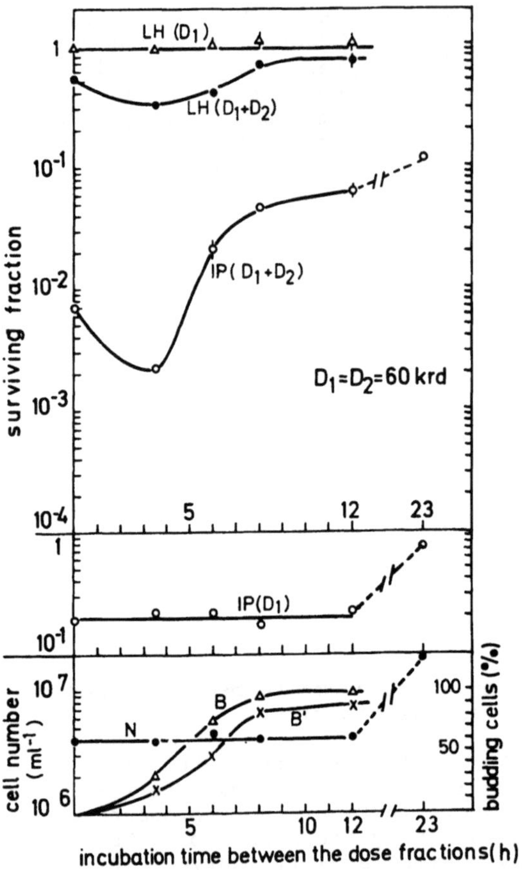

Fig. 3. Split dose survival characteristics of stationary phase cells of wild-type strain 211 for dose fractionation $D_1 = D_2 = 60$ krad. IP (D_1+D_2), cell survival after both dose fractions at immediate plating; LH (D_1+D_2), cell survival after both dose fractions after additional liquid holding treatment; IP(D_1), survival of cells irradiated with first dose and incubated in nutrient medium between dose fractions; LH(D_1), survival of cells irradiated with first dose and submitted to a further liquid holding treatment after incubation intervals indicated; N, concentration of cell culture during incubation between dose fractions; B, percentage of budding cells in culture; B', percentage of cells with bud of same size as mother cell

tration in the culture had already increased 2.5-fold, cell survival is given by shoulder curve.

The reappearance of a shoulder in the survival curves indicates that SDR occurs. The recovery process, however, does not begin simultaneously in the whole cell population. This is shown by the initial exponential part of the survival curves. With further incubation the recovery-competent fraction increases, the initial exponential part of the survival curves diminishes, and the shoulder region gets more and more pronounced.

Also the final slope of the curves decreases. The shoulder widths reached are much larger than would have been expected from the original survival curve of stationary phase cells in the case of complete SDR. For instance an analysis of the 12 h curve results in an extrapolation number of 12 for the shoulder region, compared to a value of 5 from the original curve. This can be explained by assuming a superposition of SDR and progression of the incubated cells to more radioresistant cell cycle stages. The stationary phase populations used consist mostly of G_1 cells and are therefore partially synchronized. From synchronized cultures of the same strain (de Langguth and Beam, 1973), it is known that G_1 cells are rather x-ray sensitive and that radioresistance increases toward later cell cycle stages. Survival curves obtained in a sensitive cell cycle show a small shoulder and high exponential slope, whereas in a resistant stage the shoulder is broad and the exponential slope reduced.

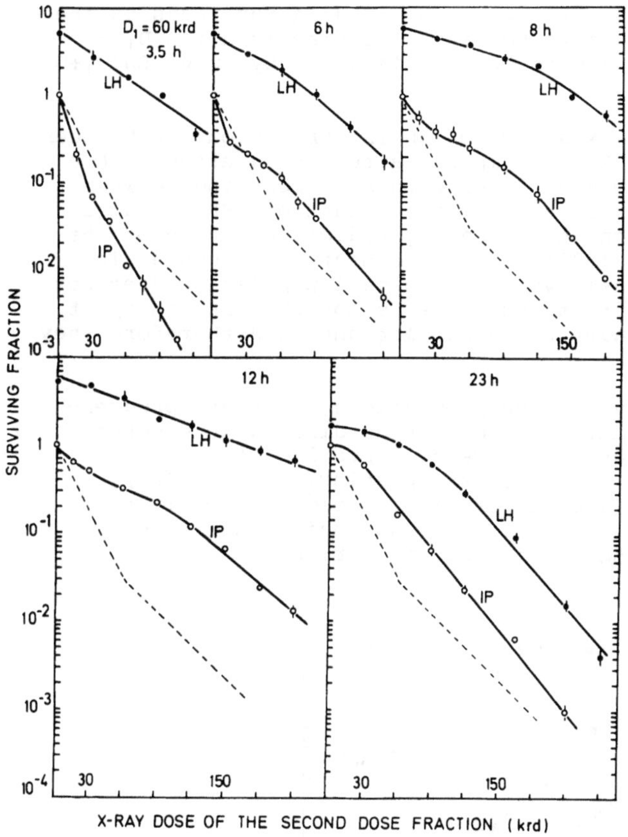

Fig. 4. X-ray survival curves of stationary phase cells of strain 211 preirradiated with D_1 = 60 krad after indicated incubation intervals in nutrient medium. IP, at immediate plating; LH, after additional liquid holding treatment. Surviving fractions plotted are normalized with respect to cell survival after first dose fraction (control C_2 in Fig. 1). Survival curve obtained if second irradiation follows first one immediately is shown as dotted line

When the preirradiated stationary population is incubated after the first dose fraction in nutrient medium, the cells progress partially synchronously through the cell cycle and reach more radioresistant stages. At a second irradiation, they behave as preirradiated cells of the momentary resistant stage having recovered more or less from sublethal damage.

The results show that a certain proportion of the preirradiated stationary phase cell population performs SDR within an incubation period of 12 h. Whether SDR is complete cannot be answered, because to do this it is necessary to know the survival curve of an unirradiated cell population in the corresponding cell cycle stage. This, however, requires experiments with synchronized cell cultures that are not yet available.

2. *Liquid Holding Recovery*

Cells irradiated with the first dose and with both dose fractions were investigated for subsequent LHR.

If the cells preirradiated with 60 krad were liquid holding-treated after different incubation intervals in nutrient medium, surviving fractions given by curve LH(D_1) in Figure 3 were obtained. During incubation in nutrient medium, the ability of the cells to perform LHR remains unchanged over an incubation period of 12 h. For longer incubation intervals LHR is masked by cell divisions occurring in the sample during the storage period.

Liquid holding of the cells after the second irradiation leads to the survival curves shown in Figure 4. For the dose fractionation $D_1 = D_2 = 60$ krad, survival after liquid holding is given by curve $LH(D_1+D_2)$ in Figure 3.

To obtain information on the existence of an interrelationship between SDR and LHR, the amount of LHR was compared after irradiation with a single dose and after fractionated doses with SDR occurring between the dose fractions. Since for strain 211 the amount of LHR measured after x-irradiation depends on the momentary radiosensitivity of the cells (de Langguth and Beam, 1973), it is necessary for such an investigation that the radiosensitivity of the cell population does not change between the two dose fractions. This is not the case in split dose experiments with stationary phase populations, and therefore they cannot be used.

In asynchronous exponentially growing populations, however, progression effects are less important. The results obtained with exponentially growing cells are shown in Figure 5. Curve 1 describes cell survival after x-irradiation for immediate plating, curve 1 (LH), after LHR. Up to doses of 48 krad after liquid holding, surviving fractions higher than 1 were obtained. This was mainly due to cell divisions occurring during the liquid holding period.

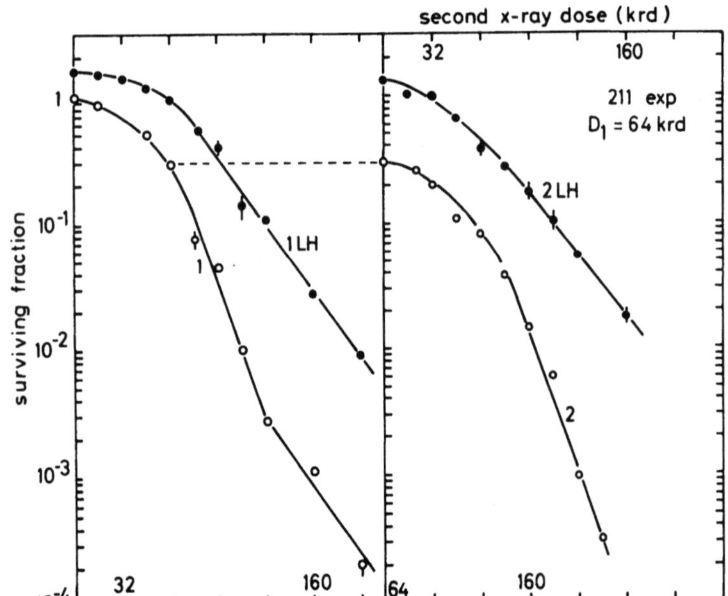

Fig. 5. Split dose recovery of exponentially growing cell cultures of strain 211 exposed to fractionated x-irradiation (dose rate 32 krad/min); left-hand side: x-ray survival of exponentially growing cells at immediate plating (curve 1) and after additional liquid holding treatment (curve 1LH); right-hand side: x-ray survival of cells pre-irradiated with $D_1 = 64$ krad after 2.5 h of incubation in nutrient medium: at immediate plating (curve 2); after additional liquid holding treatment (curve 2LH)

In split dose experiments the cells were irradiated with a first dose of 64 krad and incubated in nutrient medium. In exponentially growing cell cultures, SDR started immediately and was completed after 2.5 h, as shown by the survival curve measured at that time (curve 2 in Fig. 5). When the cells were submitted to a further liquid holding treatment after the second dose fraction, curve 2 (LH) in Figure 5 was obtained. From the intercept of curve 2 (LH) on the ordinate it can be seen that preirradiated exponentially growing cells - as stationary phase cells - do not loose their ability to perform subsequent LHR during incubation in nutrient medium. The amount of LHR measured is even higher if the preirradiated cells are incubated for 2.5 h in nutrient medium before liquid holding. This is mainly due to additional cell divisions occurring during liquid holding. By preincubation in nutrient medium about 30-50% of the cells divide during the liquid holding period, whereas no cell divisions can be detected if the liquid holding treatment is performed immediately after application of the first dose.

Since - as shown by a comparison of curves 1 and 2 in Figure 5 - the radiosensitivity of the cell population is the same at the first and second irradiations for a given survival level for immediate plating, the amounts of subsequent LHR can be compared. If the survival level is reached by a single dose, the corresponding survival level after LHR is given by curve 1 (LH); if the survival level is reached by split doses with complete SDR between them, cell survival after LHR is given by curve 2 (LH). In Table 1 for different surviving fractions on immediate plating the corresponding surviving fractions after LHR are listed. It can be seen that the amount of subsequent LHR is only determined by the value of cell survival on immediate plating, irrespective of the irradiation pattern by which it was reached. However, at fractionated irradiation with complete SDR between the dose fractions, higher dose values are needed to reach a certain survival level, than at irradiation with a single dose. If, on the other hand, corresponding dose values are considered, the amount of LHR is less after fractionated irradiation with SDR occurring between the dose fractions. LHR is affected by preceding SDR in the same way as by removal of sublethal damage during SDR: the second dose fraction becomes less efficient in inducing potential lethal damage. This, however, suggests that sublethal and potential lethal damage do not differ qualitatively but only quantitatively.

II. Fractionated X-Irradiation of the *Rad9-4* Mutant S 2095

1. SDR at Fractionated X-Irradiation

Strain S 2095 is about 4-fold more sensitive to x-ray and UV irradiation than the wild type. The experiments were again performed with stationary phase cell populations. In Figure 6 x-ray survival curves are shown for immediate plating and after liquid holding. The survival curve for immediate plating is characterized by an extrapolation number of 4 and a D_0 of 4.5 krad. The mutant shows about the same amount of LHR as the wild-type strain; for low survival levels, LHR is even greater. This suggests that a repair mechanism controlled by the RAD9 gene is not involved in LHR after x-irradiation. In split dose experiments the cells were irradiated with a first dose of 13.75 krad (which reduces cell survival to 20%) and incubated in nutrient medium. Strain S 2095 shows after irradiation with the conditioning dose almost no delay in cell budding and only a slight delay in cell division. The cells start to divide after 4.5 h (curve N in Fig. 7) whereas in unirradiated stationary phase cultures, cell division starts after 3-4

Table 1. LHR of exponentially growing diploid yeast (wild-type strain 211) after single and split doses (with complete SDR between dose fractions). IP = immediate plating; LH = liquid holding

Cell survival for IP	Doses (krad)		Cell survival after LH-treatment	
	D_1 (at irradiation with single dose)	D_2 (at irradiation with split doses)	at irradiation with single dose	at irradiation with split doses
$2 \cdot 10^{-1}$	71	99	$8.4 \cdot 10^{-1}$	$8.5 \cdot 10^{-1}$
$9 \cdot 10^{-2}$	84	127	$5.2 \cdot 10^{-1}$	$4.7 \cdot 10^{-1}$
$5 \cdot 10^{-2}$	92	141	$3.7 \cdot 10^{-1}$	$3.2 \cdot 10^{-1}$
$2 \cdot 10^{-2}$	103	155	$2.5 \cdot 10^{-1}$	$2.1 \cdot 10^{-1}$
10^{-2}	112	163	$1.8 \cdot 10^{-1}$	$1.7 \cdot 10^{-1}$
$3 \cdot 10^{-3}$	126	173	10^{-1}	$9.5 \cdot 10^{-2}$

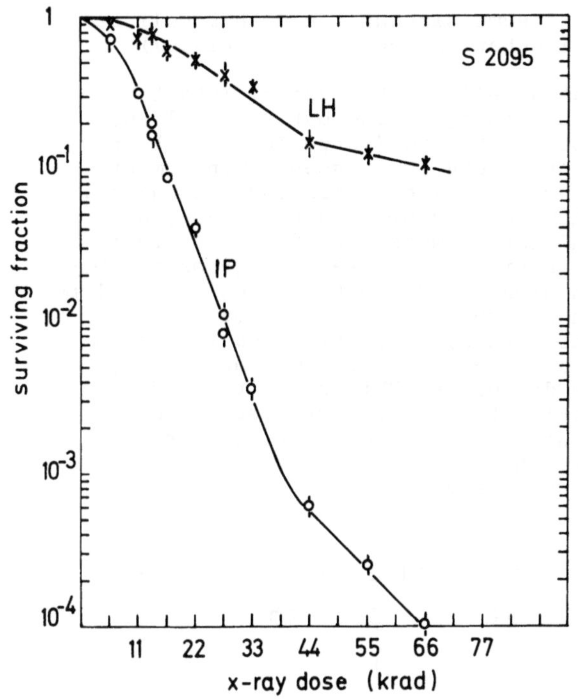

Fig. 6. X-ray survival curves of stationary phase cells of *rad9-4* mutant S 2095 at immediate plating (IP), after liquid holding treatment (curve LH)

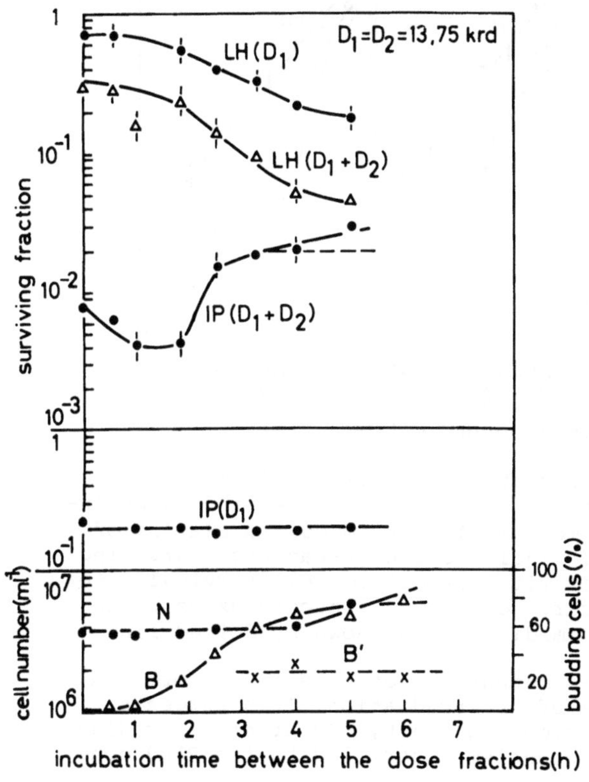

Fig. 7. Split dose survival characteristics of stationary phase cells of *rad9-4* mutant S 2095 for dose fractionation $D_1 = D_2 = 13.75$ krad. Labelling of curves corresponds to that used in Fig. 3. If curve IP (D_1+D_2) is corrected for observed increase in cell concentration N at long incubation intervals, the dotted line will be obtained

h. Also no cells with a bud of the same size as the mother cell accumulate in the cultures during incubation (curve B' in Fig. 7).

As for the wild type, preirradiated cells of the *rad9-4* mutant become more radiosensitive at the beginning of incubation. Cell survival after two dose fractions of 13.75 krad (curve IP (D_1+D_2) in Fig. 7) declines at low incubation intervals, passes a minimum at about 1 h, and shows a further increase with continuing incubation. The increase in cell survival, however, is not caused by SDR, as demonstrated by complete survival curves measured after different incubation intervals (curves IP in Fig. 8). At each stage of incubation exponential survival curves always were obtained; only the slope of the curves changes. After 1 h a higher slope was measured than that obtained without dose fractionation, indicating the increased radiosensitivity of the incubated cells. With further incubation, the slope of the survival curves decreases. This seems to be due to a progression of the cell population to more radioresistant cell cycle stages, since similar variations in radiosensitivity are observed if unirradiated stationary phase cells are incubated in nutrient medium (Wienhard, unpublished results).

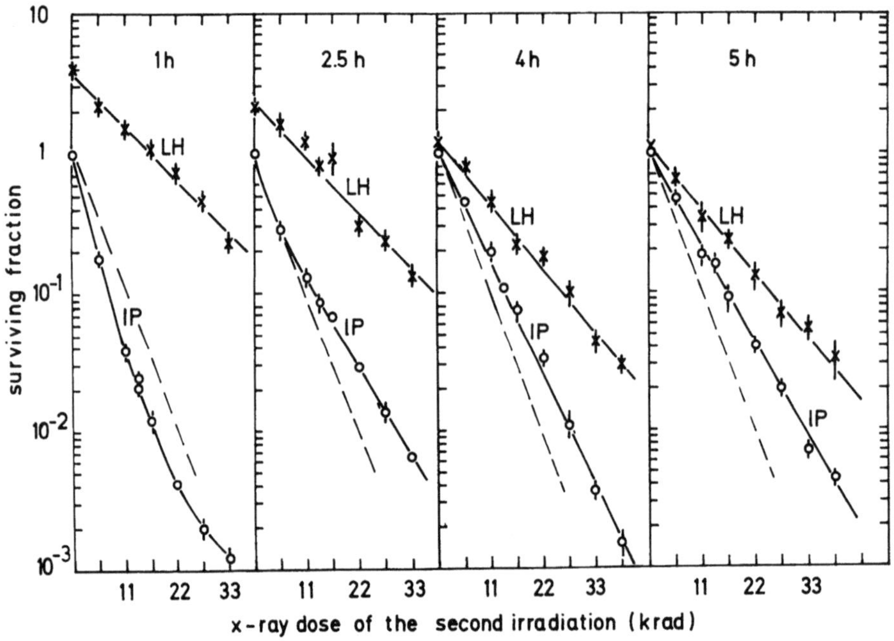

Fig. 8. X-ray survival curves of stationary phase cells of *rad9-4* mutant S 2095 preirradiated with D_1 = 13.75 krad after indicated incubation intervals in nutrient medium. IP, at immediate plating of the cells; LH, after additional liquid holding treatment. Surviving fractions plotted are normalized with respect to cell survival after first dose fraction. Survival curve obtained if second exposure to irradiation follows first one immediately is shown as dotted line

The results show that the *rad9-4* mutant is not able to perform SDR during the first division cycle after irradiation. This suggests that under optimal growth conditions in yeast, SDR after x-irradiation is dependent on a repair mechanism controlled by the RAD9 gene.

2. LHR at Fractionated X-Irradiation

Concerning LHR at fractionated x-irradiation, the *rad9-4* mutant also behaves quite differently from the wild type.

The cells irradiated with the first dose fraction completely lose their ability for subsequent LHR during incubation in nutrient medium with the first division cycle after irradiation. This is shown by the decline of curve LH (D_1) in Figure 7, which describes cell survival reached by liquid holding of the preirradiated cells after the incubation intervals indicated. When the cells are exposed to a liquid holding treatment after the second dose fraction, the LH-survival curves in Figure 8 are obtained (for the dose fractionation $D_1 = D_2 =$ 13.75 krad cell survival after LHR is given by curve LH (D_1+D_2) in Figure 7). From Figure 8 it can be seen that for a fixed survival level after fractionated irradiation (survival curves IP), the amount of subsequent LHR decreases considerably with increasing incubation intervals between the two dose fractions. A comparison of Figure 6 and Figure 8 shows that for the *rad9-4* mutant after split doses (at least for long incubation intervals between the dose fractions), the amount of LHR is considerably less than after irradiation with a single dose, leading to the corresponding survival level (normalized with respect to unirradiated controls).

D. Discussion

After x-irradiation, diploid yeast shows survival curves with an initial shoulder followed by an exponential decrease. The shoulder is generally interpreted as representing the cell's ability to tolerate a certain amount of sublethal damage before cell viability is affected. Recovery from sublethal damage can be demonstrated by the split dose technique. This phenomenon (SDR, split dose recovery) is characterized by the reappearance of a shoulder in the survival curves measured after a second exposure to irradiation.

In this report stationary phase cell cultures of a wild type and a radiosensitive strain (*rad9-4* mutant) of diploid yeast were investigated with respect to SDR after x-irradiation. The two strains behave quite differently. The wild type performs SDR within the first cell cycle after irradiation (the recovery kinetics, however, differ from those of exponentially growing cells). For the *rad9-4* mutant, SDR cannot be found within this period, although its x-ray survival curve also has an initial shoulder. This suggests that in x-irradiated diploid yeast under optimal growth conditions and within the first cell cycle after irradiation, recovery of sublethal damage is due to a repair pathway in which the gene product of the RAD9 gene is involved.

The molecular basis of the RAD9-dependent repair pathway is still unknown. It presumably involves recombinational events, since the *rad 9-4* mutant does not show radiation-induced inter- and intragenic mitotic recombination (Kowalski and Laskowski, 1975).

Another x-ray sensitive mutant of diploid yeast (the *rad51-1* mutant isolated by Nakai and Matsumoto (1967)) is also reported to be deficient in SDR after x-irradiation (Saeki et al., 1974). This mutant is also defective in radiation-induced mitotic recombination and gene conversion.

Under LHR conditions after irradiation, "potentially lethal" damage that would lead to cell death upon immediate plating is removed or bypassed. This is indicated by a higher survival probability of liquid hold cells compared to those immediately plated. Both strains show about the same amount of LHR if the treatment starts immediately after irradiation. This suggests that a RAD9-dependent repair mechanism is not involved in LHR after x-irradiation.

If, however, cells are incubated in nutrient medium after the first dose, the two strains exhibit different behavior with regard to subsequent LHR: (1) wild-type cells exposed to the first dose fraction and incubated in nutrient medium show a constant amount of subsequent LHR within the first cell cycle after irradiation. Preirradiated cells of the *rad9-4* mutant, however, completely lose their ability for subsequent LHR during incubation within this period. (2) When the wild type was examined for LHR after split dose with complete SDR between them, the extent of LHR is the same as that after irradiation with a single dose leading to the same surviving fraction upon immediate plating. (This was demonstrated quantitatively for exponentially growing cells.) On the other hand, after split doses (with long incubation intervals between them) the *rad9-4* mutant shows considerably less LHR than after irradiation with a single dose. Furthermore, the amount of LHR after fractionated irradiation decreases with increasing time intervals between the dose fractions. (Since the experiments with the *rad9-4* mutant were performed with stationary phase populations, which show additional changes in radiosensitivity during incubation, the experimental results describe the interrelationship only qualitatively.)

It was found that x-irradiated wild-type cells incubated in nutrient medium retain their ability to perform subsequent LHR within the first cell cycle after irradiation. This indicates that during this period at each stage of incubation the same amount of damage susceptible to LHR is present in the cells as immediately after irradiation. This does not mean, however, that the primary damage induced by irradiation remains unchanged during incubation. It may be modified, but this must occur in such a way that it is still susceptible to LHR. X-irradiated cells of the *rad9-4* mutant completely lose their ability to perform subsequent LHR within the first cell cycle after irradiation, which indicates that during incubation the primary potential lethal damage becomes fixed in a way in which it is no longer susceptible to LHR. Therefore, the constancy of LHR must be assumed to be linked to the RAD9.

The different behavior of the two strains at LHR after the second fractionated irradiation results from the different fates of unrepaired potential lethal damage induced by the first dose during incubation between the two fractions.

Since the wild type remains susceptible to LHR during incubation, a liquid holding treatment after the second dose fraction acts on both the potential lethal damage of the first and the second dose fractions. Therefore, at a corresponding survival level at immediate plating (determined by the total amount of potential lethal damage induced by both dose fractions), the same amount of LHR was measured as after irradiation with a single dose.

With fractionated irradiation, however, higher total doses are needed to reach the survival level, because of SDR between the dose fractions.

In the *rad9-4* mutant, potential lethal damage induced by the first dose becomes fixed during incubation between the dose fractions. Therefore only the damage caused by the second dose is modified by the following LHR treatment, if a complete fixation of potential lethal damage induced by the first dose did occur during incubation between the dose fractions.

The *rad9-4* mutant shows therefore for a corresponding survival level at immediate plating less LHR after fractionated irradiation than after a single dose, when all potential lethal damage is susceptible to LHR.

E. Summary

A wild-type strain and a radiosensitive mutant (*rad9-4*) of diploid yeast were exposed to fractionated x-irradiation and investigated with respect to split dose recovery (SDR) and subsequent liquid holding recovery (LHR). The wild type shows SDR within the first cell cycle after irradiation; no SDR was observed, however, for the *rad9-4* mutant.

Cells of the wild type exposed to the first dose fraction and incubated under optimal growth conditions retain their ability to perform LHR within the first cell cycle after irradiation, whereas the *rad9-4* mutant loses it completely. Both strains also differ in LHR after fractionated irradiation.

Acknowledgement. The authors thank Mrs. J. Textor and M.M. Hlawica for expert technical assistance. The work was financially supported by the Bundesministerium für Forschung und Technologie.

References

Averbeck, D.: Isolierung und Charakterisierung von drei UV-sensiblen Saccharomyces-Mutanten. Inaugural-Dissertation, Fachbereich Biologie, Freie Universität Berlin, 1970

Bacchetti, S., Mauro, F.: Recovery from sublethal x-ray damage in surviving yeast cells. Radiat. Res. 25, 103-114 (1965)

Elkind, M.M., Moses, W.B., Sutton-Gilbert, H.: Radiation response of mammalian cells grown in culture. VI. Protein, DNA and RNA inhibition during the repair of x-ray damage. Radiat. Res. 31, 156-173 (1967)

Elkind, M.M., Sutton, H.: Radiation response of mammalian cells grown in culture. I. Repair of x-ray damage in surviving Chinese hamster cells. Radiat. Res. 13, 556-593 (1960)

Game, J.C., Mortimer, R.K.: A genetic study of x-ray sensitive mutants in yeast. Mutation Res. 24, 281-292 (1974)

Ganesan, A., Smith, K.C.: Dark recovery processes in *E. coli* irradiated with ultraviolet light. II. Effect of uvr genes on liquid holding recovery. J. Bateriol. 97, 1129-1133 (1969)

Harm, W.: The role of host cell repair in liquid holding recovery of UV-irradiated *Escherichia coli*. Photochem. Photobiol. 5, 747-760 (1966)

Jain, V.K., Pohlit, W.: Influence of energy metabolism on the repair of x-ray damage in living cells. I. Effect of respiratory inhibitors and glucose on the liquid holding reactivation in yeast. Biophysik 8, 254-263 (1972)

Jentsch, K.D.: Isolation von DNS aus Hefezellen und quantitative Bestimmung der DNS-Synthese nach UV- und Röntgenbestrahlung. Diplomarbeit, Fachbereich Physik, Giessen, 1973
Kiefer, J.: Recovery from sublethal ultraviolet damage in diploid yeast. Int. J. Radiat. Res. 13, 399-400 (1968)
Kiefer, J.: Effect of split-dose UV-i-radiation on diploid yeast. Photochem. Photobiol. 11, 37-47 (1970)
Kiefer, J.: The importance of cellular energy metabolism for the sparing effect of dose fractionation with electrons and ultraviolet light. Int. J. Radiat. Biol. 20, 325-336 (1971)
Kiefer, J.: The effect of nucleic acid-binding metabolic inhibitors on split-dose sparing in UV- and electron-irradiation diploid yeast. Int. J. Radiat. Biol. 24, 93-97 (1973)
Kiefer, J.: The effect of caffeine on survival of UV-irradiated diploid yeast strains of different sensitivity. Mutation Res. 30, 317-326 (1975)
Kowalski, S., Laskowski, W.: The effect of three rad genes on survival, inter- and intragenic mitotic recombination in *Saccharomyces*. I. UV-irradiation without photoreactivation or liquid holding posttreatment. Molec. Gen. Genetics 136, 75-86 (1975)
Nakai, S., Matsumoto, S.: Two types of radiation-sensitive mutants in yeast. Mutation Res. 4, 129-136 (1967)
Nunes de Langguth, E.N., Beam, C.A.: Repair mechanisms and cell cycle dependent variations in x-ray sensitivity of diploid yeast. Radiat. Res. 53, 226-234 (1973)
Parry, J.M., Parry, E.M.: The effects of UV-light post-treatments on the survival characteristics of 21 UV-sensitive mutants of *Saccharomyces cerevisiae*. Mutation Res. 8, 545-556 (1969)
Patrick, M.H., Haynes, R.H.: Dark recovery phenomena in yeast. II. Conditions that modify the recovery process. Radiat. Res. 23, 564-579 (1964)
Patrick, M.H., Haynes, R.H., Uretz, R.B.: Dark recovery phenomena in yeast. I. Comparative effects with various inactivating agents. Radiat. Res. 21, 144-163 (1964)
Saeki, T., Machida, I., Nakai, S.: Split dose recovery controlled by XS1 gene in yeast. Radiat. Res. 59, 95 (1974)
Wickerham, L.J.: Taxonomy in yeast. U.S. Dept. Agr. Tech. Bull. No. 1029, 1951

9. Repair of Damage Induced by Near Ultraviolet Light Plus Furocoumarin in *Saccharomyces cerevisiae*

D. AVERBECK

Furocoumarins such as psoralen and 8 methoxypsoralen (8-MOP) and some of their derivatives are photosensitizing agents. They are known to sensitize human and guinea pig skin to near ultraviolet (UV) light and sunlight (Musajo and Rodighiero, 1972a). They are used medically in cases of vitiligo and leukoderma (El-Mofty, 1968) and also in cases of psoriasis (Walter and Voorhees, 1973; Parrish et al., 1974; Weber, 1974).

In many biologic systems the photosensitizing effects are seen by induction of lethality (Musajo and Rodighiero, 1972a). The effects are thought to be due to a covalent binding of active furocoumarins to pyrimidine bases of nucleic acids (Musajo and Rodighiero, 1972b).

Figure 1 shows the chemical structure of a typical linear psoralen derivative, 8-methoxypsoralen. Two reactive sites are seen, one at the 3, 4 and one at the 4', 5' double bond. Both are able to photoreact with the 5, 6 double bond of pyrimidines (for example, thymine) to form a C_4 cycloaddition product. One possible monoadduct of 8-MOP and thymine is seen at the bottom of Figure 1 (Musajo and Rodighiero, 1972a).

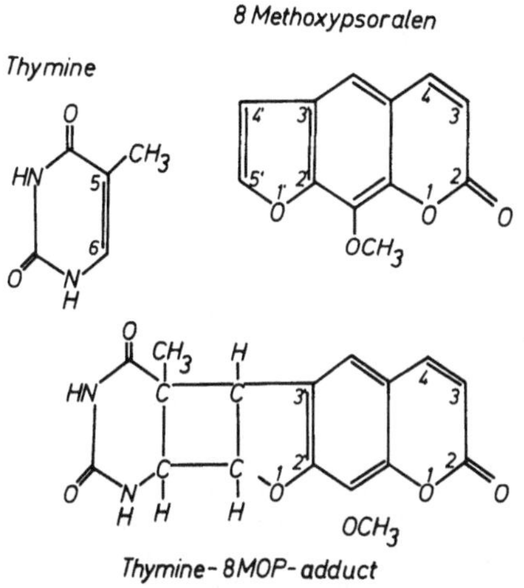

Fig. 1. Chemical structure of 8-MOP

The reaction with native DNA is thought to proceed in several steps. In the dark a loose complex of furocoumarins with DNA is formed. Treatment of the complex with 365 nm light induces covalent binding at the 3, 4 or the 4', 5' reaction site; in other words, DNA monoadducts are induced. Further illumination with 365 nm light converts the monoadducts into stable interstrand cross-links in the DNA involving the two reaction sites of the furocoumarin molecule.

In the presence of a bifunctional furocoumarin such as 8-MOP monoadducts as well as biadducts (cross-links) can be formed (in a ratio of about 3-4:1) (Musajo and Rodighiero, 1972a). However, there are so-called monofunctional furocoumarins that are able to form monoadducts only, e.g., the angular psoralen derivative, angelicin, which for steric reasons is able to form only monoadducts involving the 3, 4 double bond (Dall'Acqua et al., 1971; Rodighiero et al., 1971). Cross-link formation in DNA after psoralen-plus-light treatment can be demonstrated by the different melting and reassociation profiles during heating of DNA containing cross links (Dall'Acqua et al., 1971).

The melting and reassociation profiles during heating and subsequent cooling of yeast DNA were determined in collaboration with Drs. Reiss and Moustacchi, Orsay, France (see Averbeck and Moustacchi, 1976).

Table 1 shows the renaturation values of nuclear yeast DNA with and without treatment with 365 nm light (12600 ergs/mm^2/min) in presence of the bifunctional furocoumarin 8-MOP and the monofunctional furocoumarin 3 carbethoxypsoralen (3-CP). In comparison to the controls very little renaturation is observed after 3-CP plus 365 nm light, whereas high renaturation is observed after 8-MOP plus 365 nm light. After exposure to 365 nm light in presence of the monofunctional drug, DNA exhibits a capacity to renature that is not much different from that of untreated DNA. This indicates that no appreciable numbers of cross-links are present in the treated DNA.

Table 1. Renaturation of nuclear DNA

DNA	365 nm light exposure time (min)	% Renaturation	
			Difference
Control	0	15.0	-
8-MOP-treated (10 μg/ml)	0	11.6	0
	4	22.3	10.7
	12	42.6	31.0
3-CP-treated (10 μg/ml)	0	12.2	0
	4	15.0	2.8
	12	14.8	2.6

Up to now only bifunctional derivatives of psoralen have been used in yeast (Igali and von Borstel, 1974; Swanbeck and Thyresson, 1974; Averbeck and Moustacchi, 1975; Averbeck et al., 1975), probably because it had been observed in other biologic systems (bacteriophage λ, *Escherichia coli*, and Chinese hamster cells) that the inactivation observed in vivo was correlated to the production of furocoumarin plus light-induced cross-links in DNA (Cole, 1971; Ben-Hur and Elkind, 1973a). In the present paper we want to analyze the effects of monofunctional as compared to bifunctional furocoumarins, and we want to give evidence for the repair of the different lesions induced.

Figure 2 shows typical survival curves obtained for haploid wild-type cells of *Saccharomyces cerevisiae* in stationary growth phase after treatment with equal quantities of bifunctional furocoumarins (psoralen and 8-MOP) and the monofunctional furocoumarin angelicin followed by exposure to 365 nm light (12600 ergs mm^2 = 0.13 J/cm^2/min). It is seen that wild-type cells are about 10 times more resistant to the monofunctional than to the bifunctional agent in presence of 365 nm light. The difference in sensitivity cannot be fully understood by

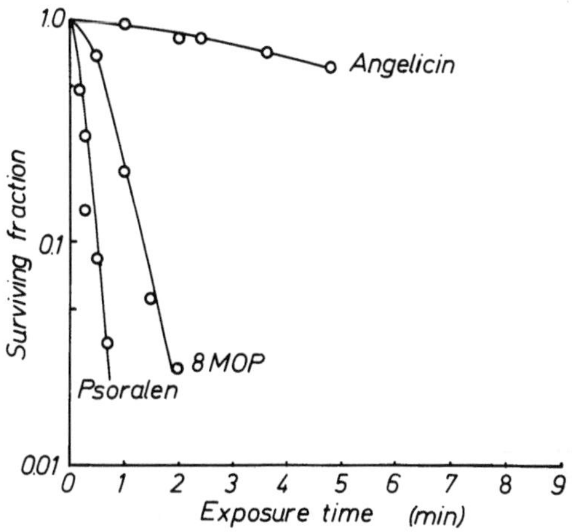

Fig. 2. Typical survival curves of haploid wild-type cells in stationary growth phase after treatment with 10 μg/ml psoralen, 8-MOP, and angelicin plus 365 nm light

the photoreactivity with native DNA. 8-MOP and angelicin are known to differ by a factor of 4 in their photoreactivity (Musajo and Rodighiero, 1972a). This suggests that the different sensitivities to the two drugs are due to different repair capacities of wild-type cells for monoadducts and cross-links in DNA.

Repair of furocoumarin plus light-induced lesions has been reported in bacteria (Igali et al., 1970; Cole, 1973), Chinese hamster cells (Ben-Hur and Elkind, 1973a, 1973b; Baden et al., 1972), and human fibroblasts. Cole demonstrated that in *E. coli* cells, the excision-resynthesis system and the postreplicative repair system are involved in the repair of furocoumarin plus light-induced interstrand cross-links in DNA (Cole, 1973).

In order to determine whether the repair pathways known in yeast are involved in the repair of monofunctional as well as bifunctional lesions, we employed repair-deficient mutants of *Saccharomyces*. We used the haploid strain S24-12, i.e., the mutant *rad2-20* (Averbeck, 1970), which is unable to excise UV-induced pyrimidine dimers (Averbeck, 1970), the haploid strain S25-13b, i.e., the mutant *rad9-4*, which affects recombinational pathways (Kowalski and Laskowski, 1975; Eckardt et al., 1975), and the double mutant *rad2-20 rad9-4* (strain S2057N1-24)(Averbeck and Moustacchi, 1975).

Figure 3 shows typical survival curves obtained after 8-MOP plus 365 nm light for the haploid wild type, the mutants *rad2-20* and *rad9-4*, and the double mutant *rad2-20 rad9-4*. The two single mutants are more sensitive than the wild type. The double mutant *rad2-20 rad9-4* is more sensitive than each of the two single mutants, but it is not supersensitive. The response of the double mutant suggests in wild-type cells additivity rather than synergism of the repair pathways blocked by the *rad2-20* and the *rad9-4* genes.

A different interaction of the repair pathways is seen when the cells are treated with a monofunctional furocoumarin such as angelicin.

Figure 4 gives typical survival curves obtained after treatment with angelicin plus 365 nm light. The two single mutants *rad2-20* and *rad9-4* are more sensitive than wild-type cells. The double mutant, however, is much more sensitive than each of the single mutants, indicating that the two repair pathways blocked by the two rad genes interact synergistically.

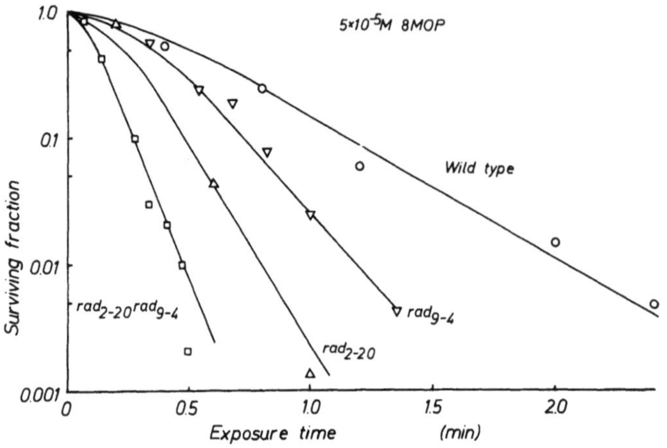

Fig. 3. Typical survival curves of haploid wild-type cells and the derived mutants *rad2-20*, *rad9-4*, and *rad2-20 rad9-4* after treatment with 5 x 10^{-5} *M* 8-MOP (bifunctional) plus 365 nm light

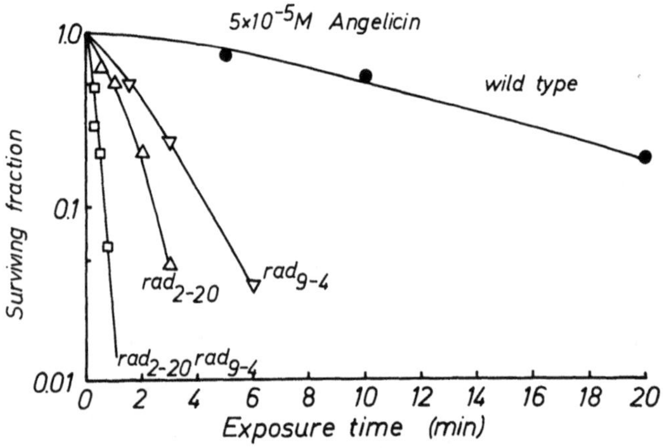

Fig. 4. Typical survival curves of haploid wild-type cells of yeast and the derived mutants *rad2-20*, *rad9-4*, and *rad2-20 rad9-4* after treatment with 5 x 10^{-5} *M* angelicin (monofunctional) plus 365 nm light

The additive interaction of repair pathways in the case of lesions induced by bifunctional furocoumarins can be understood by assuming that the two systems act on different types of lesions, i.e., monoadducts and cross-links. This would be in agreement with the repair models suggested for *E. coli*. In bacteria it is assumed that the lesion (for example, the cross-link) is altered by a first repair system (excision), this modified lesion being subsequently repaired by a second repair system (recombination repair system)(Cole, 1973).

The synergistic action of the two repair pathways of yeast in the repair of lesions induced by the monofunctional furocoumarin is in agreement with the assumption that the monoadducts induced are subject to repair by either one or the other pathway.

In order to test whether the furocoumarin plus light-induced damage is subject to delayed plating recovery, we held wild-type cells for 48 h after the treatment on non-nutrient agar before adding the growth medium.

Figure 5 shows typical survival curves of haploid and diploid wild-type cells obtained after treatment with 20 µg/ml psoralen (bifunctional furocoumarin) and 365 nm light. After immediate plating, haploid cells are more sensitive than diploid cells. Both haploid and diploid cells show a substantial increase in survival after 48 h agar holding.

The situation is similar after treatment with a monofunctional furocoumarin (angelicin, 20 µg/ml) and 365 nm light (Fig. 6). Again diploid cells are more resistant than haploid cells, and both are capable of agar holding recovery.

We interpret the fact that diploid cells are always more resistant than haploid cells as being due to a higher repair capacity of diploid cells for furocoumarin plus 365 nm light-induced lesions. The increase in survival after 48 h agar holding (dose modification a factor of about 2) seems to indicate additional repair activities during the agar holding period.

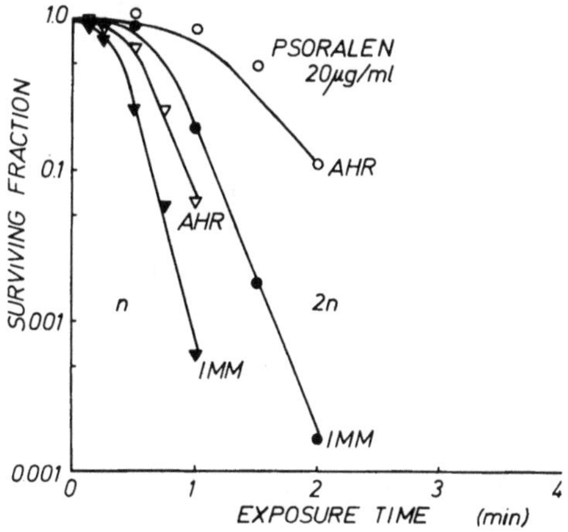

Fig. 5. Typical survival curves of haploid (n) and diploid (2n) wild-type cells after treatment with 20 µg/ml psoralen (bifunctional) plus 365 nm light. IMM, immediate plating; AHR, 48 h of agar holding

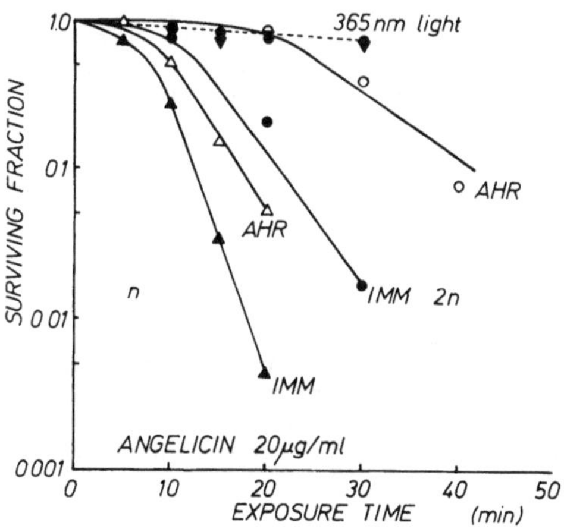

Fig. 6. Typical survival curves of haploid and diploid wild-type cells after treatment with 20 µg/ml angelicin plus 365 nm light. IMM, immediate plating; AHR, 48 h of agar holding

Now we turn to the question: are monofunctional and bifunctional lesions repaired with the same efficiency during agar holding?

Figure 7 shows the survival of haploid wild-type cells after treatment with 10 µg/ml angelicin (monofunctional) and psoralen (bifunctional) plus 365 nm light as a function of the time of agar holding. No significant differences are seen in the kinetics after treatment with the two compounds, and this is true also for diploid cells. This suggests that repair during agar holding proceeds with the same efficiency for the two compounds. Since the two drugs have the induction of monoadducts in common, it can be suggested from the similarity of the kinetics that only monoadducts are repaired during the agar holding period.

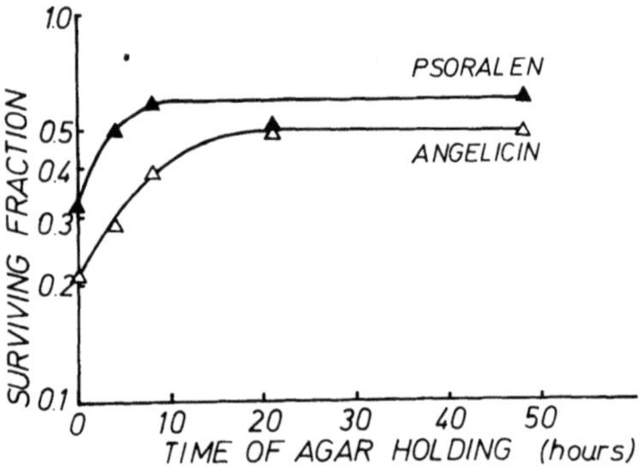

Fig. 7. Survival of haploid wild-type cells after treatment with 10 µg/ml angelicin and psoralen plus 365 nm light as a function of the time of agar holding

It is interesting to note that haploid mutant cells carrying the *rad2-20* gene, which inhibits excision repair of UV-induced pyrimidine dimers, do not show agar holding recovery after treatment with monofunctional or bifunctional furocoumarins plus 365 nm light or after treatment with 254 nm UV light (Averbeck, 1970).

In an attempt to measure the repair function working during agar holding, i.e., in dark repair conditions, we determined the binding and the release of psoralen and angelicin in conditions as near as possible to the conditions of the biologic experiments. We used tritiated compounds and the biochemical methods described by Chandra et al. (1973), which consist of precipitating the DNA of treated cells with cold trichloracetic acid and collecting it on Whatman paper discs and counting the radioactivity in a liquid scintillation counter (Intertechnique Spectromètre à scintillation liquide SL 32). We recall here the results (Averbeck et al., 1975): tritiated angelicin and psoralen are bound to DNA about linearly with the dose of 365 nm light. Per unit dose of 365 nm light, a factor of 4.3 less angelicin is bound than psoralen, reflecting the known differences of their photoreactivities with DNA (Musajo and Rodighiero, 1972a).

In conditions of dark repair (= liquid holding), the bound material decreases by about 50% for angelicin and by about 30% for psoralen after 50 h of incubation in non-nutrient medium. This corresponds to the increase of survival seen after agar holding for 50 h; i.e., the release of nucleic acid-bound material seems to be correlated to the dark recovery observed in vivo. Our data seem to indicate that more material is released per unit time in the case of angelicin than in the case of psoralen, which would agree with the idea that angelicin (monofunctional)-induced damage is repaired more easily than damage induced by bifunctional furocoumarins such as psoralen. Further biochemical studies seem to be necessary to elucidate this repair function.

In conclusion, it may be said that we have obtained biologic evidence for the repair of monoadducts and cross-links induced by monofunctional and bifunctional furocoumarins in the yeast *Saccharomyces cerevisiae*. Results obtained with rad mutants demonstrate that the excision and the recombinational repair pathways are involved in the repair. Diploid cells seem to have a higher capacity for repair than haploid cells. During agar holding further damage can be repaired. The kinetics of agar holding favor the idea that during agar holding only furocoumarin plus 365 nm light-induced monoadducts are repaired.

Preliminary biochemical data demonstrate the release of DNA-bound material during dark repair conditions. This is in agreement with the hypothesis that the induced damage might be at least in part excised from the DNA.

References

Averbeck, D.: Isolierung und Charakterisierung von drei UV-sensiblen Saccharomyces-Mutanten. Dissertation, Freie Universität Berlin, 1970

Averbeck, D., Chandra, P., Biswas, R.K.: Structural specificity in the lethal and mutagenic activity of furocoumarins in yeast cells. J. Radiat. Environ. Biol. 12, 241-252 (1975)

Averbeck, D., Moustacchi, E.: 8 Methoxypsoralen plus 365 nm light effects and repair in yeast. Biochim. Biophys. Acta (Amst.) 395, 393-404 (1975)

Averbeck, D., Moustacchi, E.: Biological effects and repair of damage photoinduced by a derivative of psoralen substituted at the 3, 4 reaction site. I. Killing effect in yeast. (Submitted to Biochim. Biophys. Acta (Amst.)) (1976)

Baden, M.P., Parrington, J.M., Delmanty, J.D.A., Pathak, M.A.: DNA synthesis in normal and XP-fibroblasts following treatment with 8 Methoxypsoralen and long wave ultraviolet light. Biochim. Biophys. Acta (Amst.) 262, 247-255 (1972)

Ben-Hur, E., Elkind, M.M.: Psoralen plus near ultraviolet light inactivation of cultured Chinese hamster cells and its relation to DNA cross-links. Mutation Res. 18, 315-324 (1973a)

Ben-Hur, E., Elkind, M.M.: DNA cross-linking in Chinese hamster cells exposed to near ultraviolet light in the presence of 4, 5', 8' Trimethylpsoralen. Biochim. Biophys. Acta (Amst.) 331, 181-193 (1973b)

Chandra, P., Biswas, R.K., Dall'Acqua, F., Marciani, S., Baccichetti, F., Vevaldi, D., Rodighiero, C.: Post-irradiation dark recovery of photo-damage to DNA induced by furocoumarins. Biophysik 9, 113-119 (1973)

Cole, R.S.: Inactivation of *Escherichia coli*, F' Episomes at transfer and bacteriophage lambda by psoralen plus 360 nm light; significance of desoxyribonucleic acid cross-links. J. Bacteriol. 107, 846-852 (1971)

Cole, R.S.: Repair of DNA containing interstrand cross-links in *Escherichia coli*. Sequential excision and recombination. Proc. Nat. Acad. Sci. USA 70, 1064-1068 (1973)

Dall'Acqua, F., Marciani, S., Ciavatta, L., Rodighiero, G.: Formation of interstrand cross-linkings in the photoreactions between furocoumarins and DNA. Z. Naturforsch. 26b, 561-569 (1971)

Eckardt, F., Kowalski, S., Laskowski, W.: The effects of three rad genes on UV induced mutation rates in haploid and diploid Saccharomyces cells. Molec. Gen. Genetics 136, 261-272 (1975)
El-Mofty, A.M.: Vitiligo and Psoralens. Oxford: Pergamon Press, 1968, pp. 1-221
Igali, S., Borstel, R.C., von: Lethal and mutagenic action of 8-Methoxypsoralen with near-ultraviolet light in *Saccharomyces cerevisiae*. 2nd Ann. Meet. Amer. Soc. Photobiol. Abstracts THAM-C2, Univ. Brit. Columbia, 1974, pp. 99
Igali, S., Bridges, B.A., Asmwood-Smith, M.J., Scott, B.R.: Mutagenesis in *Escherichia coli*. IV. Photosensitization to near ultraviolet light by 8-Methoxypsoralen. Mutation Res. 9, 21-30 (1970)
Kowalski, S., Laskowski, W.: The effect of three rad genes on survival, inter and intragenic mitotic recombination in *Saccharomyces*. Molec. Gen. Genetics 136, 75-86 (1975)
Musajo, L., Rodighiero, G.: Mode of photosensitizing action of furocoumarins. In: Photophysiology. New York-London: Academic Press, 1972a, Vol. VII, pp. 115-147
Musajo, L., Rodighiero, G.: Photo-C_4-cyclo addition reactions to the nucleic acids. In: Res. Prog. Org. Biol. Med. Chem. (Gallo, U., Santamaria, L., eds.). Amsterdam: North Holland Publ. Co., 1972b, Vol. 3, pp. 155-182
Parrish, L., Fitzpatrick, T.B., Tanenbaum, Pathak, M.A.: Photochemotherapy of psoriasis with oral methoxysalen and longwave ultraviolet light. New Engl. J. Med. 291, 1207-1211 (1974)
Resnick, M.A., Setlow, J.K.: Repair of pyrimidine dimer damage induced in yeast by ultraviolet light. J. Bacteriol. 109, 979-986 (1972)
Rodighiero, G., Dall'Acqua, F., Marciani, S., Chandra, P., Feller, H., Götz, A., Wacker, A.: Studies on the reactivation of bacteria photodamaged by an angular furocoumarin: angelicin. Biophysik 8, 1-8 (1971)
Swanbeck, G., Thyresson, M.: Induction of respiration-deficient mutants in yeast by psoralen and light. J. Invest. Dermatol. 63, 242-244 (1974)
Walter, J.F., Voorhees, J.J.: Psoriasis improved by psoralen plus black-light. Act. Dermatol. Venereol. 53, 469-472 (1973)
Weber, G.: Combined 8-methoxypsoralen and black light therapy of psoriasis. Technique and results. Brit. J. Dermatol. 90, 317-323 (1974)

10. Postreplication Repair of DNA in Mammalian Cells: A Discussion of the Mechanisms and Biological Importance

A. R. LEHMANN

A. Introduction

Postreplication repair (PRR) will refer in this paper to the processes whereby unexcised lesions in the DNA strands in damaged cells are dealt with during or after replication. As such it is very closely associated with the normal processes of replication. Many types of damage lead to disruption of the normal base pairing forces in the DNA double helix, so that special mechanisms are required for the eventual accurate synthesis of intact daughter strands. I have recently reviewed the field of PRR in mammalian cells (Lehmann, 1974) and the reader is referred to this review for a detailed account of the field and comprehensive literature list. In the present report I shall (1) summarize the main experimental results in the current literature, (2) discuss some of the controversies and discrepancies in more detail, (3) discuss the biological importance of PRR. I shall be principally concerned with ultraviolet (UV)-irradiated cells, since this is the most studied system. I should like to emphasize that much of the discussion will be speculative and will consequently reflect my personal biases and interpretations.

B. Summary of the Literature

1. In all cells studied UV decreases the overall rate of DNA synthesis, presumably because the lesions in the template strands delay the replication machinery.

2. Under many conditions, in a variety of human, rodent, marsupial, and avian cell lines, the newly synthesized DNA in irradiated cells is smaller than that in unirradiated cells. This has usually been interpreted on the bacterial model (Rupp and Howard-Flanders, 1968) as evidence for the existence of gaps in the daughter strands opposite lesions in the template strands. In some cases the size of the daughter strands approximates the distance between pyrimidine dimers in the parental strands. Direct evidence for the existence of daughter-strand gaps has recently been produced (Meneghini, 1976).

3. In some cell lines, under certain conditions, the newly synthesized DNA in irradiated cells is the same size as that in unirradiated cells. This will be discussed in greater detail below.

4. When daughter strand gaps are observed, evidence obtained from systems that are able to remove the pyrimidine dimers by photoreactivation (marsupial and avian cells) indicates that pyrimidine dimers are responsible for the formation of the majority of these gaps (Buhl et al., 1974; Lehmann and Stevens, 1975).

5. Gaps formed in daughter strands are subsequently sealed in all cells studied. (This gap formation and sealing process is often called PRR, but I shall use the term PRR in the more general sense defined in the Introduction, and the term gap-filling for the more specific process.)

6. This sealing step is inhibited by normal inhibitors of DNA synthesis, but not by inhibitors of protein synthesis (e.g., see Fujiwara, 1975). In rodent cells, but not in normal human cells, it is specifically inhibited by caffeine and a variety of caffeine analogues, which have little or no effect on DNA replication in undamaged cells (e.g., see Lehmann and Kirk-Bell, 1974). This effect of caffeine is not mediated via cyclic AMP (Ehmann et al., 1975). The mechanism of action may involve direct binding of caffeine to single-stranded regions in the DNA (Domon et al., 1970), but the recent discovery that caffeine is rapidly demethylated inside cells (Goth and Cleaver, 1976) throws doubt on this hypothesis.

7. Direct attempts to detect recombinational exchanges associated with PRR have been unsuccessful. Evidence suggests that gaps are sealed by de novo synthesis, the size of the filled-in piece being approximately 800 nucleotides. Recent work has, however, shown that a small proportion of the pyrimidine dimers are found in the daughter strands after replication. This implies that some recombinational exchanges must take place (discussed in detail below).

8. In cell lines where gaps are observed in DNA synthesized soon after irradiation, no gaps can be observed in the sections of DNA that are synthesized several hours after irradiation, despite the fact that most of the dimers remain in the template strands (Meyn and Humphrey, 1971; Lehmann and Kirk-Bell, 1972; Buhl et al., 1973).

9. Fibroblasts from patients with the variant form of xeroderma pigmentosum (XP) (normal in excision repair) are defective in PRR (discussed below).

10. Inhibition of a step in PRR by caffeine in rodent cells may give rise to a decrease in the number of induced sister chromatid exchanges, and an increase in the production of chromosome aberrations, mutations and cell killing.

C. Daughter Strand Gaps

The observation and measurement of daughter strand gaps have been made by analysis of radioactivity profiles obtained from alkaline sucrose gradient sedimentation of the DNA from cells pulse-labelled with ^{3}H-thymidine after UV irradiation. If gaps are produced in daughter DNA strands opposite pyrimidine dimers, and these gaps are only sealed some considerable time later, a stretch of growing DNA chain will contain several gaps. If these growing strands are labelled with a pulse of ^{3}H-thymidine, the Mol. Wt. distribution of the labelled DNA, as analyzed in alkaline sucrose gradients, should reflect the distribution of interdimer distances in the template strands, and the number average Mol. Wt. of the newly synthesized pieces should be equal to the average interdimer distance. Such a situation is observed in excision-deficient strains of *Escherichia coli* (Rupp and Howard-Flanders, 1968).

In mammalian cells, the situation seems to be more complex. Reports have been made that newly synthesized DNA in UV-irradiated cells is approximately the same size as the interdimer distance (Lehmann, 1972; Buhl et al., 1972b), that it is greater than the interdimer distance, and that it is the same size as in unirradiated cells (Rauth et al., 1974). This discrepancy is clearly illustrated in the work of Rauth et al. (1974). Three cell lines were irradiated with the same UV fluence to give the same frequency of pyrimidine dimers in the DNA. The newly synthesized DNA was then labelled with a pulse of ^{3}H-thymidine and the Mol. Wt. distributions analyzed. After 10 J/m^{2}, in HeLa cells the newly synthesized DNA was much smaller than in unirradiated cells, in Chinese hamster CHO cells it was only slightly smaller, and in mouse L cells it was the same size as in unirradiated cells. What are the reasons for these differences? The explanations below are developments of suggestions put forward in earlier work (Chiu and Rauth, 1972; Lehmann and Kirk-Bell, 1972; Lehmann et al., 1975).

If daughter strand gaps are formed, but the time between gap formation and gap sealing is rather short relative to the time taken to synthesize DNA between two adjacent dimers, any growing DNA strand will contain at the most only one or two gaps. The Mol. Wt. distribution of pulse-labelled DNA will then depend on a variety of factors, including the rate of DNA chain elongation, the size of the replication units, the duration of the pulse label, the number of gaps produced, and the time between gap production and gap sealing. All these factors must be taken into account, and consequently it is difficult to obtain unambiguous interpretations of the Mol. Wt. distributions of pulse-labelled DNA from UV-irradiated mammalian cells. The differences found by workers using different cell lines, different UV fluences, or different labelling conditions, may be the result of variations between one or more of the factors listed above. One obvious candidate is the time between gap formation and gap sealing (see Lehmann and Kirk-Bell, 1972). Thus the results of Rauth et al. (1974), for example, may be explained if we assume that gaps are produced opposite all dimers in all cells, but the time between gap formation and gap sealing is very short in L cells, intermediate in CHO cells, and relatively long in HeLa cells. Only in the latter cell line would gaps persist long enough to be readily observable under the experimental conditions used. A similar type of reasoning has been used (Lehmann and Kirk-Bell, 1972) to explain the observations that no gaps can be detected in DNA labelled in a short pulse given several hours after irradiation, even in cell lines in which gaps can be detected if the pulse is given immediately after irradiation, and even though the dimers remain in the parental strands (Meyn and Humphrey, 1971; Lehmann and Kirk-Bell, 1972; Buhl et al., 1973). One would postulate that the time between gap formation and gap sealing is much shorter at the late times than at the early times.

As an alternative explanation for the discrepancies discussed above, there could be two (or more) ways in which pyrimidine dimers could be circumvented. An obvious possible reason for the failure to detect daughter strand gaps in some instances is that gaps are not produced at all, i.e., that synthesis proceeds, after a delay, past the dimer in a continuous fashion rather than continuing beyond the dimer before nucleotide insertion takes place opposite the dimer. If cells contained both a discontinuous and a continuous mechanism, depending on which of these predominated, a greater or lesser number of gaps would be observed. Variation in the predominance of these two mechanisms between cell lines and at different UV fluences would give rise to different sedimentation profiles.

I do not wish to suggest that either of the two alternative explanations discussed above necessarily accounts for the experimental findings that have been reported, but to illustrate that either or both explanations could account rather simply for what appear to be contradictions in the literature.

D. Recombinational Exchanges

I have not touched in the above discussion on the mechanism whereby nucleotides are inserted opposite the dimers. Although there is no evidence in mammalian systems as to whether correct or incorrect nucleotides are inserted opposite the dimers, studies on mutation frequencies in *E. coli* indicate that the correct nucleotides are inserted opposite the majority of dimers, errors arising in only a few instances (Bridges et al., 1967). In *E. coli* the lost genetic information at the dimer/gap site is refurbished by a system of single strand recombinational exchanges, such that the piece of parental DNA from the sister duplex containing the appropriate nucleotide sequence for the gap is inserted in the gap. The discontinuity then remaining in the parental DNA, having an intact template opposite, can be filled in by normal repair replication (Rupp et al., 1971). The evidence suggests that extensive exchanges occur at each dimer/gap site (Rupp et al., 1971) with single-strand lengths of DNA of about 5×10^6 daltons being exchanged (Ley, 1973). In about half the exchanges the strand containing the dimer is exchanged, so that dimers eventually become distributed equally between parental and daughter strands (Ganesan, 1974).

In mammalian systems initial attempts to detect exchanges were unsuccessful. End-to-end attachments of parental and daughter strands have not been detected, in contrast to the bacterial situation (Lehmann, 1972; Fujiwara and Kondo, 1974a; Painter, 1974). Most gaps appear to be sealed by de novo synthesis, the piece of DNA in the filled-in gap being about 800 nucleotides in length (Lehmann, 1972; Buhl et al., 1972a). The general conclusion from these experiments was that exchanges did not seem to occur at all. However they did not rule out the possibility that exchanges only occurred at a small proportion of dimer/gap sites, or that they occurred at all dimer/gap sites but only a very limited amount of DNA was exchanged (less than 100 nucleotides). Extensive exchanges, like those found in bacteria (Rupp et al., 1971; Ley, 1973), did not seem to occur. Very recently, however, Meneghini and Hanawalt (1975) have shown that in human cells some exchanges must indeed take place, since about 20% of the dimers can be detected in the daughter DNA strands after replication. Preliminary results from our laboratory confirm these findings. Presumably the exchanges must be very small or rather infrequent and thus escaped detection in earlier experiments (Lehmann, 1972; Fujiwara and Kondo, 1974a; Painter, 1974).

Cytological evidence also indicates that genetic exchanges occur after irradiation, since UV produces an increase in the number of sister chromatid exchanges (SCE) (see Section F II below). However, as discussed by Kato (1973), it is unlikely that the SCE event is a direct outcome of exchanges such as those seen in bacteria, since the size of the exchanged DNA segments in *E. coli* (of the order of 5×10^6 daltons) is extremely small as compared with the observed length of a chromatid segment lying between two adjacent switches of silver grains in an autoradiogram (equivalent to about 10^{10} daltons of DNA). SCE's may be best explained if a very small proportion of dimer/gap sites gives rise to exchanges of very long double-stranded segments of DNA (Kato, 1973, 1974a). This would not be detectable in biochemical experiments. The production of SCE's is discussed in more detail in Section F II.

If the genetic information lost at the dimer/gap site is not regained by genetic exchanges in every case, it is necessary to postulate either (1) that a special enzyme can insert the correct (or possibly incorrect) nucleotides by recognizing the dimer (a somewhat heretical suggestion) or (2) that the correct nucleotides are inserted by using as template, not the parental strand of the same duplex, but the daughter strand of the sister duplex (which will have the same sequence but will, of course, not contain the dimer). This could occur by a kind of strand displacement mechanism similar to that suggested by Higgins et al. (1976). Interesting models for replication on damaged templates have been put forward by Meneghini (1976) and Higgins et al. (1976). Whilst, in my view, these models are not completely compatible with all the data available, it is likely that aspects of these models will be of relevance in some, if not all situations.

E. PRR in Xeroderma Pigmentosum (XP) Cells

XP is an autosomal recessive disorder, giving rise to a variety of sunlight-induced lesions in the skin, including multiple skin malignancies (Robbins et al., 1974). Cells from the majority of XP patients are unable to excise pyrimidine dimers from their DNA (see Cleaver, 1974). The genes coding for the deficient enzymes fall into five different complementation groups (designated A, B, C, D, E)(Robbins et al., 1974; Bootsma et al., 1975). Cells from a sixth class of patients have normal levels of excision repair (see Robbins et al., 1974). These are designated XP variants. We recently showed that fibroblast cells from XP variants have a defect in PRR. After a relatively low UV fluence, newly synthesized DNA in UV-irradiated normal cells is similar in size to that in unirradiated cells and is subsequently rapidly converted into high Mol. Wt. DNA. This conversion is not affected by caffeine (Buhl and Regan, 1974; Lehmann et al., 1975). In fibroblasts from XP variants, newly synthesized DNA in UV-irradiated cells is much smaller than in unirradiated cells, and conversion to high Mol. Wt. DNA takes a long time and is inhibited by caffeine (Lehmann et al., 1975). From the discussion in Section C above there are at least two possible explanations: (1) both normal and XP variant cells leave gaps opposite dimers, but normal cells seal them after a very short interval whereas the XP cells only seal them after a much longer period. Caffeine is then envisaged as a competitive inhibitor of the gap-filling enzymes (see Lehmann et al., 1975). (2) Normal cells have both a discontinuous caffeine-sensitive and a continuous caffeine-resistant mode of synthesis past dimers, the latter predominating. In XP variants the latter is missing, only the former mechanism is available. Whatever the interpretation, the results unambiguously point to a defect in PRR in XP variants. We have carried out similar experiments with skin fibroblasts from excision-defective XP patients, from patients with a large variety of sunlight-sensitive disorders, and from patients with multiple skin cancers. PRR was normal in all these cell lines except for those from XP patients. The defect described above has been found in five XP variants, and a less severe defect in PRR has been observed in fibroblasts from excision-defective XP patients from the complementation groups A, B, C, and D (see also Buhl and Regan, 1974). Cells from group E were normal.

F. Relationship to Cell Survival, Sister Chromatid Exchanges, Chromosome Aberrations, Mutagenesis and Carcinogenesis

PRR is a process that is studied at the molecular level. The biochemical procedures that are used can only give information about what happens to the majority of lesions. On the other hand biological effects, e.g., production of chromosome aberrations, are studied at the cellular level, and only a minute proportion of the lesions (perhaps 10^{-4}) will give rise to a chromosome aberration. To attempt to establish a relationship between a process at the molecular level and a biological endpoint is therefore no mean task.

In rodent cells caffeine appears to act as a specific inhibitor of the filling of daughter strand gaps after UV irradiation. No inhibition of caffeine on excision-repair of any type of damage has been detected in any mammalian cell (e.g., Regan et al., 1968; Roberts and Ward, 1973). Since PRR involves the newly synthesized DNA strands, almost by definition it is an S-phase specific process. Thus the observation that a particular biological endpoint (e.g., chromosome aberrations, cell survival) is affected by caffeine, if the caffeine is present during the S-phase of the cell cycle, provides good suggestive evidence, though not definite proof, that PRR is in some way involved in the production of the biological endpoint. Because detailed knowledge of the mechanism of PRR and of the induction of the various biological endpoints (e.g., SCE's, mutations) is very limited, we are often unable to predict whether caffeine should increase or decrease the biological parameter under observation.

I. Cell Survival

Rodent cells have only a very limited ability to excise pyrimidine dimers (Trosko et al., 1965), yet they are no more sensitive to UV than are human cells, which can excise dimers (see Rauth, 1970). Since rodent cells can survive UV fluences that produce 10^5-10^6 dimers per cell, the majority of these dimers must be handled by the PRR system, suggesting that PRR must play an important role in cell survival in rodent cells after UV irradiation. In support of this, caffeine sensitizes most rodent cell lines to the killing action of UV irradiation, and only if present during S-phase (Rauth, 1967; Domon and Rauth, 1969). Similar sensitization by caffeine has been observed following treatment with various chemical mutagens (Rauth et al., 1970; Walker and Reid, 1971; Roberts and Ward, 1973; Turnbull, 1975) so that similar conclusions can be drawn, although little is known about the details of PRR after treatment with these agents.

Normal human cells can excise pyrimidine dimers (Regan et al., 1968). The lack of this excision repair in most XP cells gives rise to a striking sensitization of the cells to UV (Cleaver, 1974; Fujiwara and Kondo, 1974b; Maher et al., 1975; Arlett et al., 1975) and to a very low ability of XP cells to reactivate UV-irradiated adenovirus-2 (Day, 1974). On the other hand the defective PRR in XP variants produces only a very small sensitization of the cells (Cleaver, 1972; Maher et al., 1975; Arlett et al., 1975) and a slight decrease in host-cell reactivation of adenovirus-2 (Day, 1975). Caffeine inhibits PRR in XP variants but not in normal human cells, and it increases the UV-induced killing of XP variants, but not of other human cells (Arlett et al., 1975; Wilkinson et al., 1970). This correlation implies that the inhibition of PRR by caffeine in XP variants sensitizes them to UV irradiation.

II. Sister Chromatid Exchanges (SCE's)

The frequency of SCE's is increased by UV light (Rommelaere et al., 1973; Kato, 1973, 1974a; Wolff et al., 1974), mitomycin C (Latt, 1974; Kato, 1974b; Kato and Shimada, 1975), 4-nitroquinoline-oxide (Kato, 1974b), and thiotepa (Kihlman, 1975). The involvement of PRR in the production of SCE's has been suggested by some of these workers (Kato, 1973, 1974b; Wolff et al., 1974) and is supported by the observations that UV-irradiated cells must pass through DNA synthesis before SCE's are produced (Kato, 1973; Wolff et al., 1974) and that caffeine decreases the number of SCE's induced by UV (Kato, 1973), mitomycin C, and 4-nitroquinoline-oxide (Kato, 1974b)(though it does not affect thiotepa-induced SCE's in plant cells (Kihlman, 1975)). Kato (1974a) has suggested that SCE's may occur where dimer/gap sites on sister duplexes overlap. This may initiate reciprocal exchanges of double-strand DNA. Bender et al. (1974) have suggested that SCE's result from exchanges of single-strand pieces of DNA, but it is difficult to conceive of a mechanism for unwinding and exchanging such enormous lengths of single DNA strands (Kato, 1974b). As discussed in Section D, SCE's are more likely to arise from double-strand exchanges.

III. Chromosome Aberrations

UV and a variety of chemical mutagens induce chromosome aberrations in plant and animal cells. Caffeine produces a dramatic increase in the number of aberrations induced by UV and most of these chemicals, and the caffeine must be present during S-phase (reviewed by Kihlman, 1974). This points to a correlation between the inhibition of PRR and the production of chromosome aberrations (Kihlman et al., 1974). In support of this hypothesis we have found a fairly good correlation between the inhibitory activities of five caffeine analogues on the gap-filling step of PRR in UV-irradiated Chinese hamster cells and their ability to potentiate UV-induced chromosome aberrations (Nilsson and Lehmann, 1975). A model for the possible conversion of dimer/gap sites into chromosomal aberrations has been put forward by Bender et al. (1973).

IV. Mutagenesis and Carcinogenesis

In rodent cells, caffeine, if present during S-phase, enhances the frequency of mutations to azaguanine-resistance following treatment with UV (Arlett and Harcourt, 1972; Morrow, 1972; Fox, 1974) and N-methyl N-nitrosourea (Roberts and Sturrock, 1973) and to ouabain-resistance after treatment with ethyl methanesulphonate (Turnbull, 1975), suggesting an involvement of PRR in mutation induction. In bacteria, UV-induced mutations are thought to be produced by an error-prone PRR pathway (Witkin, 1967), and there is probably an analogous mechanism in mammalian cells. Interference with PRR by caffeine may therefore be expected to alter the induced mutation frequency. Clearly, however, caffeine cannot act merely by inhibiting an error-prone pathway since this would lead to a decrease in the induced mutation frequency, whereas most workers have found an increase. It is necessary to postulate that PRR has an error-free and an error-prone component and that caffeine decreases the former relative to the latter. It is interesting that in *E. coli* caffeine has the opposite effect. It seems to decrease the error-proneness of PRR (Witkin and Farquharson, 1969).

Although the defective PRR in XP variants gives rise to little or no decrease in cell survival, both XP variants and excision-defective XP's are considerably more mutable by UV than are normal cells (Maher and McCormick, 1976). If mutations arise during PRR of unexcised lesions, the increased mutability of XP cells finds a ready explanation. If we assume that each unexcised dimer has a probability p of giving rise to an error during PRR, then the increased number of unexcised dimers in excision-defective XP's will give rise to an increased number of errors relative to normal cells. On the other hand in XP variants the number of unexcised dimers is the same as in normal cells, so that we must postulate that the defective PRR in these cells (Lehmann et al., 1975) gives rise to an increased value of p, i.e., PRR is more error-prone in XP variants than in normal cells. It is interesting that the completely different repair defects in the excision-defective XP's and the XP variants give rise to indistinguishable clinical symptoms.

A principal clinical feature of the xeroderma pigmentosum syndrome is the occurrence of multiple sunlight-induced skin cancers (Robbins et al., 1974). Below is a very simplified speculative summary of some factors that might affect the production of sunlight-induced skin cancers. The ideas in this summary are not particularly original, they are merely developments of similar schemes put forward by others. The main assumptions are that the production of pyrimidine dimers in the DNA of the skin gives rise to skin cancers (c.f. Setlow and Hart, 1975), that excision repair is predominantly an error-free process (Witkin, 1967), and that the skin malignancies arise by mutations in skin cells. The factors affecting the production of skin cancers by sunlight will then be:

1. The incident dose of UV light (sunlight)

2. The proportion of the incident dose absorbed by the DNA of the skin cells (greater in fair-skinned individuals and in albinos)

3. The probability that an induced pyrimidine dimer will not be excised before it is replicated, i.e., that it will be repaired by PRR rather than by excision-repair (greater in excision-defective XP's)

4. The probability that an unexcised pyrimidine dimer will give rise to an error during replication and PRR (greater in XP variants)

5. The probability that such an error occurs in a gene that is involved in the conversion of normal cells to the malignant state - this is clearly a very complex topic, and may depend on a whole variety of factors, some of which may be genetic.

In conclusion, it is clear that PRR is a process of great biological importance, and a defect in PRR has serious consequences for the affected cell, animal, or person. The exact mechanism whereby SCE's, chromosome aberrations, and mutations are induced during PRR leaves room for much speculation, and indeed several models have been put forward (e.g., Bender et al., 1973; Kato, 1974a). However, a great deal of experimentation is required before any postulated model can be validated.

References

Arlett, C.F., Harcourt, S.A.: The induction of 8-azaguanine-resistant mutants in cultured Chinese hamster cells by ultraviolet light. The effect of changes in post-irradiation conditions. Mutation Res. 14, 431-437 (1972)

Arlett, C.F., Harcourt, S.A., Broughton, B.C.: The influence of caffeine on cell survival in excision-proficient and excision-deficient xeroderma pigmentosum and normal human cell strains following ultraviolet-light irradiation. Mutation Res. 33, 341-346 (1975)
Bender, M.A., Griggs, H.G., Bedford, J.S.: Recombinational DNA repair and sister chromatid exchanges. Mutation Res. 24, 117-123 (1974)
Bender, M.A., Griggs, H.G., Walker, P.L.: Mechanisms of chromosomal aberration production. I. Aberration induction by ultraviolet light. Mutation Res. 20, 387-402 (1973)
Bootsma, D., De Weerd-Kastelein, E.A., Kleijer, W.J., Keijzer, W.: Genetic complementation analysis of xeroderma pigmentosum. In: Molecular Mechanisms for the Repair of DNA (Hanawalt, P.C., Setlow, R.B., eds.). New York: Plenum Press, 1975, pp. 725-728
Bridges, B.A., Dennis, R.E., Munson, R.J.: Differential induction and repair of ultraviolet damage leading to true reversions and external suppressor mutations of an ochre codon in *Escherichia coli* B/r WP2. Genetics 57, 897-908 (1967)
Buhl, S.N., Regan, J.D.: Effect of caffeine on postreplication repair in human cells. Biophys. J. 14, 519-527 (1974)
Buhl, S.N., Setlow, R.B., Regan, J.D.: Steps in DNA chain elongation and joining after ultraviolet irradiation of human cells. Int. J. Radiat. Biol. 22, 417-424 (1972a)
Buhl, S.N., Setlow, R.B., Regan, J.D.: Recovery of the ability to synthesize DNA in segments of normal size at long times after ultraviolet irradiation of human cells. Biophys. J. 13, 1265-1275 (1973)
Buhl, S.N., Setlow, R.B., Regan, J.D.: DNA repair in *Potorous Tridactylus*. Biophys. J. 14, 791-803 (1974)
Buhl, S.N., Stillman, R.M., Setlow, R.B., Regan, J.D.: DNA chain elongation and joining in normal human and xeroderma pigmentosum cells after ultraviolet irradiation. Biophys. J. 12, 1183-1191 (1972b)
Chiu, S.F.H., Rauth, A.M.: Nascent DNA synthesis in ultraviolet-irradiated mouse L cells. Biochim. Biophys. Acta (Amst.) 259, 164-174 (1972)
Cleaver, J.E.: Xeroderma pigmentosum: variants with normal DNA repair and normal sensitivity to ultraviolet light. J. Invest. Dermatol. 58, 124-128 (1972)
Cleaver, J.E.: Repair processes for photochemical damage in mammalian cells. In: Advances in Radiation Biology (Lett, J.T., Adler, H., Zelle, M., eds.). New York: Academic Press, 1974, 4, pp. 1-75
Day III, R.S.: Cellular reactivation of ultraviolet-irradiated human adenovirus in normal and xeroderma pigmentosum fibroblasts. Photochem. Photobiol. 19, 15-20 (1974)
Day III, R.S.: Xeroderma pigmentosum variants have decreased repair of ultraviolet-damaged DNA. Nature 253, 748-749 (1975)
Domon, M., Barton, B., Porte, A., Rauth, A.M.: The interaction of caffeine with ultraviolet-light-irradiated DNA. Int. J. Radiat. Biol. 17, 395-399 (1970)
Domon, M., Rauth, A.M.: Ultraviolet-light irradiation of mouse L cells: effects on cells in the DNA synthesis phase. Radiat. Res. 40, 414-429 (1969)
Ehmann, U.K., Gehring, U., Tomkins, G.M.: Postreplication repair: lack of involvement of adenosine-(3'5')-monophosphate (cAMP) in repair inhibition by caffeine. Radiat. Res. 61, 10 (1975)
Fox, M.: The effect of post-treatment with caffeine on survival and UV-induced mutation frequencies in Chinese hamster and mouse lymphoma cells in vitro. Mutation Res. 24, 187-204 (1974)
Fujiwara, Y.: Post replication repair of ultraviolet damage to DNA, DNA chain elongation and effects of metabolic inhibitors in mouse L cells. Biophys. J. 15, 403-415 (1975)

Fujiwara, Y., Kondo, T.: Postreplication gap-filling repair in ultraviolet irradiated mouse L cells. In: Sunlight and Man (Fitzpatrick, T.B., ed.). Tokyo: University Press, 1974a, pp. 91-102

Fujiwara, Y., Kondo, T.: Postreplication repair of ultraviolet damage to DNA in xeroderma pigmentosum, other human and mouse cells in culture. J. Radiat. Res. (Japan) 15, 81-87 (1974b)

Ganesan, A.K.: Persistence of pyrimidine dimers during post-replication repair in ultraviolet light-irradiated *Escherichia coli* K12. J. Molec. Biol. 87, 103-120 (1974)

Goth, R., Cleaver, J.E.: Metabolism of caffeine to nucleic acid precursors in mammalian cells. Mutation Res. 36, 105-114 (1976)

Higgins, N.P., Kato, K., Strauss, B.: A model for replication repair in mammalian cells. J. Molec. Biol. 101, 417-425 (1976)

Kato, H.: Induction of sister chromatid exchanges by UV light and its inhibition by caffeine. Exp. Cell Res. 82, 383-390 (1973)

Kato, H.: Photoreactivation of sister chromatid exchanges induced by ultraviolet irradiation. Nature 249, 552-553 (1974a)

Kato, H.: Induction of sister chromatid exchanges by chemical mutagens and its possible relevance to DNA repair. Exp. Cell Res. 85, 239-247 (1974b)

Kato, H., Shimada, H.: Sister chromatid exchanges induced by mitomycin C: a new method of detecting DNA damage at chromosomal level. Mutation Res. 28, 459-464 (1975)

Kihlman, B.A.: Effects of caffeine on the genetic material. Mutation Res. 26, 53-71 (1974)

Kihlman, B.A.: Sister chromatid exchanges in *Vicia faba*. II. Effects of thiotepa, caffeine and 8-ethoxycaffeine on the frequency of SCE's. Chromosoma 51, 11-18 (1975)

Kihlman, B.A., Sturelid, S., Hartley-Asp, B., Nilsson, K.: The enhancement by caffeine of the frequencies of chromosomal aberrations induced in plant and animal cells by chemical and physical agents. Mutation Res. 26, 105-122 (1974)

Latt, S.A.: Sister chromatid exchanges, indices of human chromosome damage and repair: detection by fluorescence and induction by mitomycin C. Proc. Nat. Acad. Sci. USA 71, 3162-3166 (1974)

Lehmann, A.R.: Postreplication repair of DNA in ultraviolet-irradiated mammalian cells. J. Molec. Biol. 66, 319-337 (1972)

Lehmann, A.R.: Postreplication repair of DNA in mammalian cells. Life Sci. 15, 2005-2016 (1974)

Lehmann, A.R., Kirk-Bell, S.: Post-replication repair of DNA in ultraviolet irradiated mammalian cells. No gaps in DNA synthesized late after ultraviolet irradiation. Eur. J. Biochem. 31, 438-445 (1972)

Lehmann, A.R., Kirk-Bell, S.: Effects of caffeine and theophylline on DNA synthesis in unirradiated and UV-irradiated mammalian cells. Mutation Res. 26, 73-82 (1974)

Lehmann, A.R., Kirk-Bell, S., Arlett, C.F., Paterson, M.C., Lohman, P.H.M., De Weerd-Kastelein, E.A., Bootsma, D.: Xeroderma pigmentosum cells with normal levels of excision repair have a defect in DNA synthesis after UV-irradiation. Proc. Nat. Acad. Sci. USA 72, 219-223 (1975)

Lehmann, A.R., Stevens, S.: Postreplication repair of DNA in chick cells: studies using photoreactivation. Biochim. Biophys. Acta (Amst.) 402, 179-187 (1975)

Ley, R.D.: Postreplication repair in an excision-defective mutant of *Escherichia coli*: ultraviolet light-induced incorporation of bromodeoxyuridine into parental DNA. Photochem. Photobiol. 18, 87-95 (1973)

Maher, V.M., Birch, N., Otto, J.R., McCormick, J.J.: Cytotoxicity of carcinogenic aromatic amides in normal and XP fibroblasts with different DNA repair capabilities. J. Nat. Cancer Inst. 54, 1287-1294 (1975)

Maher, V.M., McCormick, J.J.: Effect of DNA repair on the cytotoxicity and mutagenicity of UV irradiation and of chemical carcinogens in normal and xeroderma pigmentosum cells. In: Biology of Radiation Carcinogenesis (Yuhas, J.M., Tennant, R.W., Regan, J.D., eds.). New York: Raven Press, 1976, pp. 129-145

Meneghini, R.: Gaps in DNA synthesized by Ultraviolet Light-Irradiated WI38 Human Cells. Biochim. Biophys. Acta (Amst.) 425, 419-427 (1976)

Meneghini, R., Hanawalt, P.C.: Postreplication repair in human cells: on the presence of gaps opposite dimers and recombination. In: Molecular Mechanisms for the Repair of DNA (Hanawalt, P.C., Setlow, R.B., eds.). New York: Plenum Press, 1975, pp. 639-642

Meyn, R.E., Humphrey, R.M.: Deoxyribonucleic acid synthesis in ultraviolet-irradiated Chinese hamster cells. Biophys. J. 11, 295-301 (1971)

Morrow, J.: Effect of caffeine on the induction of mutations in Chinese hamster cells by ultraviolet light. Mutation Res. 14, 438-439 (1972)

Nilsson, K., Lehmann, A.R.: The effect of methylated oxypurines on the size of newly-synthesized DNA and on the production of chromosome aberrations after UV-irradiation in Chinese hamster cells. Mutation Res. 30, 255-266 (1975)

Painter, R.B.: DNA damage and repair in eukaryotic cells. Genetics 78, 139-148 (1974)

Rauth, A.M.: Evidence for dark-reactivation of ultraviolet light damage in mouse L cells. Radiat. Res. 31, 121-138 (1967)

Rauth, A.M.: Effects of ultraviolet light on mammalian cells in culture. In: Current Topics in Radiation Research (Ebert, M., Howard, A., eds.). Amsterdam: North-Holland, 1970, 6, pp. 193-248

Rauth, A.M., Barton, B., Lee, C.P.Y.: Effects of caffeine on L cells exposed to mitomycin C. Cancer Res. 30, 2724-2729 (1970)

Rauth, A.M., Tammemagi, M., Hunter, G.: Nascent DNA synthesis in ultraviolet light-irradiated mouse, human and Chinese hamster cells. Biophys. J. 14, 209-220 (1974)

Regan, J.D., Trosko, J.E., Carrier, W.L.: Evidence for excision of ultraviolet induced pyrimidine dimers from the DNA of human cells in vitro. Biophys. J. 8, 319-325 (1968)

Robbins, J.H., Kraemer, K.H., Lutzner, M.A., Festoff, B.W., Coon, H.G.: Xeroderma pigmentosum. An inherited disease with sun sensitivity, multiple cutaneous neoplasms, and abnormal DNA repair. Ann. Intern. Med. 80, 221-248 (1974)

Roberts, J.J., Sturrock, J.E.: Enhancement by caffeine of N-methyl-N-nitrosourea-induced mutations and chromosome aberrations in Chinese hamster cells. Mutation Res. 20, 243-255 (1973)

Roberts, J.J., Ward, K.N.: Inhibition of postreplication repair of alkylated DNA by caffeine in Chinese hamster cells but not HeLa cells. Chem.-Biol. Interactions 7, 241-264 (1973)

Rommelaere, J., Susskind, M., Errera, M.: Chromosome and chromatid exchanges in Chinese hamster cells. Chromosoma 41, 243-257 (1973)

Rupp, W.D., Howard-Flanders, P.: Discontinuities in the DNA synthesized in an excision defective strain of *Escherichia coli* following ultraviolet irradiation. J. Molec. Biol. 31, 291-304 (1968)

Rupp, W.D., Wilde III, C.E., Reno, D.L., Howard-Flanders, P.: Exchanges between DNA strands in ultraviolet irradiated *Escherichia coli*. J. Molec. Biol. 61, 25-44 (1971)

Setlow, R.B., Hart, R.W.: Direct evidence that damaged DNA results in neoplastic transformation - a fish story. Radiat. Res. 59, 73-74 (1974)

Trosko, J.E., Chu, E.H.Y., Carrier, W.L.: The induction of thymine dimers in ultraviolet-irradiated mammalian cells. Radiat. Res. 24, 667-672 (1965)

Turnbull, D.: Factors affecting the response of Chinese hamster cells to mutagenic alkylating agents. Ph.D. Thesis, Sussex University, 1975

Walker, I.G., Reid, B.D.: Caffeine potentiation of the lethal action of alkylating agents on L cells. Mutation Res. 12, 101-104 (1971)
Wilkinson, R., Kiefer, J., Nias, A.H.W.: Effects of post-treatment with caffeine on the sensitivity to ultraviolet light irradiation of two lines of HeLa cells. Mutation Res. 10, 67-72 (1970)
Witkin, E.M.: Mutation-proof and mutation-prone modes of survival in derivatives of *Escherichia coli* B differing in sensitivity to ultraviolet light. Brookhaven Symp. Biol. 20, 17-55 (1967)
Witkin, E.M., Farquharson, E.L.: Enhancement and diminution of ultraviolet light initiated mutagenesis by post-treatment with caffeine in *Escherichia coli*. In: Ciba Foundation Symposium on Mutation as Cellular Process (Wolstenholme, G.E.W., O'Connor, M., eds.). London: Churchill, 1969, pp. 36-49
Wolff, S., Bodycote, J., Painter, R.B.: Sister chromatid exchanges induced in Chinese hamster cells by UV irradiation of different stages of the cell cycle: the necessity for cells to pass through S. Mutation Res. 25, 73-81 (1974)

11. Radiation-Induced Mutagenesis and Mechanisms of Repair in the Yeast, *Schizosaccharomyces pombe*

A. Abbondandolo, R. Barale, S. Baroncelli, St. Bonatti, G. Bronzetti, A. Cammellini, G. Corti, N. Loprieno, A. Mazzaccaro, R. Nieri, and A. M. Rossi

Let me begin with an apology. My original intention was to present a synthesis of what is known about mutation and repair in the yeast and try to define a conclusive theory or single hypothesis consistent with the available data. When, however, I was confronted with the task of writing down on paper such coherent ideas, I realized my complete inadequacy to formulate any general theory. Consequently, I shall just tell you now my little story, containing some of the many possible approaches to the problem under discussion, in the hope that the theory, or the organic vision that is needed today by everybody, will result from the critical inspection and discussion of all available data, which is exactly the aim of meetings like the one we are attending.

The experiments I shall describe in the first part of this communication were meant to prove, or disprove, a well-defined hypothesis: the one identifying the so-called "recombinational repair" model (Rupp and Howard-Flanders, 1968) with the error-prone mechanism responsible for mutation induction by physical and possibly chemical agents.

The rationale of the experiments, already proposed and utilized by Bridges and coworkers (Bridges et al., 1970; Bridges and Mottershead, 1971) is plain: since two chromatids (or two DNA duplexes, to use the bacterial terminology) at least are required for recombinational repair to occur, induction of mutations in cells with only one DNA duplex per chromosome will invalidate the hypothesis.

The critical genetic situation was provided in our work by haploid cells of the yeast *Schizosaccharomyces pombe* synchronized in the G1 stage. To accept the logic of the preceding formulation, one has (1) to believe in the "unineme" model for the yeast chromosome (which would be questioned today by very few people) and (2) to make sure that cells treated with the mutagen in G1 do not progress through the cell cycle, in which case premutational lesions lasting long enough could be repaired by recombination-involving mechanisms in the G2 stage.

In order to satisfy the second condition, it was first demonstrated that cells synchronized in G1 of an adenine-requiring strain (*ade7-407*) are unable to duplicate their DNA when incubated on minimal medium (without adenine). The phenotypic change we are selecting for is from inability to ability for any single cell to form a colony in the absence of adenine; this presupposes, in that cell, DNA duplication that is dependent, in turn, on mutation to prototrophy: but this event requires DNA duplication (according to the initial hypothesis), so that a vicious circle is clearly defined. Results obtained with eleven different auxotrophic strains showed that mutations are induced by x-rays in G1 as in G2; this demonstrates, in our opinion, that there is no absolute requirement for recombination in radiation-induced mutagenesis. Moreover, dose-effect curves defined in the *ade7-407* strain showed that the frequency of induced mutants is approximately 2-3 times higher in G2 than in G1.

This result is not entirely explainable in terms of difference in target size, since an increase in mutation frequency was observed when cells treated in G1 were subsequently incubated in a medium permissive for DNA duplication.

Two main conclusions can be inferred from this data. The first, safe issue, is that recombinational repair has to be discarded as the sole source of mutation. The second, tentative conclusion is that an error-prone mechanism (possibly recombinational repair) operates in G2 specifically, while a different, not necessarily stage-dependent mechanism also exists and is best evidenced in the G1 situation.

Further experimental details of the work described thus far have been published elsewhere (Abbondandalo, 1975; Rainaldi and Abbondandalo, 1975).

The role of repair processes in radiation mutagenesis was also studied through a completely different approach, namely by comparing mutation induction in strains with different abilities to repair radiation-induced damage. In that case, besides reversion to prototrophy, a forward mutation system was also used allowing recovery of mutants with all possible types of molecular alterations. A more detailed description of these systems can be found in previous publications (Loprieno et al., 1974, 1975).

The relevant result in this experimental approach was the finding of a higher inducibility of mutation in the *rad*$^+$ (with normal repair capacity) compared to the *rad10-198* (with impaired repair ability) strain. The conclusion that mutations in the wild type (*rad*$^+$) are the result of an error-prone repair mechanism is enticing. A study of the molecular nature of mutations induced in the two strains indicated a higher (45%) incidence of mutations of the base-pair substitution type among the *rad*$^+$ cells as compared to the *rad10-198* (2%). This finding further supports the above conclusion and also indicates base substitutions as the end point of the postulated error-prone repair processes.

The possibility of different repair efficiency during the cell cycle was investigated by comparing induction of forward mutation in G1 and G2 cells. The G1 stage proved to be more sensitive to both the lethal and mutagenic activities of the x-irradiation. The higher mutability was particularly evident at low doses, which may lead to the hypothesis that a very efficient, or highly "misrepairing" mechanism, operates in G1, which is subsequently masked by a second, more accurate mechanism, acting in the remaining stages of the cycle.

The main results and conclusions of this work may be summarized as follows:

1. Mutations are induced by x-rays in haploid, G1 cells. This leads to two possible conclusions: either the radiation is able to modify the DNA molecule directly, or an enzymatic process exists that does not involve recombination.

2. The higher inducibility of mutations over the spontaneous background in a radioresistant compared to a radiosensitive strain clearly points to an involvement of repair mechanisms in the mutagenic process.

3. Two repair mechanisms at least seem to be involved. They correct primary lesions with different accuracy and act in different stages of the cell cycle.

4. The end point of misrepair seems to be a base-pair substitution into the double helix.

I believe that the main contribution of this communication is to make clear that the whole situation is highly confused. Perhaps part of the confusion derives from our attitude in trying to explain the relationship between repair and mutation in eukaryotes by taking *Escherichia coli* as a model. The genetic control of radiosensitivity in the eukaryotic cell, however, is by far more complex: about 30 distinct genetic loci that control sensitivity to radiation are already known in *Saccharomyces cerevisiae* (Brendel and Haynes, 1973) and almost as many have been identified in *S. pombe* (Gutz et al., 1974). This awareness does not have to discourage us, but should act as a stimulus to investigate the nature of repair mechanisms in systems that are nearer to the human cell. The recent progress made in the biochemical approach to the study of radiosensitive mutants, considered unfeasible only a few years ago, is a promising indication of what can still be done.

References

Abbondandalo, A.: Mutation and nuclear stage in *Schizosaccharomyces pombe*. I. An experimental approach to the role of recombination in mutation induction. Mutation Res. 27, 225-233 (1975)

Brendel, M., Haynes, R.H.: Interactions among genes controlling sensitivity to radiation and alkylation in yeast. Molec. Gen. Genetics 125, 197-216 (1973)

Bridges, B.A., Dennis, R.E., Munson, R.J.: Mutagenesis in *Escherichia coli*. III. Requirement for DNA synthesis in mutation by gamma rays of T4-phage complexed with *Escherichia coli*. Genet. Res. 15, 147-156 (1970)

Bridges, B.A., Mottershead, R.: RecA$^+$-dependent mutagenesis occurring before DNA replication in UV- and gamma-irradiated *Escherichia coli*. Mutation Res. 13, 1-8 (1971)

Gutz, H., Heslot, H., Leupold, U., Loprieno, N.: *Schizosaccharomyces pombe*. In: Handbook of Genetics (King, R.C., ed.). New York-London: Plenum Press, 1974, Vol. I, pp. 395-446

Loprieno, N., Barale, R., Baroncelli, S., Cammellini, A., Melani, M., Nieri, R., Nozzolini, M., Rossi, A.M.: Mutations induced by x-radiation in the yeast *Schizosaccharomyces pombe*. Mutation Res. 28, 163-173 (1975)

Loprieno, N., Barale, R., Bauer, C., Baroncelli, S., Bronzetti, G., Cammellini, A., Cinci, A., Corsi, C., Leporini, C., Nieri, R., Nozzolini, M., Serra, C.: The use of different test systems with yeast for the evaluation of chemical induced gene-conversions and gene-mutations. Mutation Res. 25, 197-217 (1974)

Rainaldi, G., Abbondandalo, A.: Mutation and nuclear stage in *Schizosaccharomyces pombe*. II. Reverse mutations induced by x-rays in the absence of recombination. Mutation Res. 27, 235-240 (1975)

Rupp, W.D., Howard-Flanders, P.: Discontinuities in the DNA synthesized in an excision-defective strain of *Escherichia coli* following ultraviolet irradiation. J. Molec. Biol. 31, 291-304 (1968)

12. Accumulation of Replicative DNA Intermediates in Response to Damage of DNA in *Tetrahymena pyriformis*[1]

O. WESTERGAARD and K. A. MARCKER

A. Introduction

The lack of mutants in DNA synthesis among eukaryotic cells has resulted in a major gap in our understanding of replication and repair of the eukaryotic chromosome. Various attempts have been made to overcome this gap. Thus, the enzymes involved in the DNA metabolism have been intensively studied and their activity and intracellular localization correlated with the physiologic conditions of the cells. Another approach has been to study the various intermediates in the DNA metabolism in order to obtain a detailed picture of the individual steps in replication and repair.

We have for some time been involved in studies of the DNA metabolism in *Tetrahymena pyriformis* and shall in the following describe how damage to DNA causes accumulation of both a particular DNA polymerase and intermediates in the DNA synthesis. In addition we have established conditions for accumulation of distinct classes of replicative DNA intermediates.

B. Induction of a Particular DNA Polymerase in Response to Damage of DNA

Radiation of *Tetrahymena* with ultraviolet (UV) light or electrons causes a considerable increase in the specific enzyme activity of total DNA polymerase activity. Figure 1 shows the variation in cell doublings and specific activity after exposure of the cultures to various doses of electron irradiation (Keiding and Westergaard, 1971). It is evident from the figure that the higher the dose, the greater is the induction of polymerase. Figure 2 shows how the activity of the polymerase increases until the cells are able to undergo divisions and thereafter declines to the normal level in untreated cells. The increase in specific activity can be completely blocked by low concentrations of cycloheximide and actinomycin D, which indicates de novo synthesis of enzyme in response to the damage (Westergaard and Pearlman, 1969).

The total DNA polymerase activity from exponentially growing cells can be separated by gel filtration on a Sephadex G-200 column into two enzyme fractions A and B (see Fig. 3a)(Westergaard, 1970). The two fractions differ considerably in their characteristics and can therefore be assayed under two different conditions. Figure 3b shows how fraction B is induced at least 40-fold in specific activity after

[1] Part of this investigation was supported under contract No. 122-74-1 Bio-DK with Euratom, C.E.C., Brussels.

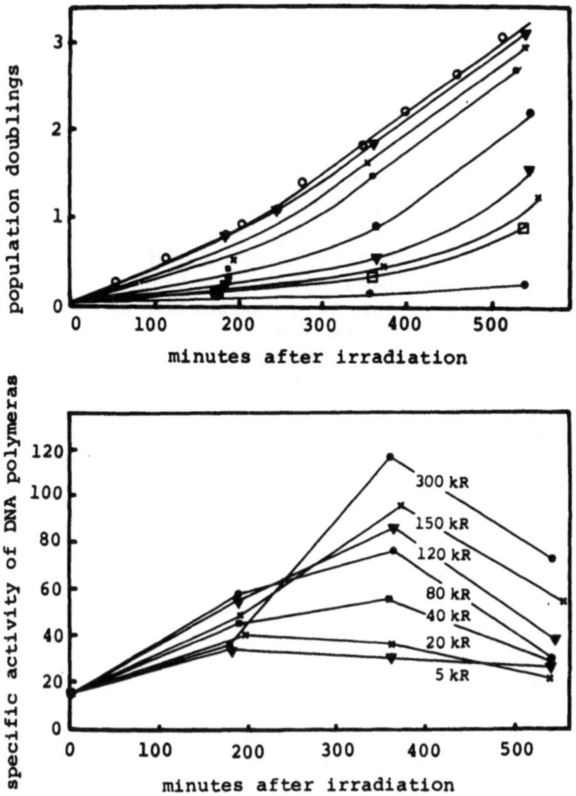

Fig. 1. Effect of electron irradiation on cell doublings and on specific activity of total cellular DNA polymerase activity. Cell number and specific activity (cpm/0.3 mg protein/20 min incubation) were determined as described by Keiding and Westergaard (1971) after irradiation with indicated doses of electrons

treatment of the cells, with a rise in total polymerase activity of 6-fold, while no or very little change is observed in the specific activity of fraction A.

Subcellular fractionation of the treated cells shows that the induced polymerase (fraction B) is localized in the mitochondria, while fraction A presumably is of nuclear origin. The mitochondrial DNA polymerase is induced also after treatment with low concentrations of ethidium bromide, which suggests that the induction of polymerase is under mitochondrial control, although the biosynthesis of the enzyme is under nuclear control (blocked by cycloheximide)(Westergaard et al., 1970). Recent investigations have demonstrated (Tabak and Borst, personal communication, 1975; Keiding, 1975) that the polymerase can be induced by treatment of the cells with chloramphenicol. This might indicate, as earlier suggested by Barath and Küntzel (1972) for other cell systems, that the activity of the nuclear polymerase is controlled by a repressor encoded in mitochondrial DNA. The synthesis or the regulation of the putative repressor may in fact be directly linked to a step in DNA synthesis. Thus thymine starvation of the cells apparently produces an immediate induction of enzyme, since an increase in polymerase activity can already be detected 20-30 min after the exposure to the treatment. However, under these conditions neither protein nor RNA synthesis is inhibited (Villadsen and Zeuthen, 1970). This clearly implies that the nucleus can indirectly recognize an effect on mitochondrial DNA synthesis without the necessity of impairment of mitochondrial protein and RNA synthesis. This hypothesis is furthermore supported by the observation that large amounts of replicative intermediates in the mitochondrial DNA accumulate under conditions, where the polymerase is induced (Upholt and Borst, 1974).

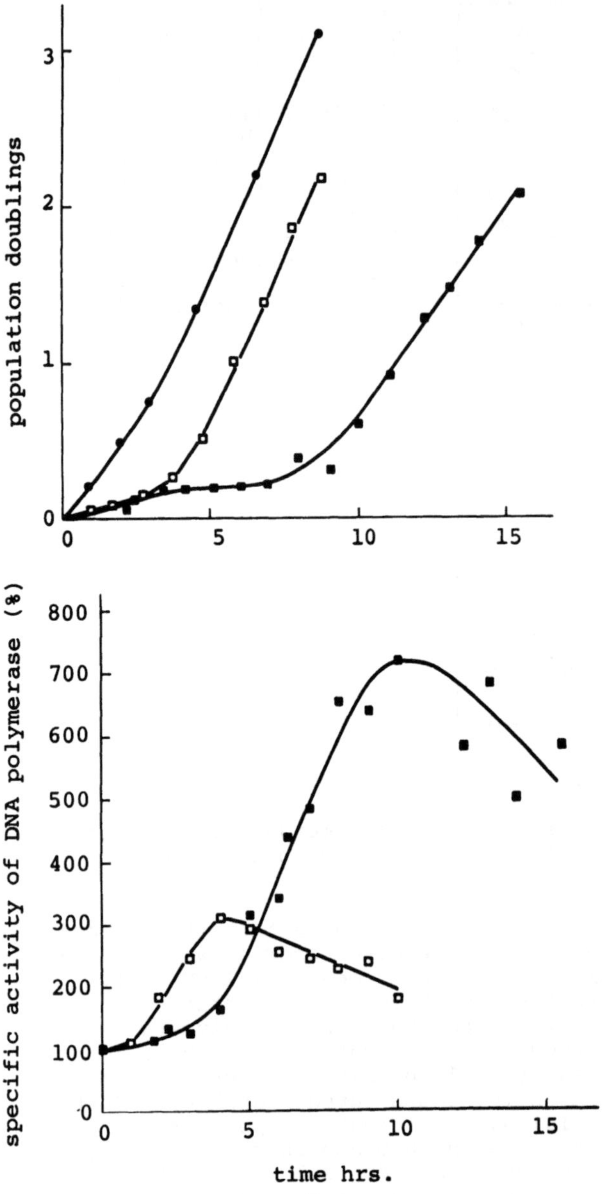

Fig. 2. Correlation between cell division and specific activity of DNA polymerase after electron irradiation. Cell number and specific activity (cpm/ 0.3 mg protein/20 min incubation) were determined as described by Keiding and Westergaard (1971) after irradiation with 0 (●——●), 100 (□——□) and 300 krad (■——■), respectively

C. Accumulation of Replicative Intermediates in Nuclear DNA

Exposure of *Tetrahymena* to excision-repairable damage of DNA results, as is also the case for mitochondrial DNA, in accumulation of replicative intermediates in the nuclear DNA within the first period of synthesis after the damage (Westergaard and Johnson, 1973; Johnson and Westergaard, 1975). Figure 4 shows how cells with DNA damage induced by different times of thymine starvation accumulate a distinct class of "interdensity" DNA (region II) in a buoyant density gradient. The

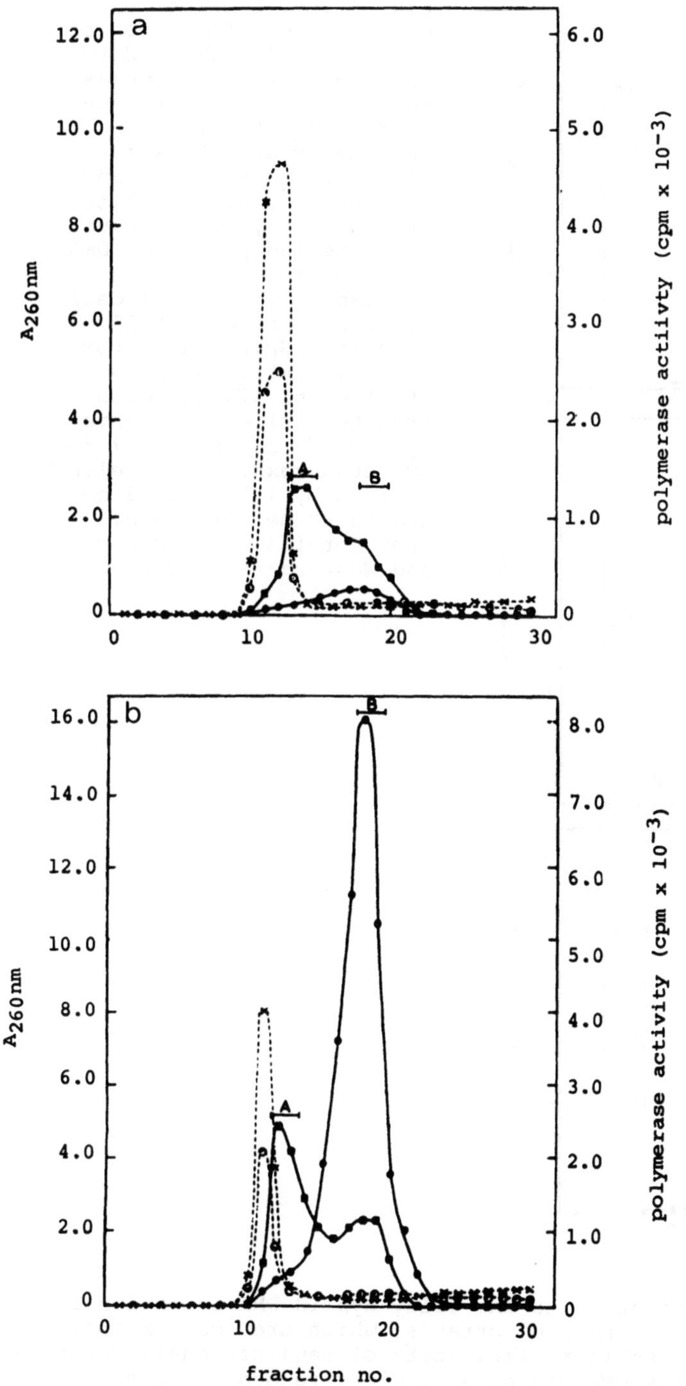

Fig. 3 a and b. Separation of total cellular DNA polymerase activity into two fractions by gel filtration on Sephadex G-200. (a) polymerase activity from exponentially growing cells. (b) polymerase activity from treated cells. (■——■) activity measured by incubation mixture with native DNA; (●——●) activity measured by incubation with heat denatured DNA; (x---x) A_{260} nm. (For details see Westergaard, 1970)

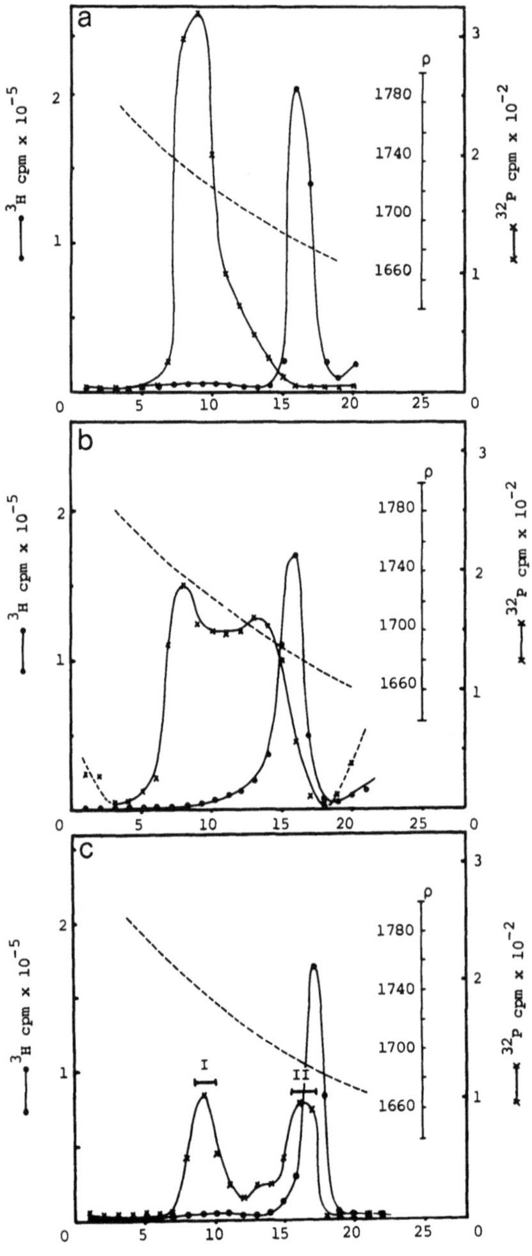

Fig. 4 a-c. Buoyant density gradient centrifugation of DNA from cells exposed to thymine starvation. Cells were prelabelled with ^{3}H-thymidine, starved of thymine by methotrexate plus uridine for 18 h and exposed to a 30-min pulse of bromodeoxyuridine plus ^{32}P-labelled phospate. DNA was isolated from cells and banded in neutral CsCl gradient for 40 h at 39 krpm in a Spinco 65 rotor. (a) cells starved by 38 μ*M* uridine and 375 n*M* methotrexate; (b) cells starved by 750 μ*M* uridine and 7.5 μ*M* methotrexate; (c) cells starved by 1.5 m*M* uridine and 15 μ*M* methotrexate. (For details see Johnson and Westergaard, 1975)

interdensity DNA is a precursor to the fully hybrid DNA in region I and contains only newly synthesized strands, which are smaller than the parental strands (≦ replicon size, units of replications). Furthermore, all the interdensity DNA contains single-stranded regions in contrast to the bulk DNA, in which only a few single-stranded regions are found. These observations have led us to believe that the DNA of region II represents replicating structures that are accumulated in front of damage sites within the individual replicons (see also Kato and Strauss, 1974; Rommelaere et al., 1974). Identical observations have been made for irradiated cells, though it has not been possible to find any accumulation of intermediates in exponentially growing cells. The described system should permit the study of the interaction between replicative and repair synthesis. In particular, it might be possible to solve the long-debated question as to how much repair synthesis occurs at growing points, compared to the relative proportion of repair occurring within the remainder of the DNA (Scudiero and Strauss, 1974).

We recently established growth conditions for *Tetrahymena* by which it is possible to isolate distinct size classes of DNA (Nymann and Westergaard, 1975). This is achieved by growing cells in nearly limiting amounts of phosphate and uridine and exposing the cells to a short cold shock, before the DNA is labelled with pulses of ^{3}H-thymidine. Sedimentation analysis on alkaline velocity centrifugation gradients of DNA labelled under such conditions reveals two distinct peaks with S values of 41 and 86, respectively. Furthermore, experiments have revealed that it is not possible to detect any intermediates between the two size classes of DNA. This is taken to imply that the 41-S fragments become directly converted into the 86-S form by ligase joining of the fragments. Because the 41-S DNA can accumulate as a distinct class of DNA and no intermediates between 41-S and 86-S can be demonstrated, we suggest that this particular DNA represents the size of a defined replicon. This interpretation is in agreement with observations made with other eukaryotic cells (Housman and Huberman, 1975; Painter, 1975).

Treatment of *Tetrahymena* cells with low concentrations (5-25 m*M*) of hydroxyurea causes accumulation of Okazaki fragments sedimenting in the range 4-8-S (Nymann and Westergaard, 1975), as was also observed for other eukaryotic cells (Magnusson et al., 1973). It is noteworthy that the synthesis of such fragments in this system continues at a nearly normal rate (inhibited 20-40%), although they cannot be joined together. In addition we have noted that the hydroxyurea also inhibits the joining of replicons into high-molecular-weight DNA. This might indicate that the same DNA polymerase (with a high K_m value for deoxyribonucleotide triphosphate) is involved in the joining of both Okazaki fragments and replicons.

The effect on DNA synthesis of the protein synthesis inhibitors cycloheximide and puromycin has also been investigated (Nymann and Westergaard, 1975). The rate of DNA synthesis in *Tetrahymena* in the presence of low concentrations of either drug is nearly normal during the first 90 min, even though protein synthesis is completely blocked under these conditions. Nevertheless cycloheximide has a specific effect on DNA synthesis in that the joining of replicons into high-molecular-weight DNA is completely blocked. This is in contrast to results obtained using puromycin, where joining of replicons occurs at a rate of 50-75% of the rate of the control. It should be emphasized that the concentrations of puromycin used (0.1-1.0 m*M*) completely block protein synthesis. In separate experiments it has been found that neither of the drugs significantly inhibits the joining of Okazaki fragments. The results presented here might indicate that the effect of cycloheximide on replicon joining is not due to inhibition of protein synthesis, but to a specific effect on the joining reaction. However, we would like to point out that although both drugs inhibit protein synthesis they do so in different ways. Cycloheximide inhibits primarily chain elongation leading to a "freezing" of polysomes, while puromycin continuously reacts with nascent polypeptides leading to dissociation of polysomes. Thus in the latter case the motions of protein synthesis are still performed to a certain extent. This would

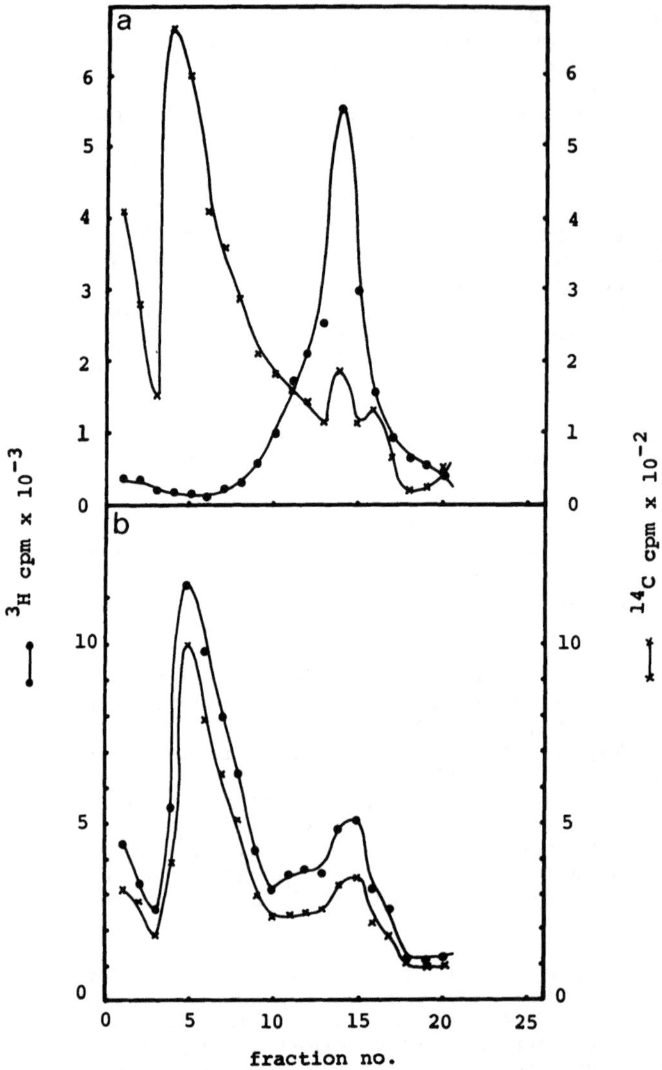

Fig. 5 a and b. Alkaline sucrose velocity sedimentation of DNA from pulse-labelled cells. Cells were prelabelled with ^{14}C-thymidine in defined medium and exposed to pulses of ^{3}H-thymidine as described by Nymann and Westergaard (1975). (a) cells exposed to a 30-min pulse of ^{3}H-thymidine; (b) cells exposed to a 30-min pulse of ^{3}H-thymidine followed by 60-min chase with cold thymidine (100 μg/ml). Gradients (5-20% sucrose on top of a 60% cushion, all in 0.2 *M* NaOH, 800 m*M* NaCl, and 1 m*M* EDTA) were spun for 2 1/2 h at 40 krpm in a Spinco SW41 rotor. λDNA, containing the deletions b515 and b519, was used as marker

indicate that joining of replicons is not dependent upon the continuous production of a particular protein, but is somehow dependent upon the continuous performance of the motions of protein synthesis. If this interpretation is correct, one would expect differential effects of protein synthesis inhibitors on DNA synthesis, which might be the reason for the conflicting results reported for such experiments in other systems (Painter, 1975).

In conclusion we have established conditions whereby intermediates of DNA synthesis can easily be obtained and accumulated. This should allow us to study in greater detail how some of the individual steps in DNA synthesis are regulated and how this process is related to the process of repair.

References

Barath, Z., Küntzel, H.: Induction of mitochondrial RNA polymerase in *Neurospora crassa*. Nature New Biol. 240, 195-197 (1972)

Housman, D., Huberman, J.A.: Changes in the rate of DNA replication fork movement during S phase in mammalian cells. J. Molec. Biol. 94, 173-181 (1975)

Johnson, B., Westergaard, O.: Accumulation of replicative DNA intermediates in *Tetrahymena* after excision repairable damage to DNA. Manuscript submitted (1975)

Kato, K., Strauss, B.: Accumulation of an intermediate in DNA synthesis by HEp.2 cells treated with methyl methane sulfonate. Proc. Nat. Acad. Sci. USA 71, 1969-1973 (1974)

Keiding, J.: Increase in specific activity of DNA polymerase in chloramphenicol treated *Tetrahymena* cultures. Exp. Cell Res. (In press) (1975)

Keiding, J., Westergaard, O.: Induction of DNA polymerase activity in irradiated *Tetrahymena* cells. Exp. Cell Res. 64, 317-322 (1971)

Magnusson, G., Craig, R., Närkhammar, M., Reichard, P., Staub, M., Warner, H.: Replication of polyoma DNA: effect of hydroxyurea and arabinosyl cytosine. Cold Spring Harbor Symp. Quant. Biol. 38, 227-233 (1973)

Nymann, O., Westergaard, O.: Manuscript in prep. (1975)

Painter, R.B.: Eukaryotic replicons. In: Handbook of Genetics (King, R.C., ed.). New York: Plenum Press, 1975, Vol. 5, pp.

Rommelaere, J., Fauris-Miller, A., Errera, M.: Isolation of replicating DNA segment from Chinese hamster cells by density equilibrium centrifugation. J. Molec. Biol. 90, 491-508 (1974)

Scudiero, D., Strauss, B.: Accumulation of single-stranded regions in DNA and the block of replication in a human cell line alkylated with methyl methane sulfonate. J. Molec. Biol. 83, 17-34 (1974)

Upholt, W.B., Borst, P.: Accumulation of replicative intermediates of mitochondrial DNA in *Tetrahymena pyriformis* grown in ethidium bromide. J. Cell. Biol. 61, 383-397 (1974)

Villadsen, J.S., Zeuthen, E.: Synchronization of DNA synthesis in *Tetrahymena* by temporary limitation of thymine compounds. Exp. Cell Res. 61, 302-310 (1970)

Westergaard, O.: Separation of two DNA polymerase fractions from *Tetrahymena* cells after excision-repairable damage to DNA. Biochim. Biophys. Acta (Amst.) 213, 36-44 (1970)

Westergaard, O., Johnson, B.: Isolation of a replication-complex from eukaryotes. Biochem. Biophys. Res. Comm. 55, 341-349 (1973)

Westergaard, O., Lindberg, B.: An induced mitochondrial DNA polymerase from *Tetrahymena*. Eur. J. Biochem. 28, 422-431 (1972)

Westergaard, O., Marcker, K.A., Keiding, J.: Induction of a mitochondrial DNA polymerase in *Tetrahymena*. Nature (Lond.) 227, 708-710 (1970)

Westergaard, O., Pearlman, R.: DNA polymerase activity in methotrexate plus uridine treated *Tetrahymena*. Exp. Cell Res. 54, 309-313 (1969)

13. Recovery from Radiation Damage Mediated by Somatic Cell Hybridisation

A. R. Kinsella, P. J. Smith, and J. M. Boyle

A. Introduction

In 1972, Harris reported that co-cultivation of a lethally irradiated Chinese hamster population with another that was unirradiated, led to the formation of hybrid clones containing functional marker alleles from irradiated nuclei. As part of a study of the radiobiology of somatic cell hybridization, we have examined further aspects of this phenomenon.

In our experiments spontaneous hybridization during co-cultivation of parent cell lines was infrequent. Therefore we used inactivated Sendai virus to mediate hybridization. The hybrid system that we used involved crosses between mutants of the purine and pyrimidine salvage pathways, that are therefore sensitive to blockage of de novo purine and pyrimidine synthesis by the drug methotrexate. Specifically we have used an interspecific cross between the Chinese hamster cell lines wg3h, deficient in hypoxanthine guanine phosphoribosyl transferase ($HGPRT^-$) and A23, deficient in thymidine kinase (TK^-). Both of these strains are derived from the DON cell line. We have also used an intrageneric cross between wg3h and mouse 3T34E(TK^-). The products of cell fusion are initially multinucleate cells among which are heterokaryons with nuclei from both parents. Some of these heterokaryons will successfully undergo nuclear fusion and mitosis, and a hybrid cell will result. Complementation of the biochemical deficiencies of the parent cell lines occurs in both heterokaryons and hybrids, and is selected for by the use of HAT medium (Littlefield, 1963) containing methotrexate.

The experiments that will be described are of two kinds. Firstly, we describe experiments of the kind performed by Harris, involving fusion of an irradiated cell line with an unirradiated cell line. The minimum requirement for hybrid production in these crosses is that the irradiated cell can provide the functional complementary marker required for HAT selection. To avoid cumbersome phraseology, in this paper we will refer to this set of conditions as a "marker rescue" experiment.

The second type of experiment involves fusions after both parents have been irradiated. Hybrid clones will only result from such crosses if complementation (in the broadest sense) has permitted the recovery of both the markers required for HAT selection, and those genes providing otherwise essential functions. We will refer to this type of experiment as a "mutual rescue" experiment.

The data to be presented will show the effect of radiation given to parent cells prior to their fusion. The end point measured is a hybrid clone in HAT medium. It is therefore convenient to describe the data as "hybrid" or "hybrid frequency" curves.

It will be demonstrated that marker rescue can occur after both intraspecific and intergeneric hybridization and that it also occurs following x-ray or ultraviolet (UV) radiation. Direct measurement of heterokaryon frequencies show that radiation-stimulated fusion cannot account for the extent of rescue observed. In crosses between parents that have received isodoses of irradiation, hybrid survival generally shows the same sensitivity (D_0) as the most resistant parent and is accompanied by an increase in zero-dose extrapolation, n.

B. Experimental Procedure

Cells were grown in Dulbecco and Vogt's modification of Eagle's minimal essential medium (Dulbecco and Vogt, 1954) supplemented with 13% foetal calf serum (13DMEM). Actively dividing cells were harvested, washed, and resuspended in Hank's balanced salt solution minus glucose (HBSS-G) at 1 x 10^6 per ml for x-irradiation. Cells for UV irradiation were washed and resuspended in phosphate buffered saline (PBS) at 1 x 10^5 per ml. Cells were irradiated with either 290 kV x-rays from a Siemens stabilipan machine operating at 12 mA, or with 254 nm light from a low-pressure mercury lamp. Dilutions were made in 13DMEM for survival assays.

The effect of hybridization on survival was investigated by fusing irradiated cells with unirradiated cells of complementary genotype for marker rescue measurements, or fusing complementary genotypes that had both received isodoses of radiation. 1 x 10^6 cells of each parental type were fused with 1500 HAU of β-propiolactone-inactivated Sendai virus for 15 min at 0°C followed by 15 min at 37°C. Dilutions of the fusion mixture were made in HAT selection medium (Littlefield, 1963) consisting of 13DMEM supplemented with hypoxanthine (1 x 10^{-4} *M*), methotrexate (4 x 10^{-7} *M*), and thymidine (1.6 x 10^{-5} *M*). Hybrid clones were scored after 10-14 days. Curves have been fitted by eye to the data points shown in the figures.

C. Results

I. Response to X-Irradiation

Figure 1 shows the x-ray survival curve of wg3h and the effect on hybridization of fusing irradiated wg3h cells with unirradiated A23. The ratio of sensitivities of the hybrid function compared to that of the parent, termed the hybrid restoration index (HRI), provides a convenient measure of the extent of rescue. In this experiment its value is 4.6 (Table 1). This result confirms those of Harris (1972) and indicates that lethally irradiated cells contribute functional genes to hybrids.

Figure 2 shows the effect of x-rays on hybridization when both parent lines are irradiated. The hybrid survival curve has a D_0 intermediate between those of the parents and shows an increased shoulder (n) (Table 1).

Figures 3 and 4 summarize several experiments with the intergeneric cross wg3h x 3T34E. Qualitatively the results are similar to those of Figures 1 and 2. Marker rescue is again observed with HRI values of 5.5 and 8.2 relative to the D_0 values of wg3h and 3T34E respectively. We do not have enough data yet to say whether rescue curves for

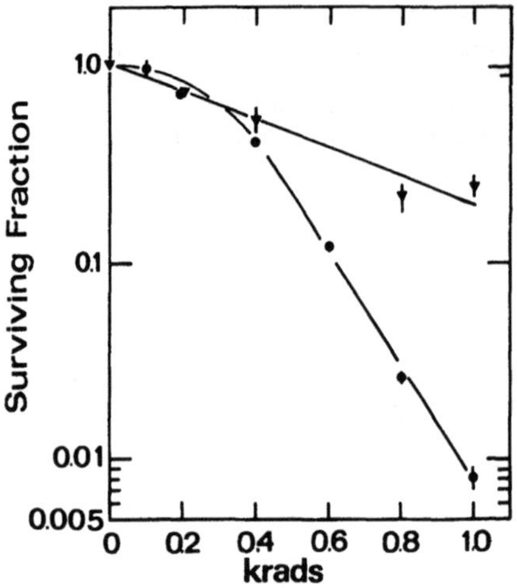

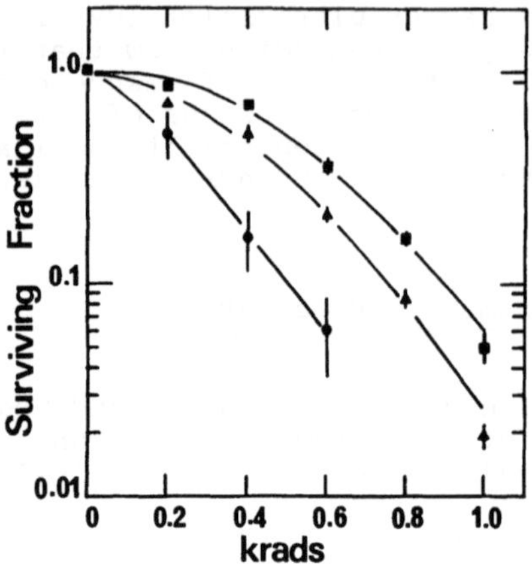

Fig. 1. Rescue of TK^+ from x-irradiated wg3h by Sendai virus-induced fusion with unirradiated A23 cells. Survival of hybrid frequency (triangles); survival of wg3h (circles). Bars are standard errors

Fig. 2. Survival of hybrids resulting from fusion of wg3h and A23 after both parents had received isodoses of x-rays. Survival of wg3h (circles); survival of A23 (triangles); survival of hybrid frequency (squares). Bars are standard errors

Table 1. Summary of survival parameters

Experiment	Survival measured	Parameters				
		n	D_q	D_o	HRI_s	HRI_d
A. Cross wg3h x 3T34E						
X-ray	wg3h	3.7	221	174	5.5	1.0
X-ray	3T34E	5.8	202	116	8.2	1.5
Marker rescue	hybrid	1.0	0.0	950		
Mutual rescue	hybrid	6.5	324	174		
B. Cross wg3h x A23						
X-ray	wg3h	6.3	274	148	4.6	
Marker rescue	hybrid	1.0	15	675		
X-ray	wg3h	1.5	70	190		1.1
X-ray	A23	2.9	245	225		0.9
Mutual rescue	hybrid	7.0	405	205		
UV	wg3h	2.1	17.7	22.0	8.7	
Marker rescue	hybrid	0.9	-20.0	192.0		
UV	wg3h	3.9	29.0	23.0		1.1
UV	A23	3.0	25.5	24.5		1.0
Mutual rescue	hybrid	7.2	50.0	25.0		

Units of D_q and D_o values are *rads* and $ergs/mm^2$ for x-ray and UV irradiations, respectively.

HRI is the hybrid restoration index $= \frac{D_o \text{ hybrid}}{D_o \text{ parent}}$

HRI_s and HRI_d refer to values obtained using hybrid D_o values from crosses involving a single parent irradiation (marker rescue) or double parent irradiation (mutual rescue).

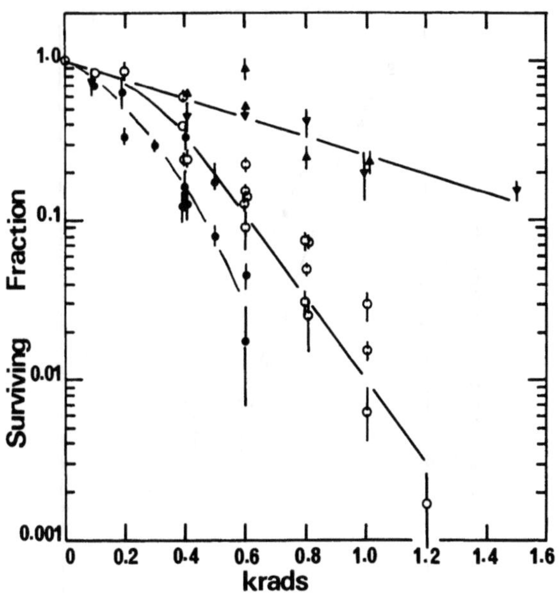

Fig. 3. Rescue of TK$^+$ and HGPRT$^+$ by hybridization of x-irradiated wg3h and mouse 3T34E with unirradiated cells of complementary genotype. Survival of 3T34E (solid circles); survival of wg3h (open circles); survival of hybrids between irradiated wg3h x unirradiated 3T34E (inverted triangles); survival of hybrids between irradiated 3T34E x unirradiated wg3h (triangles)

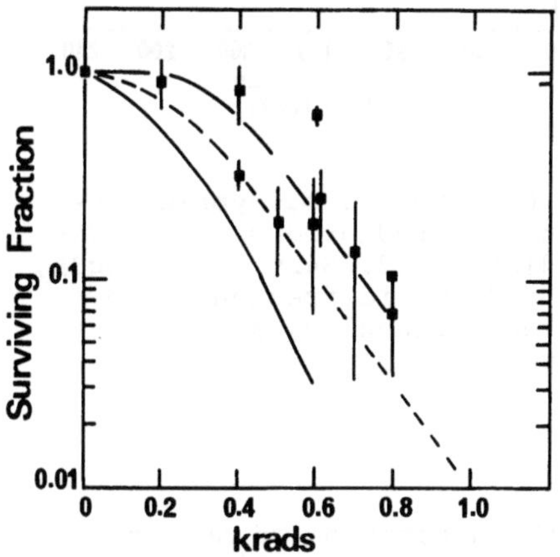

Fig. 4. Survival of hybrids between wg3h and 3T34E after x-irradiation of both cell lines. Survival of hybrid frequency (squares); curves for survival of wg3h and 3T34E (broken and solid lines, respectively) are reproduced from Figure 3

TK$^+$ and HGPRT$^+$ are different, but if they are, then the difference cannot be very large. When both parents are irradiated, an HRI close to unity is obtained with an increased value of n (Table 1).

II. Response to UV-Irradiation

To establish further the generality of the hybrid responses, we have compared the survival of hybrid frequency with the survival of the parents following UV irradiation for the cross wg3h x A23 (Figs. 5 and 6). We see the same qualitative effects as with x-rays. Quantitatively the HRI for marker rescue of TK$^+$ from irradiated wg3h is 8.7, nearly twice that obtained after x-rays. When both parents are irradiated, we see again an HRI close to unity and an increase in n.

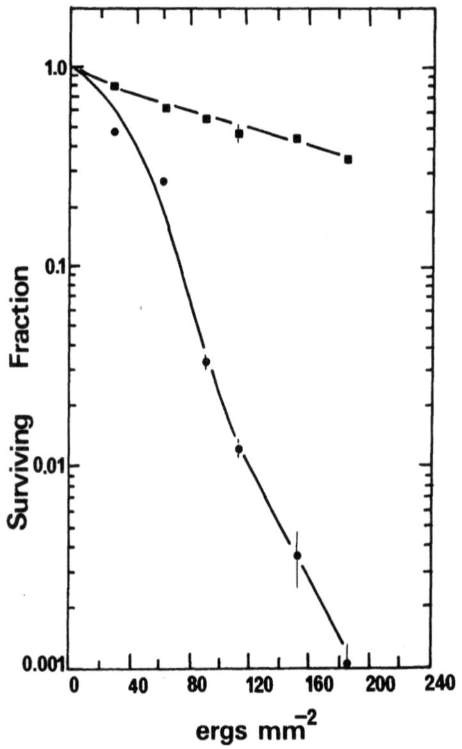

Fig. 5. Rescue of TK$^+$ from UV-irradiated (254 nm) wg3h by fusion with unirradiated A23 cells. Survival of hybrid frequency (squares); survival of wg3h (circles). Bars are standard errors

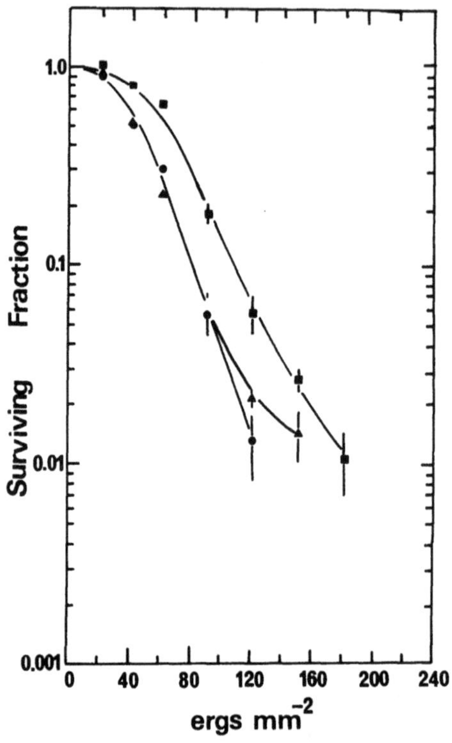

Fig. 6. Survival of hybrids between UV-irradiated wg3h and UV-irradiated A23. Survival of wg3h, A23, and hybrid frequency are circles, triangles, and squares, respectively

III. Heterokaryon Frequencies

A well-documented effect of radiation is the alteration of membrane permeability. Since alterations to membrane structure could alter the efficiency of cell fusion and hence be relevant to the interpretation of survival data, we have measured directly the effects of both UV and x-irradiation on the frequency of heterokaryon formation in the cross wg3h x A23. The nuclei of wg3h cells were labelled by growth in the presence of ^{3}H-thymidine, and these cells were fused with unlabelled A23 cells. Fusion mixtures were either sampled immediately, using a Shandon Elliott cytospin centrifuge to sediment cells onto microscope slides, or after 18 h incubation in HAT medium, by which time cells had attached to coverslips. Slides and coverslips were fixed and dip-coated with nuclear track emulsion for autoradiography. The number of labelled and unlabelled nuclei bounded by single cytoplasmic membranes were counted, and the frequency of heterokaryons was calculated (Table 2). Both methods of sampling have inherent problems. Sampling by cytospin may cause mechanical shearing and loss of cell complexes, whereas sampling after 18 h may cause some dilution of label as a result of cell division. Used together these methods probably give a reasonable reflection of heterokaryon distribution.

Table 2. Heterokaryon frequency (%) of wg3h x A23

Sampling method	Parents irradiated	Approx. numbers lethal hits (parents) 0	1	2	3	4
a) X-irradiation (*rads*)		(0)	(400)	(570)	(720)	(870)
0 hr (cytospin)	wg3h	8.2±2.5	11.0	6.6	10.8	6.2
	both		7.8	5.4	-	4.1
18 h (coverslip)	wg3h	5.6	5.4	-	-	5.7
	both		7.0	-	-	7.5
b) UV irradiation (ergs/mm^2)		(0)	(45)			(100)
18 h (coverslip)	wg3h	4.1	4.2			6.9
	both		4.3			6.0

The cytospin samples at zero time after fusion show no increase in heterokaryon formation following x-ray doses up to 4 lethal hits. Coverslip samples at 18 h suggest a marginal effect of both x-rays and UV after 4 lethal hits. This effect may result from a decrease in numbers of monokaryons after incubation of irradiated mutant cells in HAT rather than from increased heterokaryon formation. In any case the increase observed is not sufficient to explain either an HRI of >4 for marker rescue, or increases in n of >2 for crosses in which both parents are irradiated.

D. Discussion

The results summarized in Table 1 are in general agreement with those of Harris (1972). In the present study, marker rescue curves of intraspecies and intergeneric crosses, following x-irradiation or UV, all show dose responses that are very close to those expected for single hit events. The single hit character of these curves differs from those of Harris, which show a marked shoulder with D_q values approaching 700 rads. Differences in hybridization procedures used in the two studies might explain this anomaly. Our experiments used virus-induced fusion with nearly synchronous formation of heterokaryons followed by immediate HAT selection. Harris used non-synchronous fusion by cocultivation of parental cells for 24 h before HAT selection was applied. Thus the shoulder seen by Harris could result from repair of potentially lethal damage at the HGPRT$^+$ locus. The single hit kinetics seen in our studies suggests that there may only be a single locus that is functional for TK and HGPRT.

Relevant to the interpretation of these data is the observation that fusion between mitotic and nonmitotic cells causes the premature condensation of nonmitotic chromosomes (PCC) into visible structures whose morphology varies with the position of the interphase cell in the cell cycle (review by Sperling and Rao, 1974). Waldren and Johnson (1974) demonstrated with synchronized HeLa cells that x-irradiation of an interphase nucleus produced an immediate fragmentation of chromosomes when PCC occurred on fusion with an unirradiated mitotic cell. The number of fragments produced initially was linearly dose-dependent and decreased during 2 h incubation of the irradiated cells prior to fusion, suggesting that the lesions responsible for the breaks were repairable. However, this decrease was not accompanied by unscheduled DNA synthesis. A different picture emerged following UV irradiation. Here, fragmentation of chromosomes was not extensive. Instead G1 chromosomes lengthened and gave the appearance of prematurely condensed chromosomes of unirradiated S-phase cells. This change in morphology became most apparent 2 h after irradiation and was accompanied by unscheduled DNA synthesis.

Hence the repair of potentially lethal damage, suggested by the shoulders on the marker rescue curves of Harris (1972), may reflect the repair of those lesions responsible for fragmentation. Such an inference is consistent with the hypothesis of Chadwick and Leenhouts (1975) that potentially lethal lesions may be strand breaks. On the other hand, the higher HRI value obtained after UV irradiation, compared with that obtained after x-irradiation, may be related to the unscheduled DNA synthesis observed after UV but not after x-rays.

The fragmentation that occurs after PCC will cause loss of chromosomal material at mitosis and is probably the mechanism whereby radiation causes a directional loss of chromosomes (Pontecorvo, 1971; 1974).

The results of the mutual rescue experiments are characterized by HRI values close to unity and by n values that may be increased two- to three-fold over those of the parents. Experiments are in progress to determine whether a feeder layer effect, resulting from HAT-killed parent cells, is contributing to these shoulders. Our tentative conclusion at present is that these increases suggest that some mutual complementation of essential functions has occurred upon fusion after irradiation, allowing the heterokaryons to tolerate a greater amount of radiation damage (because of increased numbers of alleles) than the parent cells.

E. Summary

The rescue of TK^+ and $HGPRT^+$ alleles from lethally irradiated Chinese hamster or mouse cells by hybridization with unirradiated cells of complementary phenotype has been described. The ratio of D_o hybrids: D_o parent cells, termed hybrid restoration index (HRI), provides a measure of the extent of rescue. The HRI is greater after UV irradiation than after x-rays. Measurements of heterokaryon frequencies shows that radiation-stimulated fusion does not account for the observed rescue. The HRI observed after hybridizations involving irradiation of both parents is close to unity and is accompanied by an increase in shoulder (n). The increase in n suggests that mutual complementation occurs in heterokaryons between irradiated parent cells and is followed by the eventual reassortment of undamaged alleles in the resulting hybrid.

Acknowledgements. This work was supported by block grants provided by the Medical Research Council and the Cancer Research Campaign.

References

Chadwick, K.H., Leenhouts, H.P.: The repair of potentially lethal damage: an alternative approach. Radiat. Environ. Biophys. 11, 319-325 (1975)

Dulbecco, R., Vogt, M.: Plaque formation and isolation of pure lines with poliomyelitis viruses. J. Exp. Med. 99, 167-182 (1954)
Harris, M.: Effect of x-irradiation of one partner on hybrid frequency in fusions between Chinese hamster cells. J. Cell Physiol. 80, 119-128 (1972)
Littlefield, J.W.: The inosinic acid pyrophosphorylase activity of mouse fibroblasts partially resistant to 8 azaguanine. Proc. Nat. Acad. Sci. USA 50, 568-573 (1963)
Pontecorvo, G.: Induction of directional chromosome elimination in somatic cell hybrids. Nature (Lond.) 230, 367-369 (1971)
Pontecorvo, G.: Induced chromosome elimination in hybrid cells. In: Somatic Cell Hybridization (Davidson, R.L., de la Cruz, F., eds.). New York: Raven Press, 1974, pp. 65-68
Sperling, K., Rao, P.N.: The phenomenon of premature chromosome condensation: its relevance to basic and applied research. Humangenetik 23, 235-258 (1974)
Waldren, C.A., Johnson, R.T.: Analysis of interphase chromosome damage by means of premature chromosome condensation after x- and ultraviolet-irradiation. Proc. Nat. Acad. Sci. USA 71, 1137-1141 (1974)

14. The Effect of Liquid Holding on Survival and Mitotic Recombination of Two UV-Sensitive *Saccharomyces* Diploids

S. Kowalski and W. Laskowski

A. Introduction

In *Saccharomyces*, there is evidence for the existence of a recombination-mediated postreplication repair process (Lawrence, 1974; Eckardt et al., 1975; Kowalski and Laskowski, 1975). Of the 22 RAD genes that cause sensitivity to UV irradiation when mutated (Game and Cox, 1971), the mutant *rad5*, *rad6*, *rad9*, and *rad18* genes appear to be connected with the recombination repair pathway (Lemontt, 1971a-c; Lawrence et al., 1974). In a previous publication, we reported that a *rad9-4* diploid does not produce UV-induced recombinants, whereas a r_1^s diploid (the R_1^s gene has not yet been identified with one of the 22 RAD genes) produces UV-induced recombinants with high frequency. Both strains tested - the r_1^s and *rad9-4* diploid - are of intermediate UV sensitivity as far as survival is concerned, but show drastic differences in the UV induction of recombination events.

It is known that liquid holding (LH) affects the realization of UV-induced recombination events (Parry and Cox, 1968; Hunnable and Cox, 1971; Parry, 1971, 1972; Parry and Parry, 1972). We have investigated the effect of LH on the UV induction of recombinants in the r_1^s, *rad9-4*, and wild-type RAD diploid, in order to obtain data that allow a distinction of the recombination mechanism in these strains.

B. Materials and Methods

The strains used are listed in Table 1.

Table 1. List of diploids used

Strain number	Genotype	Referred to in text as
2126	$\frac{a}{\alpha}$ $\frac{ade2\text{-}x \quad + \quad his8}{+ \quad ade2\text{-}y \quad +}$ $\frac{lys2}{lys2}$ $\frac{ura1}{ura1}$ $\frac{RAD}{RAD}$	RAD diploid
2124	$\frac{a}{\alpha}$ $\frac{ade2\text{-}x \quad + \quad his8}{+ \quad ade2\text{-}y \quad +}$ $\frac{lys2}{lys2}$ $\frac{ura1}{ura1}$ $\frac{r_1^s}{r_1^s}$	r_1^s diploid
2125	$\frac{a}{\alpha}$ $\frac{ade2\text{-}x \quad + \quad his8}{+ \quad ade2\text{-}y \quad +}$ $\frac{lys2}{lys2}$ $\frac{ura1}{ura1}$ $\frac{rad9\text{-}4}{rad9\text{-}4}$	*rad9-4* diploid

The origin of the various markers, the media (synthetic minimal and omission media, complete medium), and buffer have already been published (Kowalski and Laskowski, 1975). Experimental procedure:

Preculturing of strains, preparation of the cell suspension, and irradiation procedure were carried out as described previously (Kowalski

and Laskowski, 1975). For the liquid holding treatment, UV-irradiated cell suspensions and the unirradiated controls (both with a titer of 10^7 cells per ml) were stored in buffer in the dark for 4 days before plating on suitable medium. The cell suspensions were diluted appropriately when necessary to yield 100 colonies per plate. All manipulations were carried out under yellow light to exclude photoreactivation.
Evaluation:

Minimal medium without adenine (MM-adenine) and complete medium (YEPD) plates were scored after 3-4 days growth at 30°C. The surviving fraction was calculated from the number of colonies growing on YEPD. The intragenic recombinants were selected for on MM-adenine. YEPD plates were replica plated onto MM and MM-histidine, and then incubated for 1-2 days at 30°C. Those colonies, half or quarter sectors growing on MM but not on MM-histidine, were scored as intergenic recombinants. One recombination event was scored per colony. More than a single sector - then smaller than a quarter sector - occurred in a single colony only in very rare cases.

The number of induced recombinants was calculated without considering the inactivation of the irradiated suspension and was corrected for the number of spontaneously occurring recombinants in the unirradiated control. Only that portion of recombinants scored after irradiation that was greater than the number of spontaneously occurring recombinants, was considered to have been induced. The UV-induced recombination frequencies (i.e., recombinants per survivors) were all corrected for the frequency of spontaneously occurring recombinants in the unirradiated control.

C. Results and Discussion

The survival characteristics of the three diploid strains following UV irradiation are shown in Figure 1. At LD_{10}, the ratio of DRF values is 1:1.8:3.3 for the RAD, r_1^S, and *rad9-4* diploids, respectively.

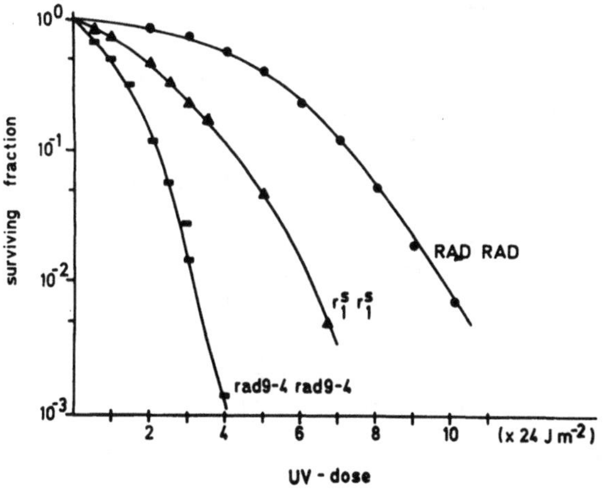

Fig. 1. UV-inactivation curves for the RAD (●), r_1^S (▲), and *rad9-4* (■) diploid. Each value is the average of at least 5 independent experiments

Table 2 lists the number of spontaneously occurring recombination events in the unirradiated controls. The curves in Figure 2 show the induction of recombinants by UV irradiation above the number of spontaneously occuring recombinants as listed in Table 2.

Table 2. Number of spontaneously occurring mitotic recombinants in the unirradiated controls of the RAD, r_1^s, and *rad9-4* diploids. Each value is the average of at least 8 independent experiments

rad diploid	RAD	r_1^s	*rad9-4*
Intergenic recombinants per 10^3 cells	6.77	25.8	0.045
Intragenic recombinants per 10^4 cells	2.23	6.66	-

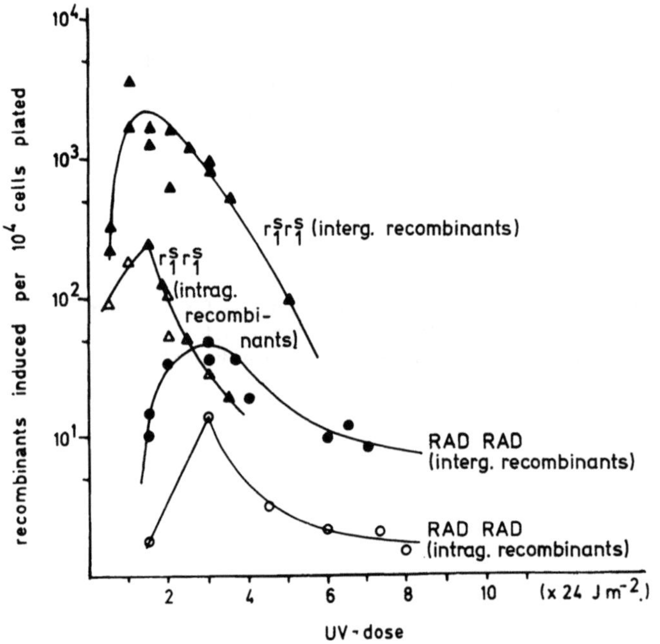

Fig. 2. Number of UV-induced inter- and intragenic recombinants per 10^4 cells for the RAD (●o) and the r_1^s (▲Δ) diploid. Solid symbols represent the intergenic, open symbols the intragenic recombinants. Each point represents an experimental value

No induction of inter- or intragenic mitotic recombinants was found in the *rad9-4* diploid. The UV induction of recombination events in the r_1^s diploid, however, is clearly higher than in the wild-type RAD strain for both the inter- and intragenic recombination events. In the wild-type RAD diploid, the maximal induction of both inter- and intragenic mitotic recombinants occurs after a UV dose of 72 J/m^2 (which corresponds to a surviving fraction of 7-8 x 10^{-1}). In the r_1^s diploid, however, the maximal induction of both inter- and intragenic mitotic recombinants occurs after a UV dose of 36 J/m^2. At this dose, the surviving fraction is 6-7 x 10^{-1}.

At the UV dose resulting in the maximal induction of intergenic recombinants in the r_1^s diploid, ca. 36% of the irradiated cells are transformed to the *his*$^-$ phenotype. These cells give rise to colonies that are, in the majority, completely *his*$^-$ (= unsectored colonies); the remaining fraction contains half or quarter sectors (= sectored colonies). The frequency of the sectored colonies among the *his*$^-$ phenotypes depends on the UV dose, as is shown in Figure 3. After irradiation with 36 J/m^2, the maximal induction of recombination events occurs in the r_1^s diploid. At the same dose, the frequency of unsectored colonies reaches a minimum, whereas the frequency of sectored colonies also attains a maximal value. With increasing UV doses, the frequency of unsectored colonies increases again, reaching 100% at 192 J/m^2. The number of UV-induced sectored and unsectored colonies are plotted versus the UV dose in Figure 4. In contrast to the r_1^s diploid, the dose-dependent frequencies of unsectored colonies in the wild-type RAD diploid show a steady decrease with increasing UV dose (Fig. 3).

In order to obtain further information on the mechanism producing *his*$^-$ cells in the RAD and the r_1^s diploids, the effect of liquid holding (LH) on the recombination induction was investigated. The effect of LH posttreatment on the survival characteristics of the three diploids is shown in Figure 5. At LD_{10}, the dose modification factors (DMF) are 1.2, 1.2, and 1.3 for the RAD, r_1^s, and *rad9-4* diploid, respectively. This shows that the three diploids tested are capable of dark repair of potentially lethal damage to a certain extent.

The effect of LH on the UV-induced recombinants is depicted in Figure 6. The *his*$^-$ phenotypes scored after LH treatment were mostly unsectored colonies in the r_1^s diploid, but were mainly sectored colonies in the wild-type RAD diploid. (A quantitative classification of these phenotypes into sectored and unsectored colonies was not performed, however.) In the r_1^s diploid, LH decreases the number of UV-induced intergenic recombinants slightly only at the maximum of UV induction; after irradiation with UV doses above ca. 50 J/m^2, LH increases the number of intergenic recombinants, as is the case in the wild-type RAD diploid at all doses tested. The effect of LH on the intragenic recombinants differs in both strains: in the r_1^s diploid, LH reduces the number of UV-induced intragenic recombinants drastically. In the wild-type RAD diploid, however, LH reduces the number of UV-induced intragenic recombinants in the lower dose range and increases it in the dose range above 100 J/m^2.

The literature on the influence of LH posttreatment on UV-induced recombination events in *Saccharomyces* (Parry and Cox, 1968; Hunnable and Cox, 1971; Parry, 1971, 1972; Parry and Parry, 1972) reports differing, partially contradicting effects. Parry and Cox (1968), as well as Parry and Parry (1972), compared the effect of LH on inter- and intragenic recombination frequencies and found that LH decreases intergenic recombination, but increases intragenic recombination in a wild-type diploid. These authors present their data as recombinants per survivors and not as the number of induced recombinants per cells plated. When our data are presented as recombinants per survivors, no significant effect of LH on the UV-induced intergenic recombination frequencies can be observed for either the RAD or the r_1^s diploid (Fig. 7). The intragenic recombination frequencies of both strains are reduced after LH in a certain dose range.

The difference between Figures 6 and 7 lies in the different reference systems used (number of recombinants per cells plated and recombinants per survivors). Our results show that LH does indeed affect the intergenic recombination process after UV irradiation, although differently in the r_1^s and the wild-type RAD diploid. In the RAD diploid, ca. 1-2%

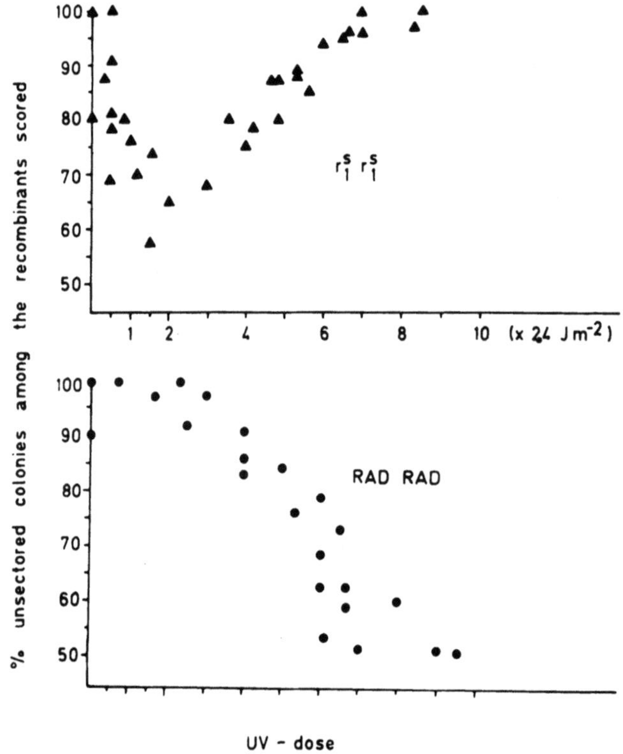

Fig. 3. Percentage of *his*⁻ phenotypes scored as unsectored colonies in the RAD (●) and r_1^S (▲) diploid plotted over the applied UV dose

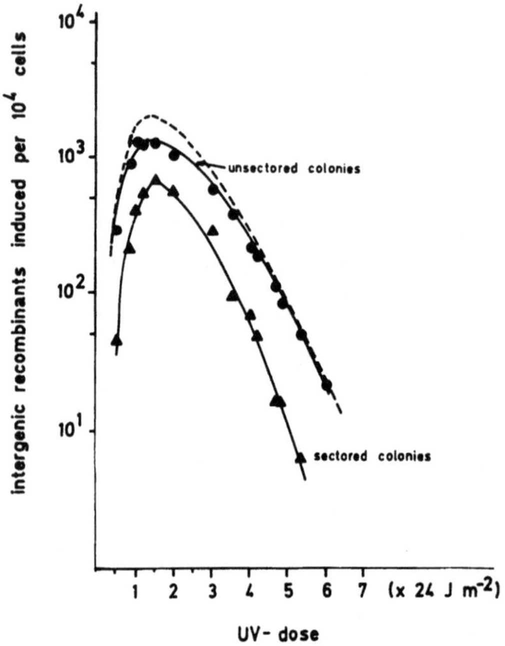

Fig. 4. Comparison of the number of UV-induced intergenic recombinants scored as unsectored (●) and as sectored (▲) colonies in the r_1^S diploid

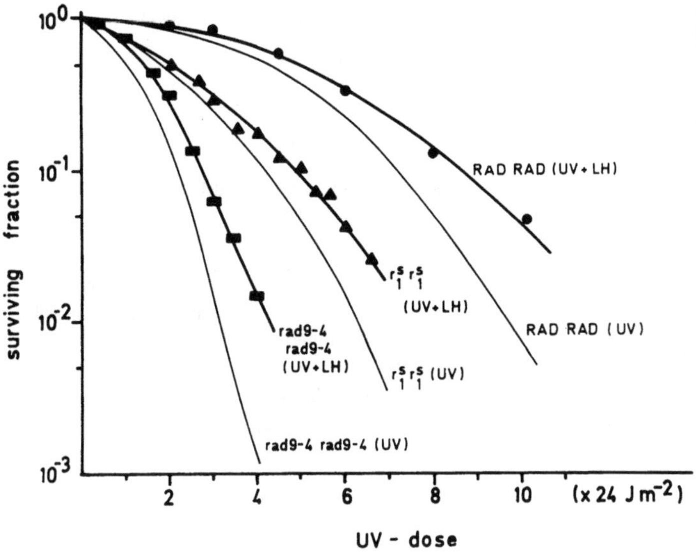

Fig. 5. Effect of LH on the UV-inactivation curves of the RAD (●), r_1^S (▲), and *rad9-4* (■) diploid. Each value is the average of at least 3 independent experiments

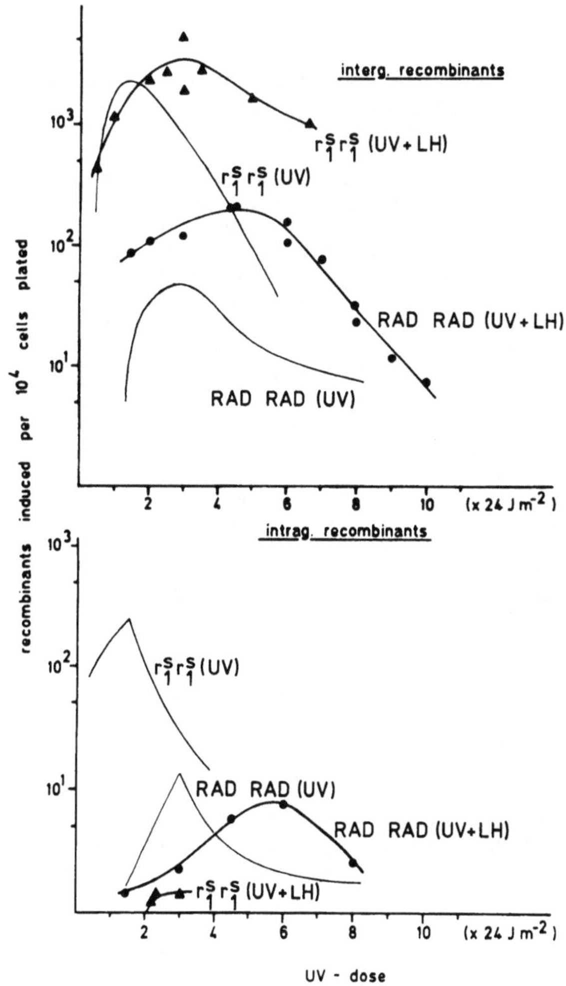

Fig. 6. Number of inter- and intragenic mitotic recombinants in the RAD (●o) and r_1^S (▲△) diploid after UV irradiation + LH. Solid symbols represent the intergenic, open symbols, the intragenic recombinants. Each point represents an experimental value

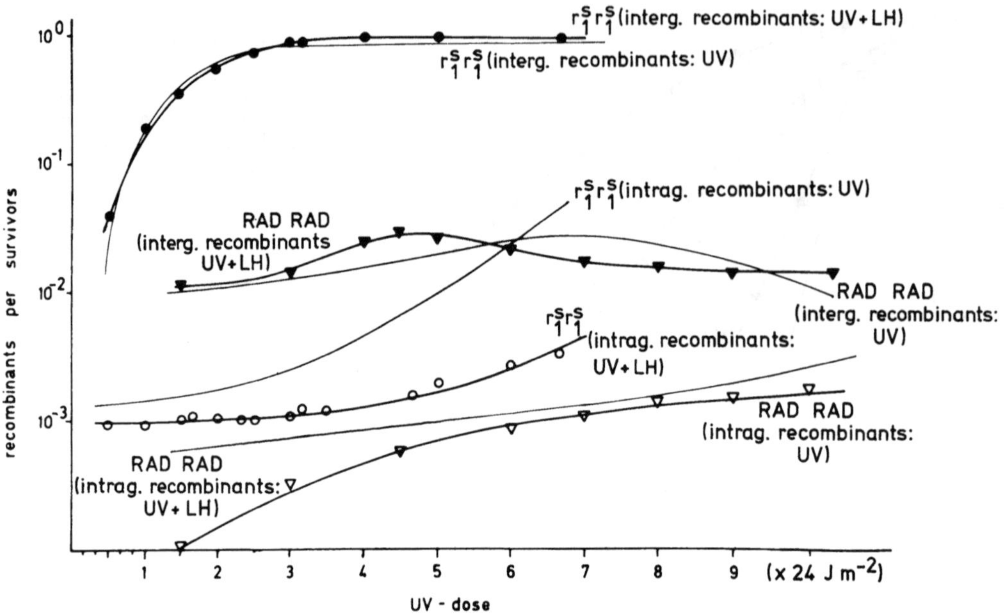

Fig. 7. Effect of LH on the UV-induced inter- and intragenic recombination frequencies of the RAD (▼▽) and r_1^S (●o) diploid. Open symbols represent the intragenic, solid symbols, the intergenic recombinants

intergenic recombinants are induced per survivors after UV irradiation as well as after UV+LH in the whole dose range studied. This means that LH posttreatment increases the number of recombinants and nonrecombinant survivors to the same extent, regardless of the UV dose applied prior to LH treatment. In the r_1^S diploid, a similar effect can be observed in the lower dose range, although the frequency of intergenic recombinants increases steadily after UV irradiation. At UV doses above 72 J/m^2, however, almost all survivors produce unsectored recombinants colonies not only after UV irradiation, but also after UV+LH. This indicates that only those potentially lethal r_1^S cells that are recombinants are reactivated during LH. We therefore assume that in this strain, a recombination-mediated repair process is the main repair mechanism operating during LH treatment. It remains to be explained, however, why the surviving cells are only of one phenotype (*his*$^-$) after UV irradiation as well as after UV+LH, since the recombination process is also expected to yield the complementary HIS$^+$ phenotype. In the attempt to explain these data, the phenomenon of lethal sectoring (James and Werner, 1966) comes to mind. If lethal sectoring occurs after UV irradiation, only one of the two cells produced by a postirradiation cell division survives. If lethal sectoring occurs in the first cell division, the resulting colonies should be unsectored. This was the case in the r_1^S strains after UV irradiation and after UV+LH. It still remains unclear, however, why the surviving cells should always give rise to colonies of the *his*$^-$ phenotype. We have no explanation for this phenomenon.

In the *rad9-4* diploid, no induced recombinants could be scored even after LH posttreatment, during which the number of induced intergenic recombinants increased in the wild-type RAD and r_1^S diploid by a factor of ca. 5.

In contrast to the results obtained for the intergenic recombinants, LH reduces the number of UV-induced intragenic recombinants drastically in the r_1^s strain. The wild-type RAD diploid shows a reduction of induced intragenic recombinants in a lower dose range and an increase in the higher dose range. This shows that inter- and intragenic recombination are affected differently by liquid holding treatment in these strains. The UV-induced intragenic recombination frequencies (recombinants per survivors) however, are reduced in both strains in different dose ranges. The different effect of LH on the UV-induced inter- and intragenic recombinants can be interpreted as evidence for the operation of two mechanisms during LH, which evidently both affect recombination.

The data we have obtained for the effect of LH on inter- and intragenic recombination in the wild-type RAD and the r_1^s diploid contradict the previously cited results of Parry and Cox and of Parry and Parry, and thus do not fit their model. These authors postulate that the activity of the REC repair system (Rupp and Howard-Flanders, 1968) is diminished by LH treatment, since the intergenic recombination frequency decreased after LH. The increase in intragenic recombination frequency is interpreted to result from the action of the EXR system, which operates during LH, leading to repaired lesions and producing intragenic recombinants.

Since the number of induced intergenic recombinants is increased after LH in our strains, and, in addition, all survivors are recombinants in the r_1^s strain, the operation of a repair process involving recombination must be promoted by LH in our strains. In contrast, the number of intragenic recombinants induced is reduced drastically in the r_1^s diploid, showing that LH leads to the removal or repair of intragenic recombinogenic lesions in our strains, which also applies to the wild-type RAD diploid in the lower dose range.

The phenomenon of liquid holding is still unclear and complex. Apparently several repair processes are operating simultaneously in a wild-type cell during LH. Perhaps some repair processes are only activated or induced during LH and develop their full activity, when the cells are transferred from saline to nutrient medium. As yet, we cannot differentiate between those processes operating during the dark storage in saline and those occurring after LH on the nutrient medium. In our opinion, more concrete information is necessary before detailed models on the effect of liquid holding posttreatment on recombination can be constructed.

D. Summary

Two mutants causing sensitivity to radiation (r_1^s and *rad9-4*) show similar survival frequencies after ultraviolet (UV) irradiation. To investigate further possible repair mechanisms that are blocked in these strains, the effect of liquid holding (LH) on survival, and inter- and intragenic mitotic recombination was studied in suitable diploids. The results were compared with those of the wild-type RAD diploid.

At LH_{10}, the ratios of dose reduction factors (DRF) for survival are 1:1.8:3.3 for the RAD, r_1^s, and *rad9-4* diploid, respectively. All strains show liquid holding recovery (LHR) or prelethal damage. The dose modification factors (DMF_{LH}) values at LD_{10} are 1.2, 1.2, and 1.3, respectively.

Inter- and intragenic recombination was measured by determining the number of recombinants induced above the spontaneously occurring level. The three diploids differed drastically in their recombination behaviour. The UV induction of recombinants is highest in the r_1^s strain (ca. 60-100 times greater than in the RAD diploid) for both inter- and intragenic recombinants. There is no UV induction of recombinants in the *rad9-4* diploid. LH increases the number of UV-induced intergenic recombinants in both the wild-type RAD and r_1^s diploids, although it has no significant effect on the UV-induced recombination frequencies. The intragenic recombination frequencies for both strains, however, are reduced by LH. In the r_1^s diploid, the number of UV-induced intragenic recombinants is reduced drastically after LH treatment, whereas the corresponding number in the RAD diploid is reduced in the lower dose range, but increases in the higher dose range ($> 72\ J/m^2$).

Acknowledgements. The authors wish to acknowledge the financial support of the Deutsche Forschungsgemeinschaft and the excellent technical assistance of Mrs. M. Brose. They also thank Dr. B.S. Cox for helpful discussions.

References

Eckardt, F., Kowalski, S., Laskowski, W.: The effects of three *rad* genes on UV-induced mutation rates in haploid and diploid *Saccharomyces* cells. Molec. Gen. Genetics 136, 261-272 (1975)

Game, J.C., Cox, B.S.: Allelism tests of mutants affecting sensitivity to radiation in yeast and a proposed nomenclature. Mutation Res. 12, 328-331 (1971)

Hunnable, E.G., Cox, B.S.: The genetic control of dark recombination in yeast. Mutation Res. 13, 297-309 (1971)

James, A.P., Werner, M.M.: Radiation-induced lethal sectoring in yeast. Radiat. Res. 29, 523-536 (1966)

Kowalski, S., Laskowski, W.: The effect of three *rad* genes on survival, inter- and intragenic mitotic recombination in *Saccharomyces*. I. UV irradiation without photoreactivation or liquid-holding post-treatment. Molec. Gen. Genetics 136, 75-86 (1975)

Lawrence, C.W.: Lecture held at the VII. International Conference on Yeast Genetics and Molecular Biology, Brighton/Sussex (1974)

Lawrence, C.W., Stewart, J.W., Sherman, F., Christensen, R.: Specificity and frequency of ultraviolet-induced reversion of an iso-1-cytochrome c ochre mutant in radiation-sensitive strains of yeast. J. Molec. Biol. 85, 137-162 (1974)

Lemontt, J.F.: Mutants of yeast defective in mutation induced by ultraviolet light. Genetics 68, 21-33 (1971a)

Lemontt, J.F.: Pathways of ultraviolet mutability in *Saccharomyces cerevisiae*. I. Some properties of double mutants involving *uvs9* and *rev*. Mutation Res. 13, 311-317 (1971b)

Lemontt, J.F.: Pathways of ultraviolet mutability in *Saccharomyces cerevisiae*. II. The effect of *rev* genes on recombination. Mutation Res. 13, 319-326 (1971c)

Parry, J.M.: The genetic effects of liquid holding recovery in ultraviolet light sensitive mutants in yeast. Molec. Gen. Genetics 111, 51-60 (1971)

Parry, J.M.: A quantitative analysis of "negative liquid holding" in some UV sensitive mutants of yeast. Molec. Gen. Genetics 118, 33-43 (1972)

Parry, J.M., Cox, B.S.: The effects of dark holding and photoreactivation on ultraviolet light-induced mitotic recombination and survival in yeast. Genet. Res. Camb. 12, 187-198 (1968)
Parry, J.M., Parry, E.M.: The genetic implications of UV light exposure and liquid-holding post-treatment in the yeast *Saccharomyces cerevisiae*. Genet. Res. Camb. 19, 1-16 (1972)
Rupp, W.D., Howard-Flanders, P.: Discontinuities in the DNA synthesized in an excision-deficient strain of *Escherichia coli* following ultraviolet radiation. J. Molec. Biol. 31, 291-304 (1968)

15. Over-Repair in *Closterium:* Increased Radioresistance Caused by an Earlier Exposure to Radiation

A. HOWARD and F. G. COWIE

A. Abstract

The unicellular green fresh-water desmid *Closterium moniliferum*, when irradiated at high dose-rate with 10 MeV electrons, yields a survival curve with a threshold to a dose of about 10 krads, followed by a shoulder and an exponential decline. Previous exposure to electrons, x-rays, or 260 nm ultraviolet (UV) causes an increase in D_T and in the width of the shoulder by a factor of up to 2, but little or no change in slope. We call this over-repair and suggest that cells normally have an efficient mechanism for repair of sublethal damage and that this, or an additional mechanism, is stimulated by the first dose of radiation and then repairs subsequent radiation damage even more efficiently. This interpretation is consistent with the large effect of dose-rate on the value of D_T and the width of the shoulder in single-dose experiments.

The effectiveness of the first dose in stimulating over-repair was small but measurable after 30 rads of x-rays and increased to an optimum between 1 and 7 krads. It was not influenced by the rate at which the first dose was given. First doses of more than 10 krads did not stimulate over-repair. After a first dose of 2 krads, measurable over-repair was seen after an interval of 1 h and increased to a plateau at 4-6 h. Over-repair was slow or absent when cells were kept cold (0.5°C) between the stimulating first dose and the second or test dose, and was suppressed by cycloheximide. Over-repair stimulated by a first dose was still apparent even when 1 or 2 cell divisions (mitoses) occurred before the test dose was given.

B. Introduction

When the desmid *Closterium moniliferum* is irradiated with fast electrons at a high dose-rate, all the cells survive doses up to a threshold of about 10 krads. After higher doses there is an exponential decline (Howard and Cowie, 1975). The threshold, and consequently large extrapolation number of the survival curve, cannot be due to cellular or nuclear multiplicity, and intranuclear target multiplicity sufficient to explain it seems very unlikely. Split dose experiments showed that with sufficient time between two doses the threshold was not only fully reconstructed but increased: that is, it was up to twice as large for cells that had been previously irradiated than for cells given only single doses. We call this phenomenon "over-repair", and we suggest that the cells normally have an efficient mechanism for repairing sublethal damage and that the same, or an additional, repair mechanism is stimulated by the first dose of radiation so that the cells are able to accumulate additional sublethal damage when exposed at a later time.

We found that there was a well-marked relationship between the size of the first dose and the amount of over-repair stimulated by it. We also found that the over-repair stimulus took time to develop, being small or absent when the interval between doses was 1 h or less, and reaching a maximum at 4-6 h (Howard and Cowie, 1976). We report here the effect of very small first doses of high dose-rate electrons and compare them with the effects of low dose-rate x-rays and of ultra-violet (UV) irradiation. We also describe experiments to explore the metabolic nature of the over-repair mechanism and the transmission of the stimulus through two cell generations.

C. Material and Methods

Details of culture media and general methods are given in Howard and Cowie (1976). We keep stock cultures of cells in suspension in nutrient medium on a light/dark cycle of 18 + 6 h. For experiments cells were plated on gridded filters in petri dishes on nutrient agar and cultured at 21-24°C with the same light cycle. Before or immediately after irradiation the number of cells on each plate was counted, and the number of colonies with 16 cells or more was determined after 11-13 days. These colonies were defined as survivors.

In the experiments reported here we gave a first (or "stimulating") dose of ionizing radiation or UV, kept the cells for an interval (usually 4 h), and then gave a graded series of second (or "test") doses to generate a survival curve. The test doses were always given with fast electrons at high dose-rate, and at the same time another set of samples, not previously irradiated, was given a series of fast-electron doses to generate a single-dose survival curve. Examples of such curves are shown in Figure 1. To express quantitatively the amount

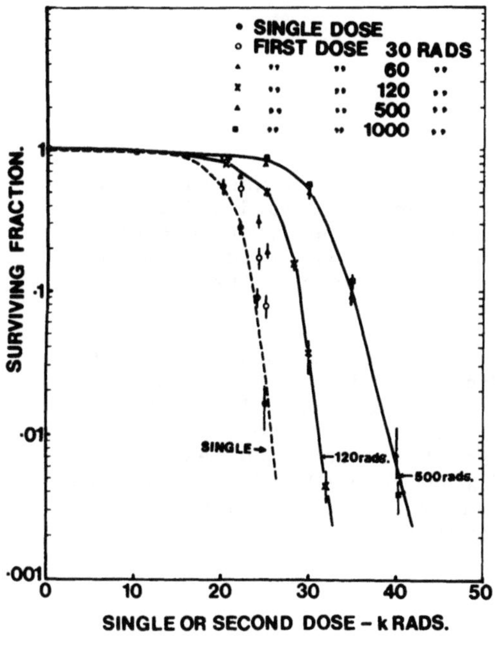

Fig. 1. Survival curves from experiment in which first or stimulating dose was given with x-rays at 100 rads/min, and second or test dose with fast electrons. Interval between doses 4 h. ●, no first dose: test dose only; o, first dose 30 rads x-rays; ▲, first dose 60 rads x-rays; x, first dose 120 rads x-rays; Δ, first dose 500 rads x-rays; □, first dose 1000 rads x-rays

of over-repair we plotted surviving fraction against the test dose of radiation (not the total dose), drew the survival curves by eye, and estimated graphically the doses required to reduce the surviving population to 10%. We called these doses D_2 and D_1 for the two-dose and single-dose curves, respectively. The ratio D_2/D_1 gave a measure of over-repair: when $D_2 = D_1$ there was full repair of sublethal damage caused by the first dose but no over-repair. When $D_2 < D_1$, less than full repair was indicated.

Cells were irradiated on plates. For exposure to fast electrons in the horizontal beam of a 10 MeV linear accelerator (dose-rate 140 rads per 5 μsec pulse or 7 krads/sec overall), the filters carrying the cells were transferred to firm non-nutrient agar surfaces and held in a vertical position; they were put back on nutrient agar immediately after irradiation. For exposure to x-rays (250 kVp, 100 rads/min) or UV light (from a mercury-vapor lamp: λ = 254 nm, dose-rate ~ 10 ergs/mm^2/sec) the cells were plated thickly (~ 5000 cells per plate), washed off the plates, and replated at appropriate densities on a series of fresh plates for the second or test dose.

To test the effect of cycloheximide (CH) between two radiation doses we gave a first dose (1 krad of fast electrons) on plates, then washed the cells off into a 0.001 *M* solution of CH where they remained for 4 h. This treatment had no effect on the viability of unirradiated cells. We then washed the cells and replated them for the test doses. The single-dose cells also had 4 h in CH before or after the test dose.

To observe the transmission of the over-repair stimulus to daughter and grand-daughter cells, we gave a stimulating dose of 1 krad of electrons; after the cells had divided once or twice, we positioned them on the gridded paper in such a way that daughter pairs and grand-daughter pairs could be identified. We thus knew the time of the test radiation dose in relation to the number of divisions that any cell had undergone since the stimulating dose.

D. Results

I. Size and Dose-Rate of First Dose

A single pulse of fast electrons (140 rads in 5 μsec) was sufficient to stimulate a significant amount of over-repair (Howard and Cowie, 1976). Figure 2 shows the effect of the size of the first dose on the amount of over-repair and allows a comparison of the results using fast electrons with new observations on the effect of first doses of x-rays given at 100 rads/min.

The x-ray experiments show, firstly, that as little as 30 rads gave an appreciable increase in shoulder (Fig. 1) and that the amount of over-repair increased with the size of the first dose. The second point demonstrated by the x-ray experiments is that their dose-effect relationship was very similar to that for fast electrons despite a difference in dose-rate by a factor of more than 4000 (Fig. 2). Thus dose-rate in this range does not apparently influence the effectiveness of the first dose in stimulating over-repair.

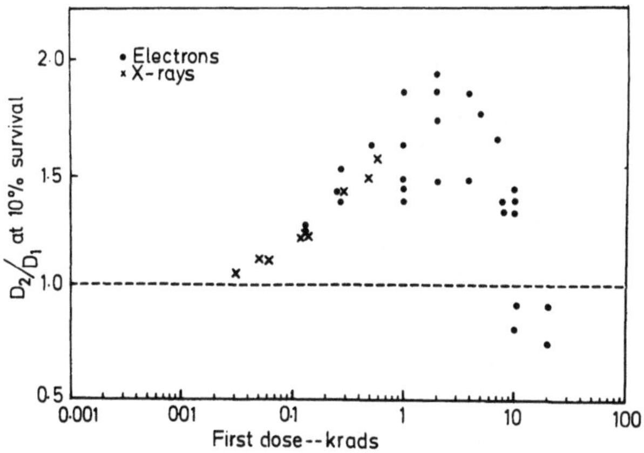

Fig. 2. Over-repair in relation to size of first or stimulating dose. Each point represents comparison of two-dose survival curve with single-dose curve obtained in same experimental run. •, fast electrons (7000 rads/sec); x, 250 kV x-rays (100 rads/min) (including data shown in detail in Fig. 1)

II. UV as a First Dose

Figure 3 shows that a first dose of 0.5 or 1.0 kergs/mm^2 of UV results in over-repair. We have the results of only one experiment, and the two-dose survival curves are incomplete, but there is no doubt that over-repair occurred and that the extrapolated two-dose curves indicate a D_2/D_1 value of at least 1.5. It is not clear whether the value would be the same or different for the two UV first doses.

III. Transmission of Over-Repair Stimulus to Later Cell Generations

We gave a first dose of 1 krad of fast electrons and a test dose after cells had performed one or two cell divisions (each of which was of

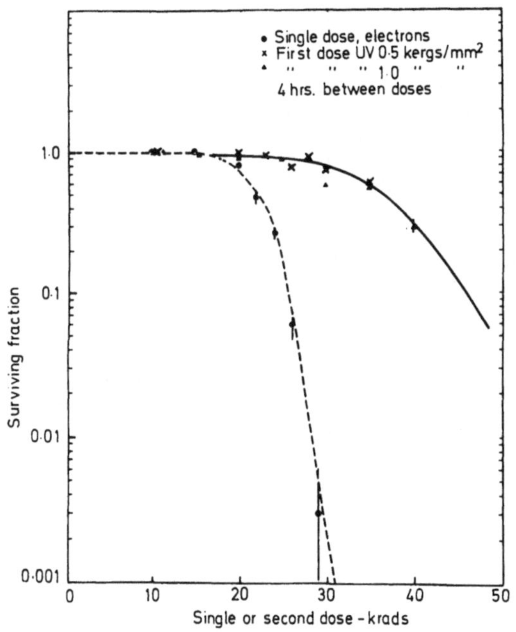

Fig. 3. Over-repair following UV irradiation. Interval between UV exposure and test dose of electrons was 4h. •, fast electrons given without previous irradiation (single doses); x, fast electrons after 0.5 kergs/mm^2; ▲, fast electrons after 1.0 kergs/mm^2

course preceded by a mitotic division of the nucleus). Each pair of daughter or grand-daughter cells was separately scored for 0, 1, or 2 surviving colonies. The results (Table 1) showed that survival after 24 krads in both these cell generations was significantly better when the original cell received 1 krad than when it was initially unirradiated: i.e., the over-repair stimulus was transmitted. The values in the last column of Table 1 suggest, however, that the increase in survival was less than in cells that did not divide before the second dose. A breakdown of the survival scores for the second generation pairs (not shown here) indicated that the over-repair stimulus was transmitted randomly to the four grand-daughters.

Table 1. Transmission of over-repair stimulus through one or two cell generations

First dose (krads)	Second dose (krads)	Number of divisions between doses	Percent survival	Increase due to first dose (factor)
0	24	-	~ 20	~ 4.5
1	24	0	~ 90	
0	24	-	27 ± 3	3
1	24	1	73 ± 2	
0	24	-	20 ± 2	2
1	24	2	52 ± 3	

IV. Effect of Cycloheximide (CH)

Preliminary tests showed that a 4 h treatment in a solution of CH of 281 μg/ml (0.001 *M*) or less had no effect on the viability of unirradiated cells. There was, however, a reduction in division rate after this treatment, as evidenced by a reduction in colony size when scored after 9-11 days' growth. This indicated that the drug entered the cells and had an effect, although a transient one, on their development.

We exposed cells to CH at 0.001 *M* for four hours either before or after single doses of electrons, and also during a 4-h interval between a first dose of 1 krad of electrons and the test doses. Figure 4 shows the results of one experiment. CH given after single doses reduced the threshold of the survival curve from about 15 krads to about 10 krads. The slope of the curve was not changed. The 10% survival dose was about 21.5 krads without the drug and about 13 krads with the drug. When we gave CH for 4 h before single-dose irradiation there was again an effect on the threshold, but the effect was a smaller one than when we gave CH after irradiation (10% survival dose about 17 krads).

In the experiment illustrated in Figure 4, D_2/D_1 = 37/22 for two-dose samples not treated with CH, giving a value of 1.7 for over-repair at the 10% survival level. The survival curve for samples treated with CH between doses is incomplete, but there is no doubt that the shoulder was greatly reduced. The survival after 20 krads was the same as in the single-dose curve without drug, and there were no survivors after 23 krads. Thus there was no over-repair.

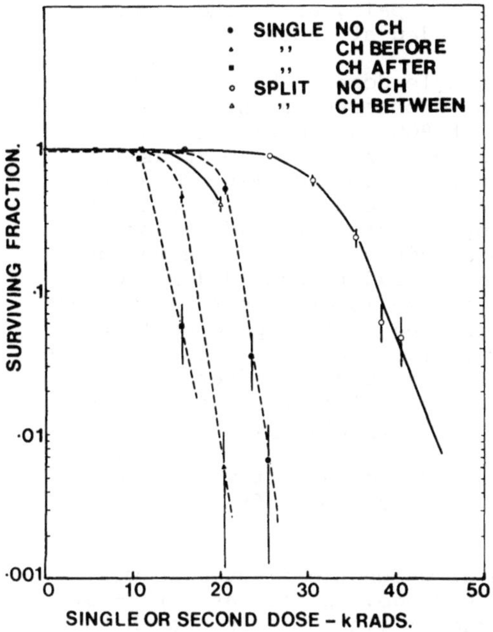

Fig. 4. Effect of 0.001 *M* cycloheximide on over-repair. All radiation doses were with fast electrons. ●, single doses, no CH; ▲, single doses, CH 4 h before irradiation; ■, single doses, CH 4 h after irradiation; o, first dose 1 krad, second doses 4 h later, water between; Δ, first dose 1 krad, second doses 4 h later, CH between

E. Discussion

There are several reasons for regarding the protective effect of a first dose against killing by a second dose as an expression of a stimulated repair mechanism in this organism. The effect is manifested as an extension of the threshold rather than as a change in the slope of the survival curve. It is not possible with our present data to exclude completely a change in slope, but it is certain that any such change must be small compared with the strikingly large effects on the shoulder. We think this justifies the view that the threshold in the single-dose curve represents an exaggerated shoulder expressing the accumulation of sublethal damage and that an earlier exposure to radiation increases the cell's ability to accumulate and repair such damage.

In spores of *Oedogonium cardiacum*, another green fresh-water alga, Horsley and Laszlo (1971) observed an unexpectedly large increase in shoulder width in the survival curve after a second exposure to x-rays. They suggested that this indicated an unexpected amount of recovery due to an "efficient recovery mechanism set into play by the first dose of radiation". We interpret our results in a similar way and use the term over-repair for this reason.

It appears in *Closterium* that to stimulate over-repair a degree of damage is required and that this damage can be caused by ionizing radiation (electrons or x-rays) or by UV. The dose-rate at which the stimulating dose of ionizing radiation was given was unimportant in the range 0.1 to 420 krads/min, but the amount of stimulation was clearly dependent on the size of the dose (Fig. 2). Up to about 1 krad there was a roughly linear relation between the effect and the log of the stimulating dose. At doses between 1 and 5 krads of electrons the effect was optimal, and at 10 or 20 krads there was little or no over-repair or even less than full repair, as if the repair mechanism had become saturated or overloaded at these doses.

When *Closterium* cells are exposed to single doses, their survival is strongly influenced by dose-rate. We have observed a threshold of 10-20 krads for survival after x-rays at 900 rads/min: this is in the same range as the thresholds seen after electrons at 420 krads/min. But when the x-ray dose-rate was reduced to 100 rads/min we observed no cell killing at 25 krads, i.e., the threshold dose was at least doubled due to a change in dose-rate by a factor of 9 (data not shown here). This observation is perhaps more understandable when we consider that at the higher dose-rate the 12 krad threshold dose took only about 15 min to deliver, whereas at the lower dose-rate it would have taken about 2 h. This consideration led us to expose cells repeatedly at 2 h intervals to 10 krad doses of the very high dose-rate (420 krads/min) electrons. We observed 100% survival after 4 such repeated doses, i.e., after 40 krads given in 6 h. Thus it appears that the cells can repair sublethal damage repeatedly. This could well be the reason why changes in dose-rate, in a range which critically affects exposure time in relation to the rate at which over-repair develops (Howard and Cowie, 1976), had a big effect on cell survival.

By contrast, changes in the dose-rate at which the first dose was given were unimportant over a very wide range. In the case of x-rays, a dose that is only 0.03% of the minimum killing dose is enough to have an observable effect. Thus a relatively small amount of damage can initiate or increase the efficiency of a repair mechanism. In addition, UV (254 nm) was effective in stimulating over-repair. It seems probable that this damage resides in the nucleus, and future work will be directed to examining this question.

The over-repair process is apparently a metabolic one. It takes time to develop and was absent or greatly reduced when the cells were kept cold between the two doses (Howard and Cowie, 1976). In addition, the protein-synthesis inhibitor cycloheximide, when present between two doses, completely suppressed over-repair (Fig. 4). A similar effect of CH was observed by Horsley and Laszlo (1973) in *Oedogonium cardiacum*, where the additional recovery was completely suppressed by CH when present between two doses. Our preliminary results indicate further that CH given before or after single doses of radiation can reduce the threshold of the survival curve. These observations support the idea that the threshold is due to a repair process that involves protein synthesis and that over-repair is due to the stimulation of the same or an additional repair process also involving protein synthesis.

Acknowledgements. This work was supported by grants from the Medical Research Council and the Cancer Research Campaign.

References

Horsley, R.J., Laszlo, A.: Unexpected additional recovery following a first x-ray dose to a synchronous cell culture. Int. J. Radiat. Biol. 20, 593-596 (1971)

Horsley, R.J., Laszlo, A.: Additional recovery in x-irradiated *Oedogonium cardiacum* can be suppressed by cycloheximide. Int. J. Radiat. Biol. 23, 201-204 (1973)

Howard, A., Cowie, F.G.: Survival-curve characteristics in a desmid. In: Cell survival after low doses of radiation: theoretical and clinical implications. T. Alper, ed. London-New York: John Wiley and Sons Limited, 1976, pp. 3-12
Howard, A., Cowie, F.G.: Induced resistance in a desmid *Closterium moniliferum*. Radiat. Res. 65, 540-549 (1976)

16. L5178Y Cell Strains Cross-Sensitive to X-Rays and UV-Light: Similarities and Differences in Recovery at the Cellular Level

M. WALICKA, I. SZUMIEL, and J. Z. BEER

A. Introduction

Most of the biologic effects of ultraviolet (UV)- or x-irradiation in bacteria can be reasonably well related to lesions observed at the molecular level, namely to DNA damage. The molecular mechanisms of the mammalian cell response to irradiation are, however, insufficiently understood.

DNA lesions have been a prime interest of radiobiologists for more than a decade, and similarities and differences between UV-induced damage and that caused by ionizing radiation have been extensively studied. Nevertheless, the manifestations of these lesions and their repair at the cellular level cannot, as yet, be fully explained in molecular terms.

Many aspects of the preceding problems can be elucidated by comparative studies on the response to x-rays and UV light of cell lines with differing radiosentivities.

Two strains of murine leukemic cells, L5178Y-R and L5178Y-S (Alexander and Mikulski, 1961; Beer et al., 1963), exhibiting cross sensitivity to UV-light and x-rays (Beer et al., 1973) present a unique eukaryotic model for such studies.

B. Material and Methods

The techniques used in growing both strains, R and S of L5178Y cells, as well as survival determinations and growth pattern characteristics were described in detail previously (Szumiel et al., 1971; Beer, 1974, 1975).

X-irradiation was carried out at a dose-rate of 382 rads/min, using a Siemens (Erlangen, GFR) Stabilipan machine operating at 200 kV and equipped with a 0.5 mm Cu filter.

UV exposures were carried out using a Philips (Eindhoven, The Netherlands) germicidal lamp, type TUV-30 with an aqueous sodium acetate filter. The exposure rate to 254 nm light was 5.35 $erg/mm^2/sec$.

DNA and protein synthesis determinations were carried out as described elsewhere (Szumiel et al., 1975) or according to Bossman and Bernacki (1970).

C. Results and Discussion

The cross sensitivity of L5178Y-R and L5178Y-S strains is characterized by the UV and x-ray dose-survival curve parameters presented in Table 1.

Table 1. X-ray and UV light dose-survival curve parameters of two strains of L5178Y murine lymphoma (Beer et al., 1973)

L5178Y strain	X-rays			UV-light		
	D_o rad	D_q rad	n	D_o erg/mm^2	D_q erg/mm^2	n
R	100	100	2.9	54	0	1
	(r e s i s t a n t)			(s e n s i t i v e)		
S	54	0	1	85	0	1
	(s e n s i t i v e)			(r e s i s t a n t)		

It is usually accepted that because of the common target molecule, i.e. DNA, cellular sensitivity to UV light parallels sensitivity to ionizing radiation. The example of L5178Y-R and L5178Y-S strains accentuates the differences between UV and x-ray damage to DNA as well as between the recovery processes.

Cells of both strains contain the same amount of DNA, and it may be assumed that no significant differences exist between the chromatin structures of L5178Y-R and L5178Y-S cells. Hence, it may be expected that radiation-induced primary lesions are quantitatively similar in the cells of both strains. The contribution of individual kinds of DNA lesions in x-ray- and UV-irradiated cells can be estimated from the literature data (for review see Painter, 1973). Results of such estimations made for an average mammalian cell are presented in Figure 1. Figure 2 shows percent survival of both L5178Y strains at the dose levels chosen for Figure 1. These data illustrate dramatic differences in ability of the cells to recover from the same DNA damages.

It is commonly understood that an unrepaired DNA damage site represents the lethal lesion. However, only a few cell lines - xeroderma pigmentosum, V7979, HeLa S3S4M, HeLa S3S2M, Yoshida S and R - can be quoted as examples of somatic cell systems in which reduced efficiency of the repair mechanisms is paralleled by increased radiation sensitivity (Cleaver, 1969, 1970; Isomura et al., 1973; Fox and Fox, 1973a). Examination of the efficiency of the repair mechanisms usually concerns repair of DNA strand breaks or elimination of UV photoproducts. However, as may be seen in Figure 1, these lesions by no means prevail in DNA of irradiated cells.

Identical DNA lesions should manifest themselves identically at the cellular level, independently of whether x-ray- or UV-induced. It is well known that although similarities between the effects of x-ray and UV light at the cellular level can be pointed out, it is even easier to indicate differences between manifestations of the actions of both kinds of radiations such as cell cycle disturbances, mode of death, chromosomal aberrations, etc. It is possible that similar phenomena observed at the cellular level reflect similar unrepaired

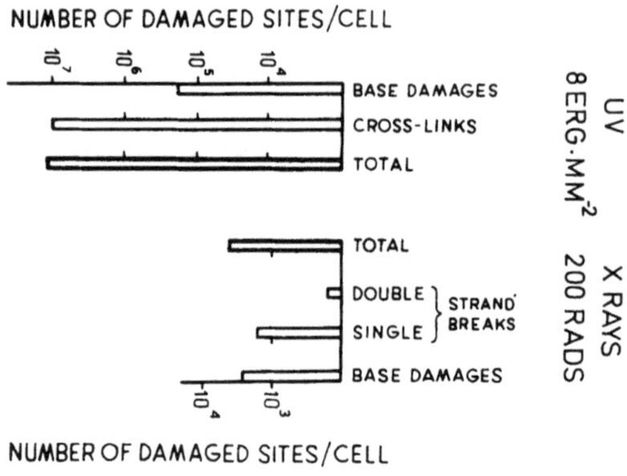

Fig. 1

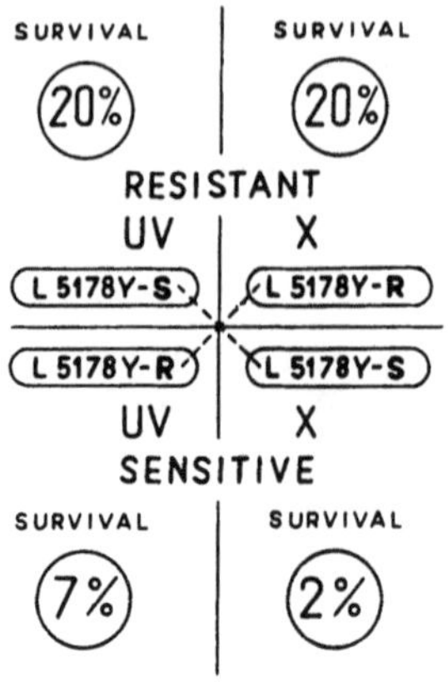

Fig. 2

Fig. 1. Types of damage and approximate numbers of damaged sites in DNA and UV- and x-irradiated cells (for references see Painter, 1973)

Fig. 2. Survival of L5178Y-R and L5178Y-S cells irradiated with 200 rads of x-rays or 8 erg/mm^2 of UV light. Data for UV light were obtained after exposure of 2-mm-thick layers of culture in Fischer medium; doses were calculated taking into account 94.5% absorption of 254 nm UV light in the medium (Beer et al., 1973; Walicka, 1975)

DNA structure distortions whereas discrepancies between behavior of x-ray- and UV-damaged cells are connected with differences in DNA lesions. Alternatively, structures other than DNA may be considered as the targets, especially in the case of randomly absorbed ionizing radiation.

I. Cell Cycle Changes Induced by UV and X-Irradiation in L5178Y-R and L5178Y-S Cells

One of the typical early effects of x-irradiation in mammalian cells is blockage of progression through the G2 phase of the cell cycle. UV, on the other hand, causes primarily disturbances in DNA replication, thus extending the S phase. The respective changes established by us for the two strains of L5178Y cells are summarized in Figure 3. The exceptional features here include an extremely long x-ray-induced G2 block in L5178Y-S cells (reported also by Ehmann et al., 1974) and no UV-induced block in S phase in L5178Y-R cells (Walicka, 1975).

II. Susceptibility of DNA Synthesis in L5178Y Cell Strains to UV Light and X-Rays

Both UV light- and x-ray-induced damage disturbs DNA replication as measured by ^{3}H-TdR incorporation: (1) UV photoproducts and possibly DNA-protein cross-links, by distorting the double helix and thus slowing down the replicase progression along the replicon; (2) x-ray lesions, by a not yet understood mechanism inhibiting initiation rather than elongation processes (Makino and Okada, 1974).

Our observations on ^{3}H-TdR incorporation into irradiated L5178Y-R and L5178Y-S (Fig. 4) cells derive from experiments in which various proportions of the populations survived. Nevertheless, it can be seen that in the case of x-rays, cell sensitivity was paralleled by the DNA synthesis susceptibility to x-irradiation (Szumiel et al., 1975), while the reverse relationship could be seen for UV-exposed cells: a

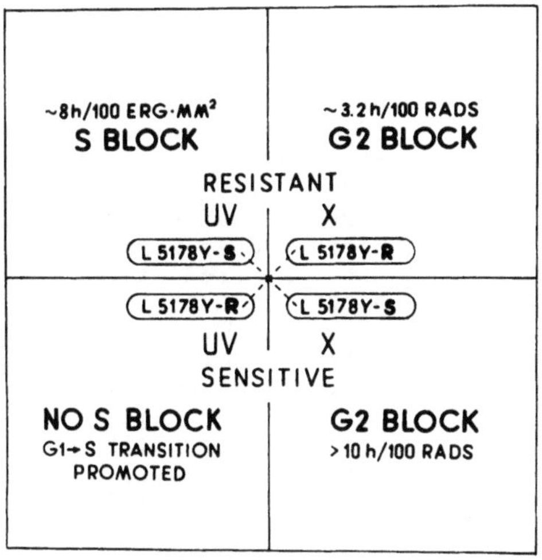

Fig. 3. Cell cycle disturbances induced by UV- and x-irradiation in L5178Y cells (Walicka, 1975)

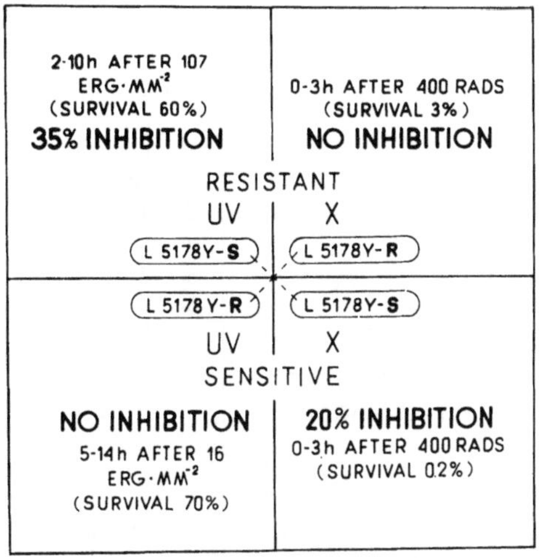

Fig. 4. Available data on DNA synthesis inhibition in UV- and x-irradiated L5178Y cells (Walicka, 1975; Szumiel et al., 1975)

typical S block was observed in the more resistant L5178Y-S strain while no block was found in the sensitive L5178Y-R strain (Walicka, 1975).

III. Effects of Split-Dose UV and X-Irradiation

The most striking differences in the behavior of L5178Y-R and L5178Y-S cells were observed in split-dose experiments. Results of these experiments, presented in Figure 5, cannot be fully interpreted at the moment, because of lack of data on cell age dependence of x-ray and UV sensitivity. Nevertheless, a clear correlation between cell sensitivity and the presence or lack of sparing effect of split-dose irradiation can be seen.

1. UV Split-Dose Effects

Although the nature of the lesions repaired during the interval between UV exposures is unknown, it is widely understood that the repair processes are S phase-dependent (for review see Painter, 1973). Available data suggest the existence of a relation of this kind in L5178Y cells: the increase of relative survival of UV-resistant L5178 Y-S cells observed for 7-15 h intervals between the exposures (cf. Fig. 5) overlapped with a block in S phase. On the other hand, in UV-sensitive L5178Y-R cells the minimum relative survival values corresponded with the normal rate of DNA synthesis. Moreover, it was found that the relative survival could be enhanced by storing L5178Y-R cells between UV exposures at 25°C (Walicka, 1975). This temperature causes a considerable slowing-down in DNA synthesis (Makino and Okada, 1974). Thus, it seems that in UV-irradiated L5178Y cells the split-dose sparing effect is dependent on the rate of DNA synthesis: slow replication facilitates recovery from the damage.

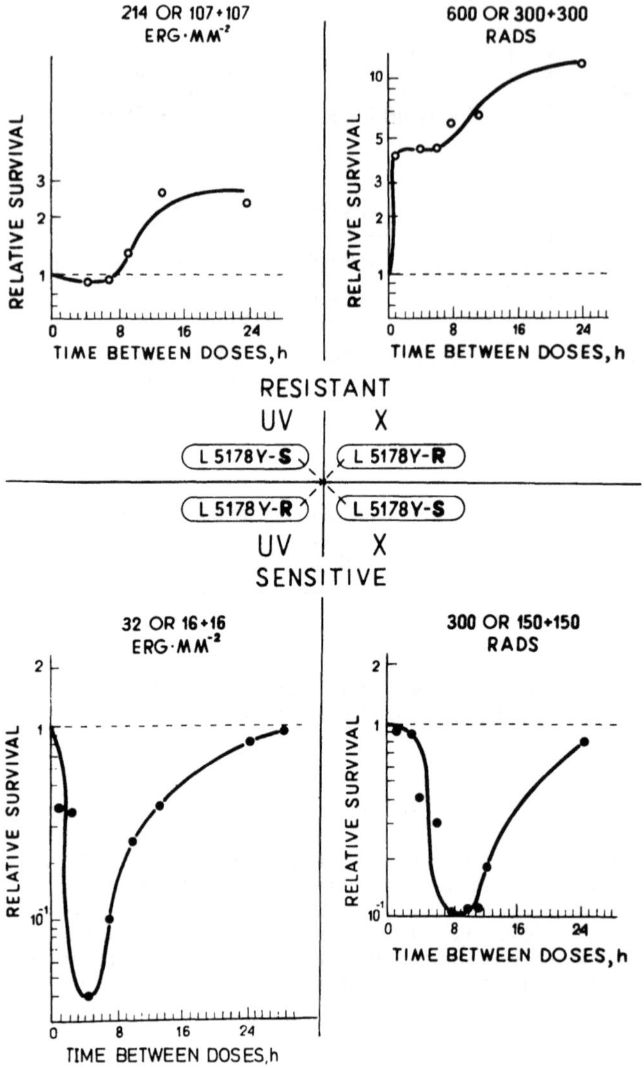

Fig. 5. Relative survival (i.e., survival related to that observed after single dose) of L5178Y cells after split-dose UV- and x-irradiation. Survivals after cumulative doses were, after x-irradiation: 0.02% for L5178Y-R cells and 0.25% for L5178Y-S cells; after UV irradiation: 22.5% for L5178Y-R cells and 4.6% for L5178Y-S cells (Beer et al., 1973; Walicka, 1975)

2. X-Ray Split-Dose Effects

The x-ray split-dose sparing effects are usually attributed to the repair of sublethal damage during the time interval between doses (Elkind et al., 1961; Lange, 1970). Several authors (see Bryant, 1975) suggested that decreased sensitivity of cells after split-dose recovery was connected with a more efficient repair induced by the first exposure. Lange et al. (1973), who associate repair of sublethal damage with the repair of double-strand breaks, have also postulated that a first dose of radiation can reduce effects of the second dose by inducing synthesis of certain enzymes taking part in the repair processes (Lange, 1971). These suggestions find a strong support in Bryant's (1975) experiments showing that protein synthesis is a prerequisite for split-dose sparing effect.

From Table 2 it can be seen that the behavior of x-irradiated L5178Y-R and L5178Y-S cells provides further indication that protein synthesis is involved in mechanisms underlying split-dose sparing effects.

Table 2. X-ray split-dose recovery and protein synthesis in two strains of L5178Y cells

L5178Y strain	Split dose sparing effect	Protein synthesis inhibition
R	+	-
S	-	+

The usual responses to split-dose irradiation consist either of an increase in the relative survival (sparing effect) or of no change in relative survival.

As can be seen in Figure 5, L5178Y-S cells are clearly more sensitive to the second x-ray dose than to the first. In contrast to the repair+ progression model of recovery from sublethal damage (Elkind et al., 1961; Lange, 1970), hypotheses predicting radiation-stimulated induction of repair processes provide possible explanations of this phenomenon; it can be postulated that decrease of the relative survival is related to a lowered activity of repair enzymes after the first exposure.

The reason for the discrepancies between Bryant's observations and those of other authors, who claim that protein synthesis inhibitors do not affect the sublethal damage repair (for review, see Terasima, 1973; Gautschi et al., 1973) is not clear. It is possible that the sources of these discrepancies are to be sought in individual characteristics of the cell lines, concentrations of the inhibitors, and time of treatment, because these and similar factors can be of basic importance in studies on the mechanisms modifying radiation sensitivity.

D. Conclusions

Assuming that the same type of radiation damage at the molecular level causes the same phenomena at the cellular level, the response of two strains of L5178Y cells to UV and x-irradiation can be interpreted as being indicative of lethal lesions of distinctly different character induced by two kinds of radiations. Thus, formation of single-strand breaks in DNA of UV-irradiated cells, due to uncompleted repair mainly of the postreplication type, may be of less importance than usually assumed. The data presented suggest that other types of UV damage deserve more attention, in particular DNA protein cross links, as studied by Han et al. (1975).

The results presented point also to some problems concerning intracellular mechanisms determining x-ray and UV sensitivity. In the search for an explanation of different radiosentivities in closely related cell lines, the most obvious approach is to look for differences in efficiency of DNA repair processes. Surprisingly, the efficiency of replication repair in a similar pair of x-irradiated cell lines, studied by Fox and Fox (1973b), is the same in both resistant and sensitive lines. However, as proposed by these authors, the quality of repair may differ in sensitive and resistant cells. Misrepaired lesions are in fact "fixed" lesions and produce lethal effects. Thus, the extent of lethal damage would be determined by (1) quantity of the primary lesions, (2) efficiency of the repair mechanisms, and (3) ratio of the rate at which imperfect repair ("lesion fixation") processes proceed, and the rate of perfect repair processes leading to restoration of the original structure of the DNA molecule. The data presented permit the postulation that the DNA replication rate presents a factor that may influence this ratio in irradiated L5178Y cells.

E. Summary

Among mammalian cells, murine leukemic lymphoblasts L5178Y-R and L5178Y-S show the unique feature of being cross-sensitive to x-rays and ultraviolet (UV) light. The mean lethal doses for x-rays are 100 and 54 rads and for UV light 54 and 85 erg/mm^2.

In addition to lower x-ray sensitivity, L5178Y-R cells also exhibited the dose-fractionation sparing effect (relative survival (RS) > 5). No such effect, or even potentiation, was observed in the more x-ray-sensitive L5178Y-S cells (RS ≙ 0.1). Similarly, UV-resistant L5178Y-S cells were susceptible to the sparing effect of UV dose fractionation (RS ≙ 2.5), whereas UV-sensitive L5178Y-R cells were sensitized by the first fraction (RS ≙ 4 x 10^{-2}).

Preliminary information on the mechanisms underlying these differences was obtained from experiments on the effects of x-ray and UV exposure on DNA and protein synthesis. Irradiation of L5178Y-S cells with 400 rads of x-rays inhibited ^{3}H-TdR incorporation by 20% and nuclear protein synthesis by 25%. Neither of these processes was inhibited in L5178Y-R cells by x-rays. Thus, in L5178Y cells, there exists a relation between inhibition of macromolecular synthesis and x-ray sensitivity.

On the other hand, ^{3}H-TdR was incorporated into UV-irradiated L5178Y-R cells at the normal rate, whereas L5178Y-S cells exposed to 107 erg/mm^2 incorporated only 65% of ^{3}H-TdR in comparison to the controls. This may indicate that enhanced UV resistance depends on slowed-down DNA replication, thus increasing the efficiency of the repair processes.

References

Alexander, P., Mikulski, Z.B.: Mouse lymphoma cells with different radiosensitivities. Nature (Lond.) 192, 572-573 (1961)

Beer, J.Z.: Late post-irradiation phenomena in mammalian cell populations. I. Radiation-induced heritable changes in x-irradiated L5178Y-S cells in vitro. Institute of Nuclear Research, Warsaw, Report No. 1484/X/B/A, 1974

Beer, J.Z.: Late post-irradiation phenomena in mammalian cell populations. II. Intraclonal recovery in sublines isolated from x-irradiated L5178Y-S cell populations. Institute of Nuclear Research, Warsaw, Report No. 1501/X/B/A, 1975

Beer, J.Z., Lett, J.T., Alexander, P.: Influence of temperature and medium on the x-ray sensitivities of leukaemia cells in vitro. Nature (Lond.) 199, 193-194 (1963)
Beer, J.Z., Szumiel, I., Walicka, M.: Cross-sensitivities to UV-light and x-rays of two strains of murine lymphoma L5178Y cells in vitro. Studia Biophys. (Berlin) 36/37, 175-182 (1973)
Bossman, H.B., Bernacki, R.J.: Glycosidase activity. Glycosidase activity in L5178Y mouse leukaemic cells and the activity of acid phosphatase, β-galactosidase, and β-N-acetylgalactosaminidase and β-N-acetylglucosaminidase in a synchronous L5178Y cell population. Exp. Cell Res. 61, 379-386 (1970)
Bryant, P.E.: Decrease in sensitivity of cells after split dose recovery: evidence for the involvement of protein synthesis. Int. J. Radiat. Biol. 27, 95-102 (1975)
Cleaver, J.E.: DNA repair of Chinese hamster cells of different sensitivities to ultraviolet light. Int. J. Radiat. Biol. 16, 277-285 (1969)
Cleaver, J.E.: DNA repair and radiation sensitivity in human (xeroderma pigmentosum) cells. Int. J. Radiat. Biol. 18, 557-565 (1970)
Ehmann, U.K., Nagasawa, H., Petersen, D.F., Lett, J.T.: Symptoms of x-ray damage to radiosensitive mouse leukaemic cells: asynchronous populations. Radiat. Res. 60, 453-473 (1974)
Elkind, M.M., Sutton, H., Moses, W.B.: Postirradiation kinetics in mammalian cells grown in culture. J. Cell. Comp. Physiol. 58 (Suppl. 1), 113-134 (1961)
Fox, M., Fox, B.W.: Repair replication after UV-irradiation in rodent cell-lines of different sensitivity. Int. J. Radiat. Biol. 23, 359-376 (1973a)
Fox, B.W., Fox, M.: DNA single strand rejoining in two pairs of cell lines showing the same and different sensitivities to x-rays. Int. J. Radiat. Biol. 24, 127-135 (1973b)
Gautschi, J.R., Young, B.R., Cleaver, J.E.: Repair of damaged DNA in the absence of protein synthesis in mammalian cells. Exp. Cell Res. 76, 87-94 (1973)
Han, A., Korbelik, M., Ban, J.: DNA-to-protein cross-linking in synchronized HeLa cells exposed to ultraviolet light. Int. J. Radiat. Biol. 27, 63-74 (1975)
Isomura, K., Nikaido, C., Horikawa, M., Sugahara, T.: Repair of DNA damage in ultraviolet sensitive cells isolated from HeLa S3 cells. Radiat. Res. 53, 143-151 (1973)
Lange, C.S.: On the relative importance of repair and progression in Elkind recovery as measured in synchronous HeLa cells. Int. J. Radiat. Biol. 17, 61-79 (1970)
Lange, C.S.: The organization and repair of mammalian DNA. FEBS Letters 44, 153-156 (1971)
Lange, C.S., Liberman, D.F., Sheck, L.E.: The repair of radiation induced double-strand breaks in the DNA of exponentially growing L5178Y cells. Radiat. Res. 55, (Cc-9) (1973)
Makino, F., Okada, S.: Comparative studies on the effects of carcinogenic and antitumor agents on the DNA replication of cultured mammalian cells. Mutation Res. 23, 387-394 (1974)
Painter, R.B.: Similarities and differences between UV and ionizing radiation death. In: Advances in Radiation Research (Duplan, J.F., Chapiro, A., eds.). New York-Paris-London: Gordon and Breach, 1973, Vol. I, pp. 411-419
Szumiel, I., Rosiek, O., Budzicka, E.: Nuclear and cytoplasmic protein synthesis in two strains of L5178Y cells of different radiosensitivities. Studia Biophys. (Berlin) 51, 193-200 (1975)
Szumiel, I., Ziemba-Zak, B., Rosiek, O., Sabliński, J., Beer, J.Z.: Harmful effects of an irradiated cell culture medium. Int. J. Radiat. Biol. 20, 153-161 (1971)

Terasima, T.: DNA break repair and sublethal damage recovery. In: Advances in Radiation Research (Duplan, J.F., Chapiro, A., eds.). New York-London-Paris: Gordon and Breach, 1973, Vol. I, pp. 421-429

Walicka, M.: Krzyżowa wrażliwość komórek L5178Y-R i L5178Y-S na promieniowanie X i światło UV (Cross-sensitivity of L5178Y-R and L5178Y-S cells to x-rays and UV-light). Ph.D. Thesis, Institute of Nuclear Research, Warszawa, 1975

17. Detection of Hybrid DNA Formed during Mitomycin C-Induced Sister Chromatid Exchange in Chinese Hamster Cells

P. D. Moore

A. Introduction

Most cells appear to possess several repair mechanisms capable of removing the damage produced in their DNA by mutagens such as γ-rays, ultraviolet (UV) light, and mitomycin C. The most widely studied of these is that involving excision of the damage and patching of the DNA by repair synthesis (Howard-Flanders, 1973). However, it is now known that other systems of repair involve a process of recombination. In particular, this second type of mechanism is necessary for the repair of damage involving both strands of the DNA molecule. Such damage occurs, for example, when UV-irradiated cells replicate their DNA and gaps are left in the newly synthesized strand opposite pyrimidine dimers in the old strand, or when cells are treated with mitomycin C, an agent that causes cross links to be formed between the two strands (Iyer and Szybalski, 1963). Genetic recombination is greatly stimulated by agents causing this type of damage.

In bacteria the gaps in DNA newly synthesized after UV irradiation are repaired by the insertion of DNA from a sister chromosome by a recombinational process (Rupp et al., 1971). In mammalian cells, however, the gaps appear to be repaired by de novo synthesis rather than by recombination (Lehmann, 1972). Nevertheless, the induction of cytologically detectable sister chromatid exchanges by agents such as x-rays (Gatti and Olivieri, 1973), UV (Kato, 1973), and mitomycin C (Latt, 1974) strongly indicates that some sort of recombinational repair occurs in mammalian cells. Since it is widely believed that hybrid DNA is an essential intermediate in genetic recombination between homologous chromosomes, it is very likely that it also occurs in sister chromatid exchange. This hybrid DNA would be formed from the annealing of complementary homologous strands from two sister chromatids. In this study a density labelling technique is used to detect a DNA fraction that is probably hybrid DNA formed during sister chromatid exchange in Chinese hamster cells.

B. Methods

I. Cell Culture

Chinese hamster cells (CHO) were grown either as a monolayer or in suspension culture in Eagle's basal medium, supplemented with 10% dialysed foetal calf serum, non-essential amino acids, and antibiotics (penicillin 100 units/ml; streptomycin 100 μg/ml). Twenty-four hours prior to the beginning of an experiment, the cells were split and tylosin (60 μg/ml) was added to the growth medium to suppress any mycoplasma (Grüneisen et al., 1975).

II. Radioactive Labelling

The growth medium was removed from the actively dividing cells and replaced with one containing fluorodeoxyuridine (FUdR), 0.25 μg/ml and bromodeoxyuridine (BUdR), 3 μg/ml. After 30 min ^{3}H-thymidine, 15 Ci/mmole was added to a concentration of 20 μCi/ml and incubation continued for a further 6 h. The cells were then removed from the radioactive medium, washed, and then replaced in normal growth medium containing unlabelled thymidine, 5 μg/ml to compete out any remaining BUdR or radioactive label and then incubated for 5 h with mitomycin C treatment as described in the text. Where mitomycin C treatment for less than the full 5 h is indicated, the drug was given at the beginning of the 5 h. Precautions were taken throughout to avoid exposing cells labelled with BUdR to illumination containing long wavelength UV light. The use of undiluted high specific activity ^{3}H-thymidine under these conditions does not detectably alter the density shift in the DNA caused by the BUdR substitution (Davidson and Bick, 1973).

III. Extraction of DNA for Caesium Chloride Gradients

After harvesting, the cells were washed and then resuspended in 0.9 ml of 50 m*M* Tris-HCl (pH 7.5), 10 m*M* EDTA at 0-4°C, lysed by the addition of 0.1 ml of 20% Sarkosyl, and stored overnight at -20°C. Samples were thawed, treated with pre-digested protease (Sigma type VII) 0.5 mg/ml for 1 h at 37°C, extracted three times with equal volumes of chloroform/amyl alcohol (24:1), and then dialysed for 36 h against 10 m*M* Tris-HCl (pH 7.5), 1 m*M* EDTA. Before analysis on CsCl gradients the DNA was sheared by being forced four times through a gauge 27 (0.4 mm diameter) needle in 250 m*M* NaCl. The molecular weights, as determined by neutral sucrose velocity gradients, were 9 x 10^7 and 1.4-1.8 x 10^7 daltons for untreated and sheared DNA, respectively.

The sample volume was adjusted to 4 ml with buffer and 5.55 g CsCl were added. The solution was then centrifuged for 60 h at 33,000 rpm in a Beckman 65 fixed angle rotor. The tubes were punctured and 7 drop fractions were collected. Aliquots were precipitated onto filter paper with ice-cold 5% TCA, washed in ethanol, dried, and counted in a liquid scintillation counter. Densities across the gradients were monitored either by weighing or refractometry. Fractions to be rebanded were first pooled and dialysed for 48 h against 10 m*M* Tris-HCl (pH 7.5), 1 m*M* EDTA.

IV. Measurement of Sister Chromatid Exchanges

The method was based on that of Korenberg and Freedlander (1974). Cells were grown in medium containing FUdR, 0.25 μg/ml and BUdR, 3 μg/ml for 30 h (i.e., about 2 cell generations). Six h before the end of this period, mitomycin C was added as required, and then 5 h later colchicine was added to a concentration of 0.8 μg/ml. At the end of the labelling the cells were harvested, washed, resuspended in hypertonic solution (10 m*M* trisodium citrate, 30 m*M* KCl), and incubated at 37°C for 20 min. The cells were then fixed in methanol/acetic acid (3:1), dropped onto cold wet slides, and air dried. Differential staining of sister chromatids was achieved by incubating the slides in 1 *M* phosphate buffer (pH 8.0) at 88°C for 10-12 min, rinsing in distilled water, and staining for 6 min in 3% Giemsa in Gurr's pH 6.8 buffer. Chromatids containing BUdR in only one strand of the DNA stained darkly, while chromatids with BUdR in both strands remained unstained. At least 50 chromosomes were examined for each determination of the frequency of sister chromatid exchange.

C. Results

I. Experimental Design

The basic rationale behind the experiments is shown in Figure 1. The cells are grown in BUdR together with a radioactive label for less than one complete round of DNA replication. At the end of the labelling the radioactivity is present only in the heavy strand of DNA molecules containing one heavy and one light strand (HL DNA), and for each pair of sister chromatids the radioactivity will be in strands of opposite polarity. On returning the cells to medium containing neither BUdR nor radioactive label, all further synthesis, whether replicative or repair, can only result in either completely light (LL) DNA or in HL DNA. On the other hand, hybrid DNA formed by recombination between sister chromatids can result in two types of molecule, completely light (LL) or completely heavy (HH), the latter of which is also radioactive. When DNA from these cells is extracted and banded on neutral CsCl equilibrium gradients, the bulk of the radioactivity will be found in a peak corresponding to half heavy DNA, while hybrid DNA will be detected at the fully heavy density position. A similar technique was developed independently by Rommelaere and Miller-Faurès (1975).

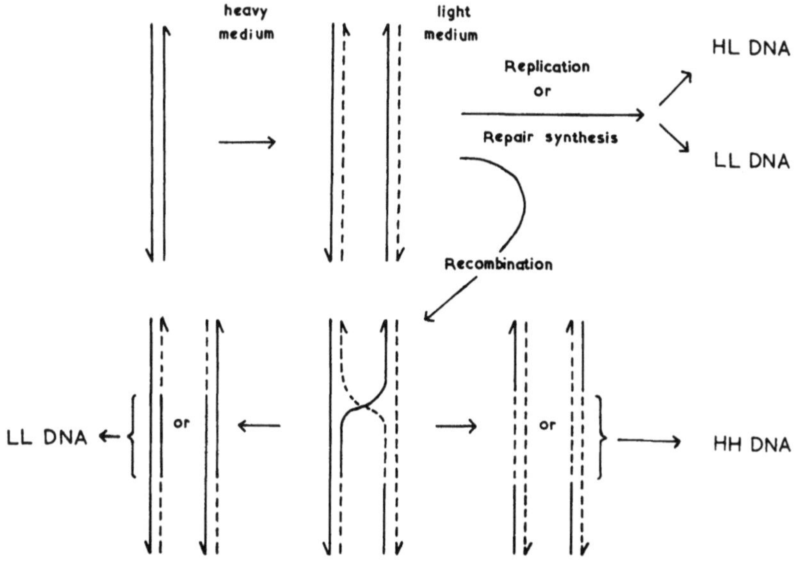

Fig. 1. Rationale behind the experimental design; for details see text. For simplicity, the recombination model used is that proposed by Holliday (1964b) for meiotic recombination, although similar models for recombinational repair of cross-linked DNA (e.g., Cole, 1973) would give essentially the same result. ---DNA containing BUdR and (^{3}H); —unlabelled DNA

In order to show that HH DNA formed in these circumstances represents an intermediate in recombinational repair, it is necessary to show that treatments leading to an increase in recombination also lead to an increase in HH DNA. The agent chosen for this was mitomycin C, since (1) it is a potent recombinagen (Holliday, 1964a; Latt, 1974),

(2) it is a cross-linking agent (Iyer and Szybalski, 1963) that produces double-strand damage without the requirement for further DNA synthesis, and (3) it was necessary to avoid UV irradiation of cells containing BUdR in their DNA.

II. Detection of Hybrid DNA

Figure 2 shows the density profile on a neutral caesium chloride gradient of DNA extracted from cells prelabelled as described in Methods and then treated with 2 μg/ml of mitomycin C for 5 h. The vast majority of the radioactivity is found at the position of HL DNA; however, a small amount, in this experiment about 0.4%, bands at a density close to that of HH DNA. If the DNA from this latter region of the gradient is isolated and rerun on a second neutral caesium chloride gradient, the radioactivity again bands much closer to the position of HH DNA than to that of HL DNA (Fig. 3). The material from this peak is degraded by pancreatic DNase I, but is completely resistant to the single strand specific deoxyribonuclease of *U. maydis* (Holloman, 1973). The possibility that this DNA represents a discrete fraction of HL DNA differing in base composition from the average DNA can be ruled out, since such AT-rich fractions are not displaced further than the heavy side of the main peak (Rommelaere and Miller-Faurès, 1975). It is clear, therefore, that the material banding at densities approaching that of HH DNA consists of double-stranded DNA, mostly HH

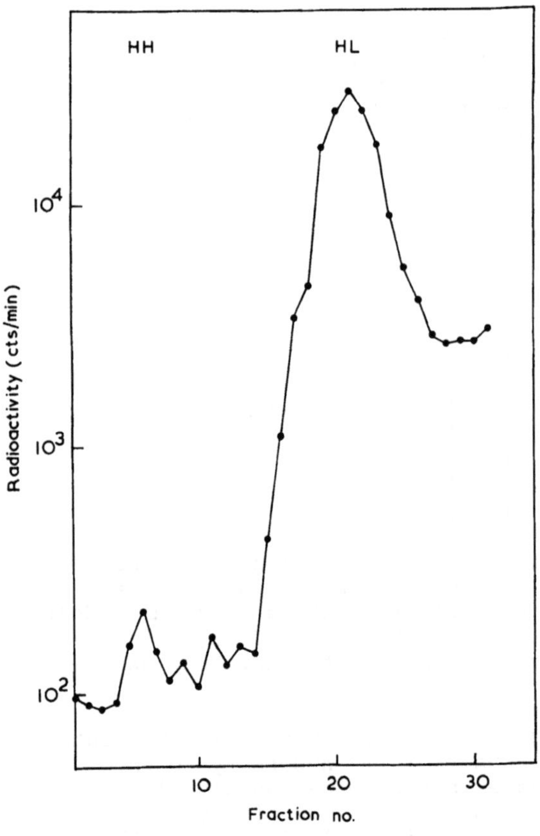

Fig. 2. Density distribution in neutral CsCl gradient of DNA from cells grown for 6 h in medium containing BUdR and then treated with mitomycin C (2 μg/ml) for 5 h in normal medium

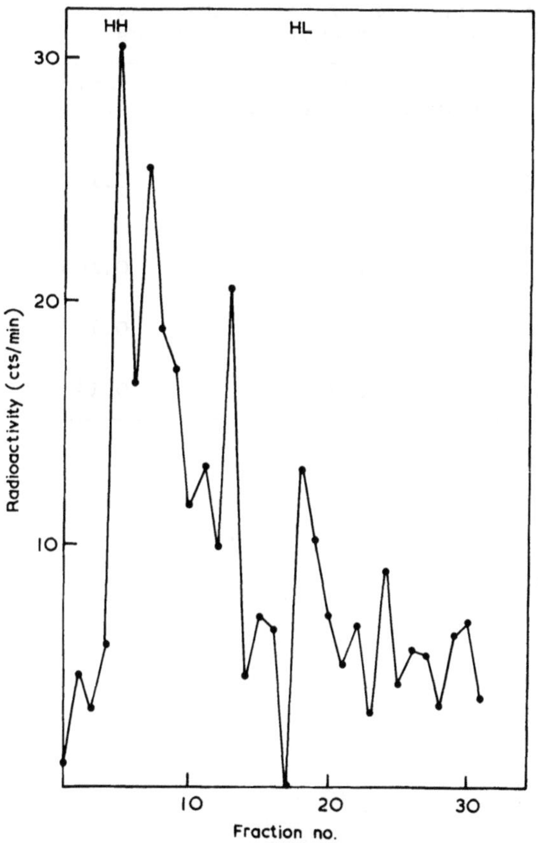

Fig. 3. Recentrifugation in neutral CsCl of DNA isolated from fractions 5-13 of gradient shown in Fig. 2

but containing some regions of HL. The relative proportions of HH and HL in these molecules, and consequently their actual density, depend on the length of the regions of HH DNA in the chromosomes and the degree of shearing during preparation, which is somewhat variable. This material will be referred to as the heavy fraction of the DNA.

In similar experiments in which the cells were not treated with mitomycin C the amount of radioactivity banding between HL and HH is much lower, but when sufficient material from this region is rebanded the presence of a heavy fraction is revealed. This is in good agreement with the results of Rommelaere and Miller-Faurès (1975). The size of the heavy fraction found with different mitomycin C treatments is shown in Table 1.

III. Relationship between HH DNA and Sister Chromatid Exchange

An independent measure of the amount of mitotic recombination occurring in the cells is the frequency of sister chromatid exchange. This was measured by the method of Korenberg and Freedlander (1974). For this technique the cells are labelled with BUdR for two rounds of replication before treatment with mitomycin C, so that for each pair of sister chromatids one is fully substituted with BUdR (HH) and the other only half substituted (HL). The two chromatids (HH and HL) can then be differentially stained, so that metaphase chromosomes have a "harlequin" appearance (Fig. 4), and sister chromatid exchanges (SCEs) are clearly visible.

Table 1. Effect of mitomycin C on the fraction of DNA containing regions of HH and on the frequency of sister chromatid exchanges

Mitomycin treatment	DNA in heavy fraction (%)	A mean	B exchanges/chromosome	Ratio A/B
0	0.11 0.05 0.19	0.117	0.62	0.189
1 µg/ml 1 h	0.26 0.09	0.175	1.67	0.105
1 µg/ml 5 h	0.34	0.340	2.42	0.140
2 µg/ml 5 h	0.40 0.40	0.400	2.75	0.145

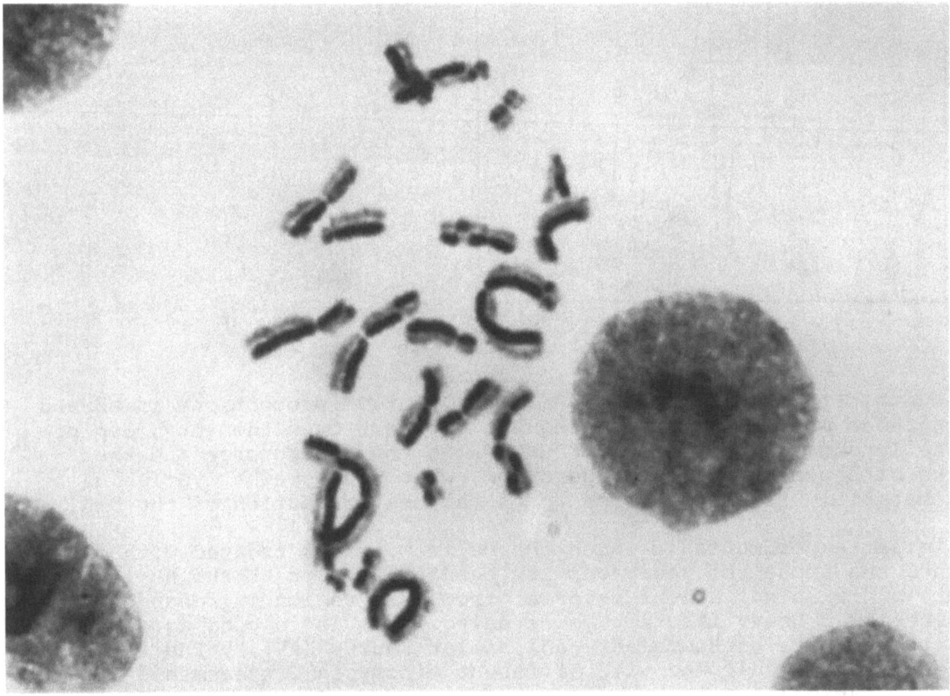

Fig. 4. Chinese hamster chromosomes showing sister chromatids differentially stained with Giemsa. Cells were labelled with BUdR (3 µg/ml) for 30 h before collection. X 2,000

The effect of mitomycin C on frequency of SCEs is shown in Table 1. It is clear that increasing doses of mitomycin C result in increases in both SCEs and the size of the heavy fraction of DNA on CsCl gradients. Further, there appears to be a fairly constant relationship between frequency of SCEs and the size of the heavy fraction. These results strongly suggest that the DNA in the heavy fraction is an intermediate in recombination arising from the formation of hybrid DNA.

D. Discussion

Hybrid or heteroduplex DNA is generally accepted as being an intermediate in genetic recombination; however, direct evidence for this has only previously been obtained in bacteriophages (Broker and Lehman, 1971; Benbow et al., 1975). In eukaryotes the formation of hybrid DNA has only been inferred from genetic data. The preceding experiments provide direct evidence that hybrid DNA is formed in Chinese hamster cells after treatment with the cross-linking agent mitomycin C, probably as a result of recombinational repair.

The formation of HH DNA after only one round of synthesis in BUdR-containing medium has been found by Rommelaere and Miller-Faurès (1975). They also showed that UV irradiation prior to growth in BUdR resulted in an increase in the amount of HH DNA. In this report it has been demonstrated that treatment with mitomycin C after the cells have been removed from the BUdR-containing medium increases the fraction of the DNA-containing regions of HH. The correlation between the increase in the heavy fraction and the increase in SCE is strong circumstantial evidence that the HH DNA is hybrid DNA formed during a process of recombination.

This correlation also allows an estimate to be made of the amount of hybrid formed per exchange. Because the (^{3}H) label is in the heavy strand of the DNA, HH DNA obviously has twice the specific activity of HL. However, since an equal amount of LL and therefore undetected hybrid might be expected to occur and since the DNA in the heavy fraction is usually at least 2/3 HH, the size of the heavy fraction in Table 1 is probably only a slight overestimate of the proportion of hybrid in the DNA as a whole. Thus if 0.4% heavy fraction is found when there are 2.75 SCEs/chromosome (ca. 50 SCEs/cell), the amount of hybrid per exchange totals about 700,000 base pairs. The length of the regions of HH DNA can also be estimated and appears to be in the range 15,000-30,000 base pairs (Rommelaere and Miller-Faurès, 1975; and preliminary results in our laboratory). From these figures it would appear that there could be between 20 and 50 lengths of hybrid per SCE. Two of these could be accounted for if hybrid was always formed reciprocally (see Fig. 1). However, there are several other possible explanations for an excess of hybrid DNA: (1) a high frequency of non-reciprocal, gene conversion-type events, such as those found for induced allelic recombination in fungi (Roman and Jacob, 1958), (2) a higher frequency of exchanges in labelling experiments than in SCE experiments due to the (^{3}H) in the DNA, (3) more than one process leading to the formation of hybrid DNA, (4) errors in measuring the amount of DNA in the heavy fraction. Further experiments are necessary to distinguish between these possibilities.

E. Summary

Hybrid or heteroduplex DNA is generally accepted as being an intermediate in genetic recombination, although such an intermediate has not yet been demonstrated in an eukaryotic organism.

Recombinational or post-replication repair should also involve a hybrid DNA intermediate, and a method of detecting this by using a density-labelling technique has been developed. The DNA of Chinese hamster cells labelled for one round of replication with BUdR has a half heavy (HL) density. Hybrid DNA formed during subsequent recombinational repair between sister chromatids would have a density of either HH or LL.

A small amount of DNA with HH density has been detected on neutral CsCl gradients. The amount of this HH DNA is increased by treatment with mitomycin C after the cells have been removed from the BUdR-containing medium. Increasing doses of mitomycin C produce similar increases in both the amount of HH DNA and the frequency of sister chromatid exchanges measured cytologically. This correlation provides evidence that the HH DNA is an intermediate in recombinational repair.

Acknowledgements. I would like to thank Dr. Robin Holliday for originally suggesting the experiments and for many useful discussions. I am also grateful to Gill Tarrant for providing the cells and instruction in the art of mammalian tissue culture. This work was supported in part by a grant from the Cancer Research Campaign.

References

Benbow, R.M., Zuccarelli, A.J., Sinsheimer, R.L.: Recombinant DNA molecules of bacteriophage X174. Proc. Nat. Acad. Sci. USA 72, 735-739 (1975)

Broker, T.R., Lehman, I.R.: Branched DNA molecules: Intermediates in T4 recombination. J. Molec. Biol. 60, 131-149 (1971)

Cole, R.S.: Repair of DNA containing interstrand crosslinks in *Escherichia coli*: Sequential excision and recombination. Proc. Nat. Acad. Sci. USA 70, 1064-1068 (1973)

Davidson, R.L., Bick, M.D.: Bromodeoxyuridine dependence: a new mutation in mammalian cells. Proc. Nat. Acad. Sci. USA 70, 138-142 (1973)

Gatti, M., Olivieri, G.: The effect of x-rays on labelling pattern of M_1 and M_2 chromosomes in Chinese hamster cells. Mutation Res. 17, 101-112 (1973)

Grüneisen, A., Rajewsky, M.F., Remmer, I., Uschkoreit, J.: Inhibition of (^{3}H)-thymidine incorporation by hydroxyurea. Atypical response of mycoplasma-infected cells in culture. Exp. Cell Res. 90, 365-373 (1975)

Holliday, R.: The induction of mitotic recombination by mitomycin C in *Ustilago* and *Saccharomyces*. Genetics 50, 323-335 (1964a)

Holliday, R.: A mechanism for gene conversion in fungi. Genet. Res. 5, 282-304 (1964b)

Holloman, W.K.: Studies on a nuclease from *Ustilago maydis*. II. Substrate specificity and mode of action of the enzyme. J. Biol. Chem. 248, 8114-8119 (1973)

Howard-Flanders, P.: DNA repair and recombination. Brit. Med. Bull. 29, 226-235 (1973)

Iyer, V.N., Szybalski, W.: A molecular mechanism of mitomycin action: linking of complementary DNA strands. Proc. Nat. Acad. Sci. USA 50, 355-362 (1963)

Kato, H.: Induction of sister chromatid exchanges by UV light and its inhibition by caffeine. Exp. Cell Res. 82, 383-390 (1973)

Korenberg, J.R., Freedlander, E.F.: Giemsa technique for the detection of sister chromatid exchanges. Chromosoma 48, 355-360 (1974)

Latt, S.A.: Sister chromatid exchanges, indices of human chromosome damage and repair: detection by fluorescence and induction by mitomycin C. Proc. Nat. Acad. Sci. USA 71, 3162-3166 (1974)

Lehmann, A.R.: Post-replication repair of DNA in UV irradiated mammalian cells. J. Molec. Biol. 66, 319-337 (1972)
Roman, H.L., Jacob, F.: A comparison of spontaneous and UV induced allelic recombination with reference to the recombination of outside markers. Cold Spring Harbor Symp. Quant. Biol. 23, 155-160 (1958)
Rommelaere, J., Miller-Faurès, A.: Detection by equilibrium centrifugation of recombinant-like DNA molecules in somatic mammalian cells. J. Molec. Biol., in press (1975)
Rupp, W.D., Wilde, C.E., Reno, D.L., Howard-Flanders, P.: Exchanges between DNA strands in ultraviolet irradiated *Escherichia coli*. J. Molec. Biol. 61, 25-44 (1971)

18. Dependence of Photoreactivating Enzyme - Substrate Complex Formation in *Saccharomyces* on Liquid Holding Treatment

A. Yasui and W. Laskowski

A. Introduction

Photoreactivating enzyme (PRE) repairs pyrimidine dimers that have been formed in DNA irradiated with ultraviolet (UV) light, according to the following reaction scheme (Rupert, 1964):

$$E + S \underset{k_2}{\overset{k_1}{\rightleftharpoons}} ES \xrightarrow[\text{light}]{k_3} E + P$$

where E is the PRE, S the pyrimidine dimer (substrate), ES the PRE-substrate complex, and P the repaired product. Light flash experiments have recently proved to be very useful in studying photoenzymatic repair (Harm et al., 1968, 1971). By using a short intense light flash, only those substrates can be repaired that are complexed with PRE at the moment of the flash.

The aim of this paper is to apply this technique to yeast cells for an investigation of the time dependence of the formation of PRE-substrate complexes.

B. Materials and Methods

The following *Saccharomyces* strain was used:

Haploid: MB1030-1A (α, *rad1-18*, *rad2-17*, *rad18-2*). The symbols in brackets indicate mating type alleles and mutated genes causing radiation sensitivity (Game and Cox, 1971). *rad1* and *rad2* mutants are known to block the excision repair mechanism (Unrau et al., 1971) whereas *rad18* blocks another not clearly identified dark repair mechanism (Game and Mortimer, 1974). The strain was obtained from M. Brendel.

Medium: Yeast extract peptone medium (YEP) containing 1% yeast extract (Difco), 0.5% peptone (Merck), 2% glucose, and 2% agar (Merck) was used in all experiments.

Growth conditions and cell preparation: The strains were stored on YEP agar slants at 4°C. Cell samples were inoculated on fresh YEP agar slants and incubated for 48 h at 30°C. The cells were then suspended in buffer (0.05 *M* KH_2PO_4; pH 4.8), washed three times in buffer for further treatment. After irradiation and photoreactivation appropriate cell samples were plated on YEP medium and incubated for 72 h at 30°C.

UV irradiation: The UV source was a low pressure mercury lamp (Osram HNS 12) with a maximum emission at 254 nm wavelength. The dose rate was 3 erg/sec/mm^2. The dose-rate was measured by a photon flux counter (Schaarschmidt, 1970).

Photoreactivation (PhR): For photoreactivating light flashes an electronic flash unit (Rollei E 36 RE) was used, delivering a light flash of 1/800 sec duration. This light flash was filtered by a Jenaer Filter Glass WG 320, transmitting only wavelengths longer than 320 nm. UV-irradiated cell suspensions, placed in a small glass dish, were positioned 5 cm from the surface of the flash unit and exposed to one or more light flashes. With the help of a photodiode it was ascertained that by using smaller distances between the cell sample and the surface of the flash unit more photons reach the cells. The amount of photoreactivation, however, did not increase. Maximum photoreactivation was obtained by 100 light flashes delivered in 30-sec intervals. More flashes brought no essential increase in PhR. This maximum was identical with one obtained by continuous illumination with photoreactivating light of an Osram-L-20W-70 lamp.

Liquid holding treatment: After UV irradiation or UV irradiation plus flash PhR, the cells were kept in buffer for various times in the dark at 30°C. Unirradiated control cells were also kept under the same conditions. Appropriate samples were then plated on YEP medium and incubated for 72 h at 30°C. All experiments were performed under yellow light to prevent uncontrolled PhR, at room temperature of approximately 22°C. The results depicted in the figures are average values obtained from three to five independent identical experiments.

The cell concentration during UV irradiation, photoreactivation, and liquid holding was 5 x 10^6 cells per ml buffer. Before plating, the cell suspensions were diluted appropriately to obtain from 100-400 colonies per plate.

Quantitative Evaluation of the Photoreactivation Effects by the Dose Decrement. In the theoretical example illustrated in Figure 1 the lower curve represents the number of lesions in a cell produced by UV irradiation immediately before application of one light flash. Since effects of dark repair on the number of lesions in a cell cannot be excluded completely, even in the UV-sensitive strains used in our experiments, the slope of the lower curve is dependent on the time interval between UV irradiation and the light flash. In UV-irradiated cells a single flash may decrease the number of lesions in the cell from a → c (lesion decrease, LD), resulting in a dose decrement c → b. If dark repair is effective in cells representing b and c to the same extent, it leads to the observed survival rates d and e. We shall call f → e lesion decrease after dark repair (LD*).

This analysis permits investigation of the effects of liquid holding treatment on photorepairability of UV-irradiated cells.

C. Results

Figure 2 shows the survival of the haploid strain after UV irradiation (left curve) and the results of PhR by single light flashes at various times after UV irradiation (right curves). These data show that 70% of the maximum survival can be obtained by one flash as early as 1 min after UV irradiation, but that about 20 min are necessary to achieve maximum survival.

Figures 3a and 3b show the effect of liquid holding on cell survival after UV irradiation and after UV irradiation plus photoreactivation (PhR). Strain MB1030-1A always exhibits negative liquid holding re-

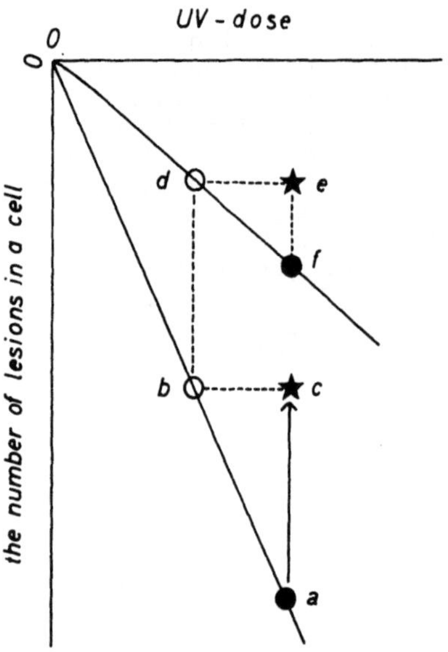

Fig. 1. For explanation see text

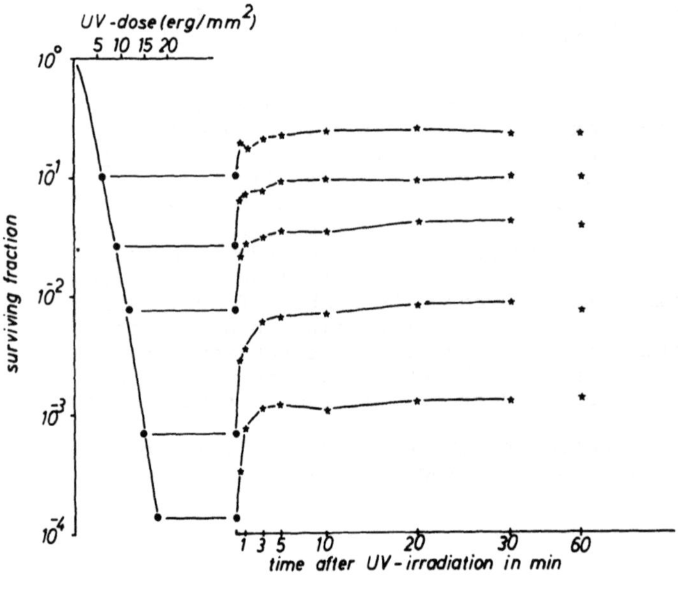

Fig. 2. Survival of strain MB1030-1A after UV irradiation (left curve) as well as after UV irradiation and application of a single light flash at various times after UV irradiation (right curves)

covery. The curves obtained from samples that were either UV-irradiated or UV-irradiated and photoreactivated by a single light flash 40 min after UV irradiation, and then kept in the dark in buffer 1-6 days at 30°C, are almost parallel to each other. Figure 3b depicts an increase in survival with increasing liquid holding periods longer than 2 days. It has not been determined whether this increase in survival is caused by a residual dark repair process, which becomes effective after 2 days of liquid holding, or by multiplication of surviving cells

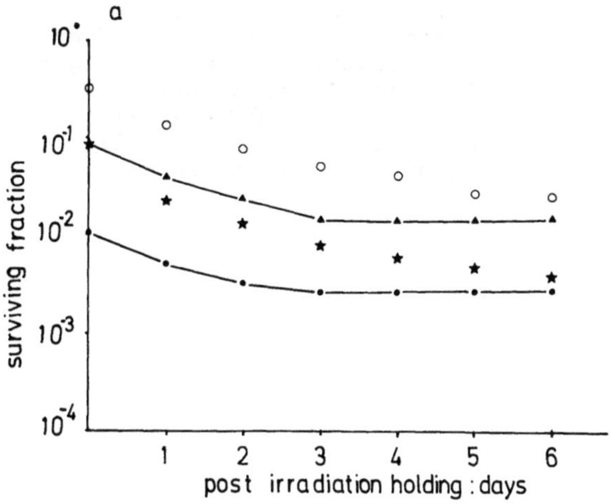

Fig. 3 a and b. Effect of liquid holding treatment on photoreactivability of strain MB1030-1A. ●, UV irradiation + liquid holding; ▲, UV irradiation + PhR with a single light flash 40 min after UV irradiation + liquid holding; ★, UV irradiation + liquid holding + PhR with a single light flash; o, UV irradiation + liquid holding + maximum PhR

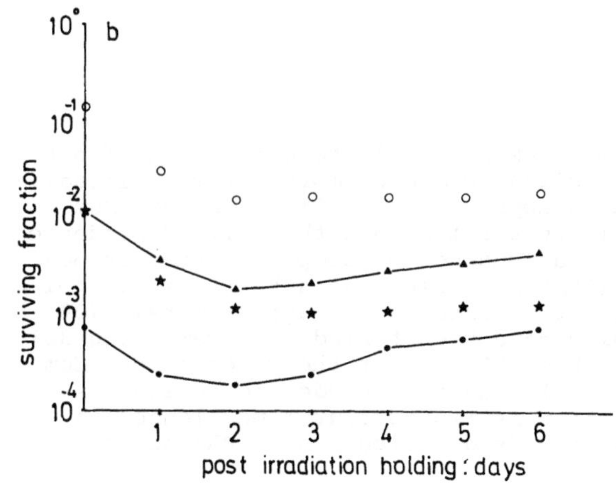

due to material released from dead cells. The fact that all the data obtained with photoreactivated cells do not show an increase in survival with increasing liquid holding time does not support the assumption of cell multiplication (Fig. 3b).

During the liquid holding period of up to 6 days, the survival of the controls decreased 10% per day on the average. The data for each day of liquid holding have been calculated by using the surviving controls as reference.

In order to test the photoreactivability after increasing days of liquid holding, samples were taken daily from the UV-irradiated suspensions and photoreactivated with a single light flash as well as 100 light flashes for maximum PhR.

Figure 4 gives results taken from Figure 3a and 3b as well as results obtained after UV irradiation with 6 erg/mm^2. The proportions of single flash PhR to maximum PhR, which are determined in these figures by dividing the distance between closed circles and stars by the corresponding distance between closed circles and open circles, decrease with increasing LHR-duration from 0.6 - 0.7 to 0.2 - 0.3.

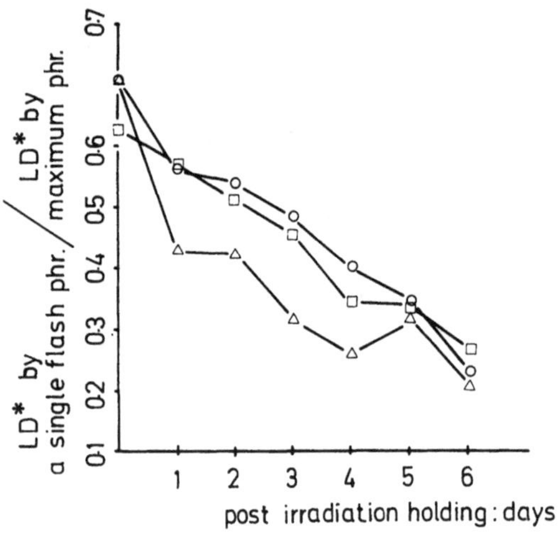

Fig. 4. Change of ratio LD* (lesion decrease by PhR after dark repair) by a single light flash to LD* by maximum PhR in UV-irradiated MB1030-1A cells. Δ, 6 erg/mm^2; o, 12 erg/mm^2; □, 15 erg/mm^2

D. Discussion

It is shown in Figure 2 that a maximum survival increase by one flash is obtained, when the flash is applied 20 min or more after UV irradiation. One min after UV irradiation approximately 70% of the maximum survival increase is obtained. It is assumed that this time dependence of survival is caused by the time dependence of complex formation between photoreactivating enzyme (PRE) molecules and substrates. It follows, therefore, that one min after UV irradiation approximately 70% of the maximum number of complexes have been formed. The remaining 30% PRE-substrate complexes require about 20 min for their formation. Compared to *E. coli* (Harm et al., 1968), the time needed to obtain maximum PRE-substrate complex formation is two to four times longer in *Saccharomyces*. This may indicate a wide distribution of PRE molecules in the cell.

The strain used in these experiments is incapable of excision repair (because of the *rad1*, *rad2* mutation) and is blocked in a still unknown dark repair process (*rad18* mutation). There are, however, additional dark repair processes still effective, such as recombination repair, which has been shown to exist in *Saccharomyces* (Kowalski and Laskowski, 1975). The effect of the remaining dark repair processes on the photoreactivability was studied by using extended time intervals between UV irradiation and photo flash. The cells were kept in buffer for these periods (liquid holding). As may be seen in Figures 3a and 3b, liquid holding exerts similar effects on cells that have been either UV-irradiated or UV-irradiated and photoreactivated with a single flash 40 min after UV irradiation. This indicates that PhR is without effect upon negative liquid holding recovery (NLHR), which is characteristic for this strain. This interpretation agrees with that of Parry (1972), who concluded that NLHR results from the reduced activity of an enzyme system involved in the dark repair of nonphotoreactivable UV-induced lesions. If, however, the UV-irradiated cells are incubated in buffer in the dark before photoreactivation the photoreactivability (maximum photoreactivation as well as photoreactivation by a single flash) of these cells decreases steadily. In *E. coli* a similar decrease of maximum photoreactivability after UV irradiation was observed in uvr$^+$ but not in uvr$^-$ (i.e., excision repair defective) cells (Moss and Davies, 1974). From these data it was concluded that in *E. coli* the excision repair mechanism must be undisturbed to cause a decrease of photoreactivability during liquid holding. If the same holds true for *Saccharomyces*, one has to assume that despite the presence of the mutated genes *rad1* and *rad2* there is some activity of the excision repair process in these cells (i.e., these mutants are leaky), or the changes of the substrate for the PRE molecules can be caused in *Saccharomyces* by another dark repair mechanism. The decrease of the photoreactivability by one flash during liquid holding is more rapid than the decrease of maximum photoreactivation (Fig. 4). The decrease of single flash photoreactivation may be explained either by a change in the activity of the PRE molecules or by a change in the substrate. Since control cells, kept under liquid holding conditions for 6 days and then UV-irradiated and photoreactivated by a single flash, always exhibited the same photoreactivability as cells without liquid holding treatment (Yasui, unpublished results), a change in the activity of the PRE molecules can be discarded as a possible explanation. Therefore changes of the substrate are considered to cause the observed rapid decrease in single flash photoreactivation and the slower decrease in maximum photoreactivation.

E. Summary

Complex formation between photoreactivating enzyme and pyrimidine dimers (substrate) after UV irradiation has been studied in *Saccharomyces* by using photo flashes. Under the experimental conditions used, the time necessary for maximum complex formation was about 20 min. By extending the time between UV irradiation and photo flash to up to 6 days (liquid holding), the survival increase produced by one light flash decreased more rapidly than the survival increase obtained by maximum photoreactivation. This effect is interpreted as being indicative of a change in the substrate, since it could be shown that the efficiency of the photoreactivating enzyme remains unchanged in control cells.

Acknowledgements. The authors thank Dr. W. Harm for discussions on the problem of flash photoreactivation and the Deutsche Forschungsgemeinschaft for financial support.

References

Game, J.C., Cox, B.S.: Allelism tests of mutants affecting sensitivity to radiation in yeast and a proposed nomenclature. Mutation Res. 12, 328-331 (1971)

Game, J.C., Mortimer, R.K.: A genetic study of x-ray sensitive mutants in yeast. Mutation Res. 24, 281-292 (1974)

Harm, W., Harm, H., Rupert, C.S.: Analysis of photoenzymatic repair of UV lesions in DNA by single light flashes. In vivo studies with *Escherichia coli* cells and bacteriophage. Mutation Res. 6, 371-385 (1968)

Harm, W., Rupert, C.S., Harm, H.: The study of photoenzymatic repair of UV lesions in DNA by flash photolysis. In: Photophysiology (Giese, A.C., ed.). New York-London: Academic Press, 1971, Vol. I, pp. 279-324

Kowalski, S., Laskowski, W.: The effect of three rad genes on survival, inter- and intragenic mitotic recombination in *Saccharomyces*. Molec. Gen. Genetics 136, 75-86 (1975)

Moss, S.H., Davies, D.J.G.: Interrelationship of repair mechanisms in ultraviolet-irradiated *Escherichia coli*. J. Bacteriol. 120, 15-23 (1974)

Parry, J.M.: A quantitative analysis of "Negative Liquid Holding" in some UV sensitive mutants of yeast. Molec. Gen. Genetics 118, 33-43 (1972)

Rupert, C.S.: Photoreactivation of ultraviolet damage. In: Photophysiology (Giese, A.C., ed.). New York-London: Academic Press, 1964, Vol. II, pp. 283-327

Schaarschmidt, B.: Eine einfache Methode zur Bestimmung des Quantenflusses von Bestrahlungslampen im Wellenlängenbereich 2400 A - 6000 A. Z. Naturforsch. 25b, 330-331 (1970)

Unrau, P., Wheatcroft, R., Cox, B.S.: The excision of pyrimidine dimers from DNA of ultraviolet irradiated yeast. Molec. Gen. Genetics 113, 359-362 (1971)

19. Evaluation of the Action Spectrum of Yeast Photoreactivation in vivo by Means of Pulsed Dye Lasers

A. ANDERS, A. YASUI, H. ZACHARIAS, I. LAMPRECHT, and W. LASKOWSKI

A. Introduction

The normally obtained photoreactivation (PhR) by continuous illumination raised three questions that could not be answered by the classical instrumental set-up:

1. How many molecules of the photoreactivating enzyme exist per cell?
2. What are the kinetics of PhR?
3. What is the action spectrum of PhR?

The first two problems could be solved by means of photo flashes (see the preceding paper of Yasui and Laskowski) and the third with the aid of monochromators and very strong light sources (e.g., mercury arc lamps). An ideal combination of the two methods may be realized by a pulsed tunable dye laser operating in the desired wavelength range and with sufficient intensity. As far as we know, this is the first time that lasers have been used in such experiments. Before reporting some results of our experiments we want to give a short description of the instrumental set-up.

B. Instrumentation

Two different tunable dye lasers were used in our experiments, a flash light pumped laser (Zeiss, FL 3 B) and a nitrogen-laser pumped dye laser (Molectron, DL 300, UV 1000). Some characteristic data are compiled in Table 1. The lower border of wavelength of the laser-pumped laser ranges to 365 nm, of the FL 3 B, to 430 nm. But frequency doubling by means of KDP crystals opens the ultraviolet (UV) region for the latter one, so that the FL 3 B was used within the full spectrum of PhR, the Molectron only for the upper end. The laser dyes used, together with their solvents, the fundamental wavelengths, the wavelengths after frequency doubling, and the respective intensities are given in Table 2.

Because of the small diameter of the laser beam it was not useful to use the normal irradiation dishes of some centimeter diameter and a content of some milliliters. Instead a petri dish was prepared in the following way: on a diameter of the dish 8 small blocks of agar were distributed at equal distances. These blocks are approximately 3 mm high and 3 mm in diameter and fit well to the agar surface (Fig. 1). The petri dishes are fixed to a vertical frame in an exactly reproducible manner, and the single blocks are moved stepwise into the laser beam. Each gets one pulse; when the line is finished the dish is exchanged and irradiated again after a prechosen interval. By applying different laser wavelengths the flash technique is usable for spectroscopic analyses. Figure 2 contains the total instrumental set-up.

Table 1. Characteristic properties of two dye lasers and photo flash

	Molectron	Zeiss	Rollei
Directly accessible wavelength	360-740 nm	425-750 nm	320-770 nm
Pulse power	5-150 kW	5-150 kW	< 4 kW
Pulse energy	25-1000 μJ	1-50 mJ	< 4 J
Pulse duration	5 ns	200-500 ns	> 1 ms
Repitition rate	5-50 Hz	5 Hz	< 0.1 Hz
Diameter of beam	3 mm	3 mm	-
Line width	0.02 nm	0.1 nm	-

Table 2. Dyes used in the laser experiments

Laser	Dye	Solvent	Wavelength fundamental	Wavelength after frequency doubling	Energy per pulse
Molectron DL 300 UV 1000	PPO	ethanol/ toluene	370 nm	-	30 μJ
			380 nm	-	10 μJ
	PBBO	ethanol/ toluene	397 nm	-	70 μJ
Nitrogen pump laser UV 1000	-	-	337 nm	-	8000 μJ
Zeiss FL 3 B	coumarine 2	methanol	428 nm	-	155 μJ
	coumarine 102	methanol	460 nm	-	1300 μJ
	Nile blue perchlorate + rhodamine B	methanol	(710)	355 nm	5 μJ

Comparing the effectivity of photo flash versus laser pulse, one obtains for the Rollei E 36 RE flash an electric energy per flash smaller than 80 J; the light energy is less than 4 J, distributed over approximately 150 cm^2. The flash spectrum shows that only 30% of its energy is photoreactivating. With these figures the active photo flash intensity amounts to 8 mJ/cm^2 or 1.1×10^{-9} J per cell. These data compare to 2.2 mJ/cm^2 or 0.28×10^{-9} J per cell for the Zeiss laser pulse (428 nm) and to 1.1 mJ/cm^2 or 0.14×10^{-9} J per cell for the Molectron dye laser pulse (437 nm).

Because of the extremely small divergence of the beam, the distance between laser and dish is of no importance. In most experiments the time interval between two pulses was 2 or 3 min. With the Molectron dye laser a few experiments were run with a pulse frequency of 25 Hz and up to 1000 pulses. These results will be published elsewhere.

The wavelength region of PhR is as yet very unfavorable for lasers, because it is at the lower end of the range of action for one (Molectron) and after frequency doubling at the upper end for the other (Zeiss).

Fig. 1. Supporting frame for petri dishes and device for reproducible shift of dishes. Eight agar blocks containing cells are seen in dish

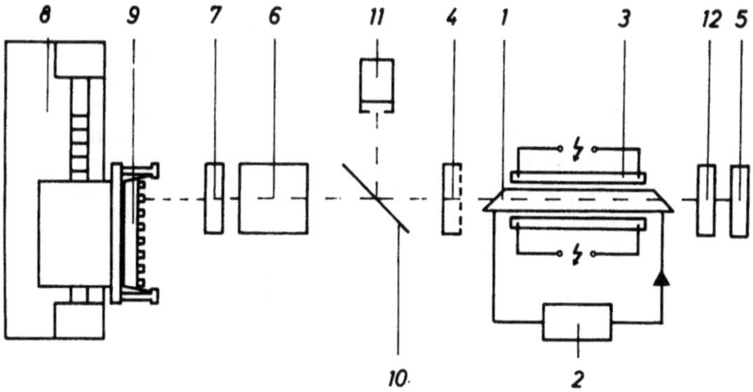

Fig. 2. Schematic drawing of total instrumental set-up: (1) dye laser cuvette; (2) circulation system with pump for dye; (3) optical pumping flash lamps; (4) output mirror; (5) totally reflecting mirror; (6) KDP crystal for frequency doubling; (7) filter absorbing visual light (fundamental wavelength); (8) supporting and transporting frame; (9) petri dish with agar blocks; (10) glass plate; (11) light meter; (12) interference filter

C. Biologic Materials and Methods

The experiments were performed with the haploid strain MB1030-1A of *Saccharomyces cerevisiae*. The strain was stored, precultured, and handled as described by Yasui and Laskowski (preceding paper). After UV irradiation, leading to 0.1% survivors, 5 µl of a suspension containing 8 x 10^5 cells/ml were transferred onto each agar block. The petri dishes with 8 agar blocks were kept for 40 min at room temperature until laser irradiation. The agar blocks were then suspended in buffer, appropriate cell samples plated on nutrient medium, and incubated for 3 days at 30°C. The number of colonies was counted and the increase in survival calculated. By this method the dose decrement (Harm et al., 1971) obtained by 10 laser pulses was determined. By comparison with the dose decrement produced by one photo flash, the relative photoreactivating efficiency was computed.

All results were referred to data obtained from non-UV-irradiated control samples, which were otherwise handled in the same way, i.e., irradiated by photo flash or laser.

D. Results

Figure 3 compares the PhR obtained by photo flash and by pulses of two different laser wavelengths. It is obvious that because of the higher total energy per photo flash maximum PhR is gained by a smaller number of photo flashes. Even with 150 laser pulses of both wavelengths no saturation was obtained. These wavelengths are of course at the edges of the PhR spectrum (see Fig. 4), but the laser intensity at these frequencies is higher than in the maximum of PhR. Therefore these wavelengths were chosen.

Until now the spectrum of PhR for yeast was measured only in vitro (Harm and Rupert, 1970), whereas in vivo the whole spectrum (white light) was used. Figure 4 gives the wavelength dependence of the PhR for *Saccharomyces*. Because the laser intensity is quite different for the various wavelengths, PhR was normalized to an energy unit per pulse. The spectrum shows a broad maximum between 370 and 410 nm and a steeper decline to shorter wavelengths than to longer ones. Irradiation of cells with 355-nm pulses without preceding UV irradiation demonstrates that at this frequency there is a simultaneous inactivation and reactivation of the yeast cell. With changing wavelength this ratio of reactivation/inactivation changes significantly. At 330 nm and even more at 306 nm no reactivation is left. Jagger et al. (1969) observed for *E. coli* that for the highest PhR doses at 313 nm 50% of the cells were killed and no PhR existed at 302 nm.

E. Discussion

With these experiments the action spectrum of photoreactivation in whole yeast cells could be established. It fits to in vitro results that have been obtained by Harm et al. (1971). It is also comparable to the action spectrum of PhR in integer *E. coli* cells (Jagger et al., 1969). Moreover, for the first time spectral research and pulse technique could be combined by means of different dye lasers. They appear to be suitable and promising instruments in the field of photoreactivation. Until now, the spectral region of PhR in microorganisms was hardly accessable with dye lasers. But an interesting field for future research with lasers has been opened by the results of Sutherland and Sutherland (1975) on human photoreactivating enzyme. These authors have shown that, in contrast to former assumptions, such an enzyme exists. In this case maximum photoreactivation is obtained around 410 nm with an action spectrum extending from 300 to approx. 600 nm. These longer wavelengths are easily realized using dye lasers

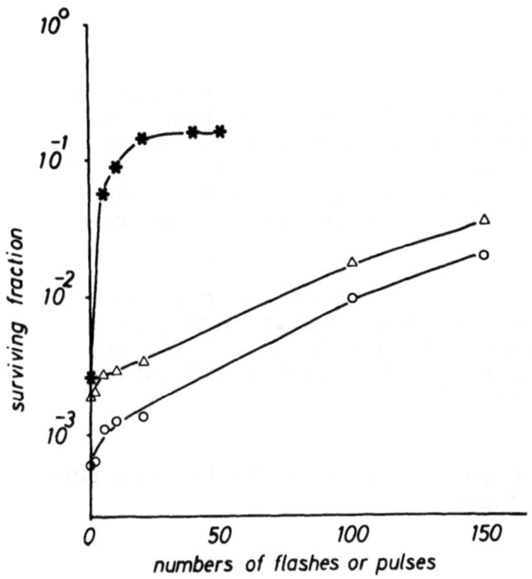

Fig. 3. Comparison between photoreactivation by photo flashes (*) and by laser pulses of two different wavelengths (Δ, 460 nm; o, 367 nm)

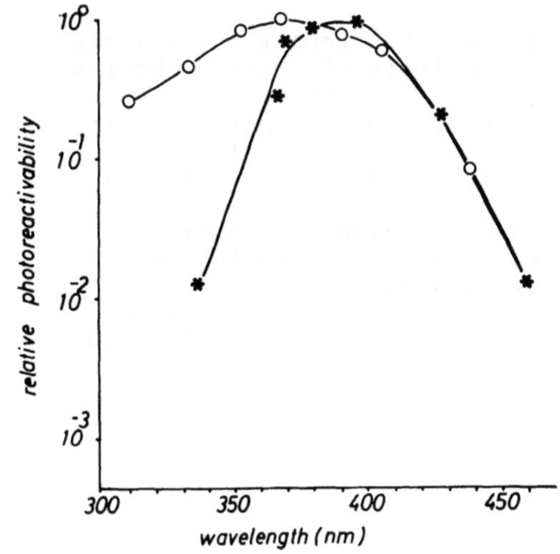

Fig. 4. Action spectrum of photoreactivation of *Saccharomyces* by ten laser pulses (*) compared to action spectrum of isolated photoreactivating enzyme from *Saccharomyces* (o) (adapted from Harm et al., 1971)

with a sufficiently high intensity so that the lasers become at least equivalent to the photo flash.

Because of the extremely short duration of the laser pulses (Table 1), one may use them for kinetic experiments with very small time constants and for the investigation of processes with two photons by dividing the beam and introducing a delay line into the optical pathway. By this method it should be possible to obtain pulses with a temporal interval of some nanoseconds.

F. Summary

By means of two different dye lasers the dependence of photoreactivation on the wavelength of the reactivating light used was measured for a haploid yeast strain in vivo. A broad maximum between 370 nm and 410 nm was found, as compared to a maximum in the 355-385 nm region obtained in vitro experiments by Harm and Rupert (1970). At wavelengths shorter than 350 nm a simultaneous inactivation and reactivation of the cells was obtained.

The instrumental set-up for the laser irradiation, the energy and power per pulse, and their influence on the reactivation are discussed as well as the future applications of tunable pulsed lasers for kinetics and for more photon processes in the photoreactivation of microorganisms and mammalian (human) cells.

Acknowledgement. W. Laskowski acknowledges the support by the Deutsche Forschungsgemeinschaft, Bonn-Bad Godesberg.

References

Harm, H., Rupert, C.S.: Analysis of photoenzymatic repair of UV lesions in DNA by single light flashes. VII. Photolysis of enzyme-substrate complexes in vitro. Mutation Res. 10, 307-318 (1970)

Harm, W., Rupert, C.S., Harm, H.: The study of photoenzymatic repair of UV lesions in DNA by flash photolysis. In: Photophysiology (Giese, A.C., ed.). New York: Academic Press, 1971, Vol. VI, pp. 279-324

Jagger, J., Stafford, R.S., Snow, J.M.: Thymine-dimer and action-spectrum evidence for indirect photoreactivation in *Escherichia coli*. Photochem. Photobiol. 10, 383-395 (1969)

Sutherland, J.C., Sutherland, B.M.: Human photoreactivating enzyme. Action spectrum and safelight conditions. Biophys. J. 15, 435-440 (1975)

Chapter III
Cell Division and Progression

1. Effects of UV- and Ionizing Radiation on Mitosis and Enzyme Regulation in the Synchronous Nuclear Division Cycle of *Physarum polycephalum*[1]

W. SACHSENMAIER and E. DWORZAK

A. Introduction

Macroplasmodia of the Myxomycete *Physarum polycephalum* exhibit unique features suitable for radiobiologic studies on cell cycle-dependent events. Nuclear divisions are naturally synchronous due to the syncytial nature of the plasmodium (Howard, 1932). Disc-shaped surface cultures containing millions of nuclei in a single mass of protoplasm may be grown under axenic conditions up to several centimeters in diameter (Fig. 1). This multinuclear state provides the organism with an extraordinary radioresistance as far as survival is concerned. On the other hand, the perfect synchrony of the nuclear mitotic cycle offers ideal conditions to study cycle-dependent radiation effects other than mere cell killing, as for example mitotic delay, interference with nucleic acid metabolism, and enzyme regulation. This report reviews briefly experiments with ultraviolet (UV) light and soft ionizing radiation dealing with effects on the timing of synchronous nuclear mitosis and on the temporal pattern of certain "step" enzymes in *Physarum* plasmodia.

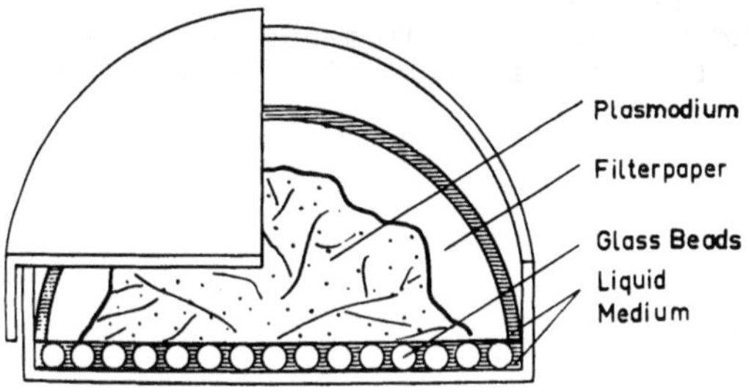

Fig. 1. Surface culture (macroplasmodium) of *Physarum polycephalum*

B. Methods

Stationary plasmodia of *Physarum polycephalum* were cultivated axenically in petri dishes (Ø = 9 cm) on filter paper (Schleicher & Schuell Nr. 595) as described elsewhere (Daniel and Baldwin, 1964; Sachsenmaier and Rusch, 1964). Mitotic stages of the nuclei were observed in ethanol-fixed smear preparations with a phase contrast microscope (Guttes et al., 1961). During mitosis more than 95% of the nuclei in a single plasmodium divide within five to ten minutes. Entire plasmo-

---

[1] Dedicated to Prof. Dr. O. Hoffman-Ostenhof on his 60th birthday.

dia or segments were irradiated with UV light (Mineralight UVS-11, Ultraviolet Products Inc., San Gabriel, Cal.; λ_{max} 253.7 nm, 120 erg/ mm^2/sec at a distance of 10 cm) or with ionizing radiation (Siemens "Dermopan" x-ray generator, 29 kV, 25 mA, 0.3 mm Al-filter, dose rate 870 R/min). Incorporation of radioactive uridine into RNA was measured by pulse labeling with 0.2 µCi/ml ^{14}C-uridine (62 mCi/m*M*) for 30 min in the presence or absence of actinomycin C. If actinomycin was present, pulse labeling was started 30 min after the drug was added to the medium. Specific radioactivity of RNA was determined by liquid scintillation counting (Sachsenmaier et al., 1967). Enzymes were assayed in the high-speed supernatant of plasmodial homogenates: thymidine kinase (Sachsenmaier and Ives, 1965; Sachsenmaier et al., 1970a); thymidine-monophosphate kinase (Coutsogeorgopoulos et al., 1965); thymidine-diphosphate kinase (Chiga et al., 1963); deoxycytidine kinase, analogous to thymidine kinase using ^{14}C-deoxycytidine as substrate; glucose-6-phosphate dehydrogenase (Sachsenmaier and Ives, 1965).

C. Results and Discussion

I. Cycle Dependent Variation of Radiation-Induced Mitotic Delay

Figure 2 demonstrates the change of radiosensitivity during the synchronous nuclear cycle with regard to mitotic delay. It is clear that different mechanisms are responsible for the retarding effects of UV light and ionizing radiation on mitosis. A single dose of UV is almost equally effective at any stage during a long period of the cycle including the S and part of G2 phase. Ionizing radiation, on the other hand, acts most efficiently during mitosis (M-1), which precedes the delayed mitosis (M) by one complete cycle and is followed immediately by the onset of the S period. A second maximum of radiosensitivity is observed at the end of the G2 period if high doses (10 kR) are applied.

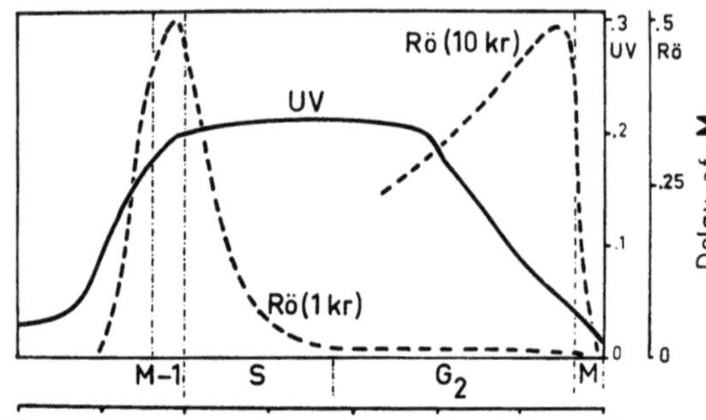

Fig. 2. Variation of radiosensitivity during synchronous nuclear division cycle. Plasmodial segments (2-6 cm^2) were irradiated at various stages of mitotic cycle in absence of medium with UV light (2300 ergs/ mm^2) or x-rays (1 kr or 10 kr) and subsequently returned to fresh medium. Abscissa: time of irradiation. Ordinate: delay of mitosis (M) relative to nonirradiated control plasmodia. Mitosis (M-1) precedes delayed mitosis (M) by one complete cycle. Length of normal mitotic cycle was 8 to 10 h (one unit on time scale). (Sachsenmaier et al., 1970a,b)

The mechanisms leading to mitotic delay by either type of radiation probably involve primary damage of nuclear DNA. This is suggested by observations that the retarding effect of UV light on mitosis may be enhanced by prior incorporation into DNA of 5-bromo-2'-deoxyuridine or partially reversed by visible light (Sachsenmaier, 1966; Rupff, 1967). Although irradiation with UV light during the S phase decreases DNA synthesis, this effect cannot explain the extension of the radiosensitivity plateau into a substantial part of the G2 period. Interference with the transcription and/or translation of genetic messages essential for the onset of mitosis may therefore be more likely causes for UV-induced division delay.

Ionizing radiation induces DNA single-strand breaks as primary lesions which in turn may interfere with the transcription of division specific messages. This would explain why the peak of radiosensitivity immediately precedes the onset of DNA replication. Single-strand breaks induced during the G2 period are rapidly repaired in *Physarum* plasmodia (Brewer and Nygaard, 1972). Similar lesions induced during the S phase, however, may be "fixed" as the replicating fork passes the gap before it has had a chance to be repaired. The position of the sensitivity maximum suggests that early replicating genes are more important for the timing of the next mitosis than late replicating genes; this follows from the sharp drop of radiosensitivity during the first half of the S period.

The second sensitivity maximum for ionizing radiation at the end of the G2 period probably reflects a different action mechanism leading to mitotic delay. The slope of the sensitivity curve suggests that a slowly repaired primary radiation damage (e.g., DNA double-strand breaks) or a "setback" process analogous to that proposed for temperature shock-induced division delay (Zeuthen and Rasmussen, 1972), is involved.

II. Permanent Phase Shifts of the Mitotic Cycle by UV Irradiation

The time course of consecutive nuclear divisions following the first (delayed) mitosis after UV irradiation exhibits a peculiar pattern (Fig. 3). The second and third postirradiation cycles are considerably shorter than normal. This advancement of the division schedule not only compensates for the initial delay of the first mitosis but actually causes subsequent mitoses to occur earlier than the corresponding normal mitoses of the control plasmodium. This phenomenon most likely is related to the partial degradation of irradiated nuclei observed during the first S period after UV treatment (Sachsenmaier et al., 1970b). According to a model of division control discussed elsewhere (Sachsenmaier et al., 1972), the timing of mitosis in multinuclear plasmodia is controlled by a critical ratio of cytoplasm/nuclei or mass/DNA (Fig. 4). This ratio steadily increases between divisions and drops again as the nuclei divide and replicate their DNA. UV irradiation reduces the number of viable nuclei after the first postirradiation mitosis without concomitantly decreasing the overall growth rate; thus the critical DNA mass ratio is reached earlier than normal. The original balance between growth and division is approached during one or two shortened intervals following the first delayed mitosis, resulting, however, in a permanent negative phase shift of the mitotic schedule.

Plasmodia treated with ionizing radiation do not exhibit this effect. DNA synthesis and the number of nuclei are not changed significantly by radiation doses in the order of 1 kR (Sachsenmaier et al., 1970a).

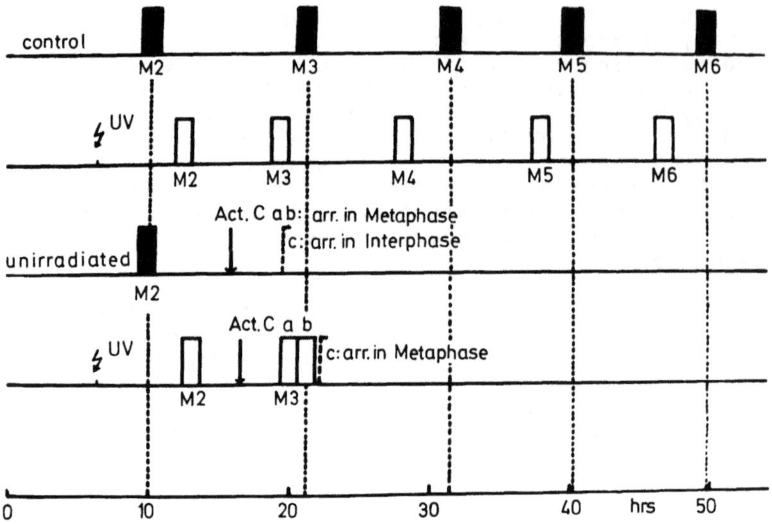

Fig. 3. Timing of consecutive mitoses in UV-treated plasmodia; effects of actinomycin C. Macroplasmodium (Ø 4 cm) was dissected into two equal parts 3-5 h prior to mitosis (M2). One half was irradiated immediately with UV light (7500 ergs/mm^2). At the time indicated by the straight arrow (↓), small pieces (~ 2 cm^2) of each moiety were transferred onto medium containing actinomycin C (a: 100 µg/ml; b: 250 µg/ml; c: 500 µg/ml). Control and experimental media were changed every 12 h. Black and white bars indicate time of complete nuclear division in unirradiated and irradiated plasmodia, respectively. In normal plasmodia treated with 100 and 250 µg/ml actinomycin C (a, b), and in UV-irradiated plasmodium, treated with 500 µg/ml actinomycin C, most nuclei were arrested in metaphase after entering mitosis at the time indicated by (⊤). Nuclei of unirradiated plasmodium, treated with 500 µg/ml actinomycin C, were arrested in interphase and exhibited condensed chromatin granules and diminished nucleoli

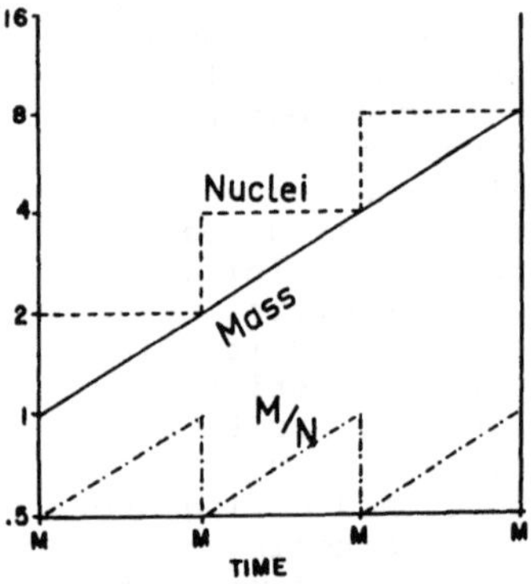

Fig. 4. Schematic diagram of change of total mass (——) and of number of nuclei (- - -) in exponentially growing synchronous plasmodium. Ratio of total mass per nuclei (-.-.-). M: synchronous nuclear mitoses. (Sachsenmaier et al., 1972)

Nuclear divisions following the first delayed postirradiation mitosis tend to "catch up" but do not overtake the corresponding control mitoses (Scholze, 1965). Again this would be expected if the timing mechanism of nuclear division is normally linked to the DNA/mass ratio and becomes transiently perturbed by ionizing radiation; the original rhythm is approached during postirradiation cycles as the system reverts to normal.

III. Radiation-Induced Resistance to Actinomycin C

Preparation for mitosis requires synthesis of RNA throughout the major part of the division cycle in *Physarum* as well as in most other proliferating cell systems. The onset or progress of nuclear mitosis in normal synchronous *Physarum* plasmodia is blocked by actinomycin C for up to about one hour prior to the expected mitosis (Sachsenmaier and Becker, 1965; Sachsenmaier et al., 1967). Interestingly, plasmodia treated with UV or ionizing radiation become partially resistant to this drug (Figures 3 and 10). Substance concentrations of 100 or 250 mg/ml added to the medium of an unirradiated plasmodium during mid-interphase cause mitotic nuclei to be arrested in metaphase, whereas 500 mg/ml prevent nuclei from entering prophase. UV-irradiated cultures, however, undergo normal mitosis in the presence of 100 and 250 mg/ml actinomycin C, and the highest concentration tested (500 mg/ml) only blocks nuclei in metaphase. Table 1 indicates that RNA synthesis, as measured by pulse labeling with ^{14}C-uridine, also become less sensitive to actinomycin C after UV treatment. This explains easily the loss of antimitotic activity of the drug and discounts the supposition (Devi and Guttes, 1972) that the onset of postirradiation mitoses in UV-treated *Physarum* plasmodia no longer depends on the prior synthesis of new RNA. Ionizing radiation similarly decreases the sensitivity of RNA synthesis and nuclear division to actinomycin C (Fig. 10). The reason for the development of actinomycin resistance after irradiation with UV or x-rays is not yet clear. Perhaps, the binding capacity of DNA for the antibiotic is reduced during DNA repair synthesis.

Table 1. Effect of actinomycin C on RNA synthesis in normal and UV-irradiated plasmodia

	Actinomycin C	Specific radioactivity of RNA			
		cpm/mg RNA		% of control	
	µg/ml	exp. A	exp. B	exp. A	exp. B
Control	0	10.1	9.9	100	100
	100	1.9		19	
	250	1.3		13	
	500		1.3		13
UV irradiation	0	8.6	5.6	100	100
	100	4.6		54	
	250	5.8	6.2	68	110
	500		3.4		61

Macroplasmodium (Ø = 4 cm) was divided into two equal parts 3.5 h prior to mitosis (II) and one segment was irradiated immediately with UV light (7500 ergs/mm^2). Small pieces (2 cm^2) of each moiety were transferred onto medium containing various concentrations of actinomycin C 6 h (exp. A) or 7 h (exp. B) after control mitosis (II). RNA synthesis was measured by pulse labeling with ^{14}C-uridine 30 min after exposure to actinomycin C as described under Methods

IV. Radiation Effects on Enzyme Regulation

Certain enzymes are produced discontinuously during the cell cycle in various systems (Mitchison, 1971). Figures 5 and 9 demonstrate the periodic enzyme activity patterns of thymidine kinase and deoxycytidine kinase in synchronous *Physarum* plasmodia. These two enzymes are induced at the end of the G2 period and exhibit a distinct maximum in

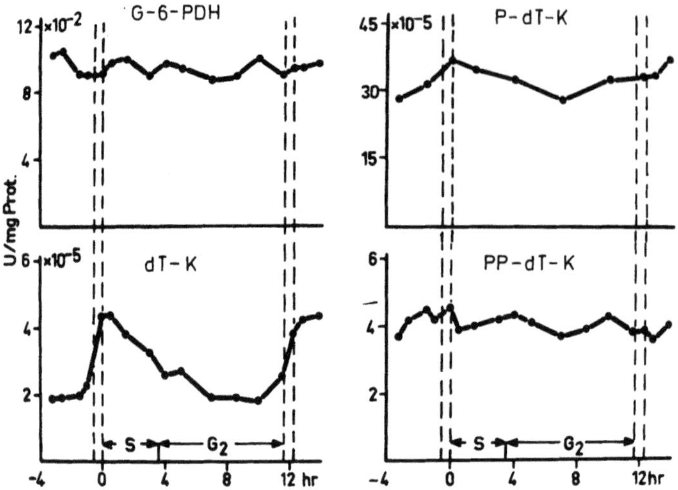

Fig. 5. Specific enzyme activities in high-speed supernatants of plasmodial homogenates prepaired at various points during mitotic cycle. G-6-PDH: glucose-6-phosphate dehydrogenase; dT-K: thymidine kinase; P-dT-K: thymidine-monophosphate kinase; PP-dT-K: thymidine-diphosphate kinase

early S phase. Other enzymes, including those catalyzing subsequent steps of the thymidine triphosphate biosynthetic pathway, exhibit constant specific activities throughout the cycle, suggesting that these enzymes are constitutively produced parallel to the bulk of soluble proteins. Irradiation with UV light or x-rays under conditions that cause a delay of nuclear mitosis also delays the appearance of the enzyme activity maximum of thymidine kinase (Fig. 6). This suggests that the induction of certain "step" enzymes, like thymidine kinase, is functionally linked with the initiation of mitosis or with the onset of DNA replication, which immediately follows mitosis in *Physarum*. Perhaps the close temporal association of enzyme induction, mitosis, and onset of DNA replication results from a common trigger event acting at the end of the G2 period. This event may be identical with or related to the phosphorylation of H1-histones. The phosphate content of H1-histones reaches a maximum at exactly the same time that the enzyme activities of the pyrimidine deoxynucleoside kinases start to increase (Bradbury et al., 1973).

Ionizing radiation strikingly enhances the increase of thymidine kinase (Fig. 7) and deoxycytidine kinase (Fig. 9) during the delayed induction period.

This stimulatory effect is only seen, however, if irradiation is administered during the same narrow period of the cycle, i.e., mitosis (M-1), when x-rays induce a maximum delay of the following mitosis (M) (Fig. 8). When irradiation is administered during other parts of the cycle, even if mitotic delays are induced with a high dose (10 kR), no excess production of thymidine kinase occurs (Fig. 7). This clearly indicates that "superinduction" of thymidine kinase is not a mere consequence of mitotic delay. The coincidence of the most sensitive periods for the effects on mitosis and enzyme production (Fig. 8), however, points to a common mechanism responsible for both effects; in particular, the findings would agree with the assumption that unrepaired DNA lesions that are derived from single-strand breaks by

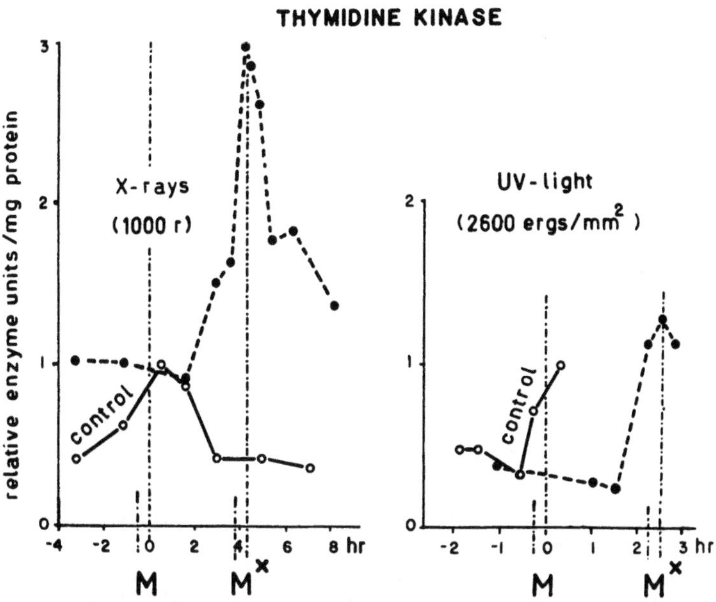

Fig. 6. Specific activity of thymidine kinase in controls (——) and in plasmodia treated with x-rays or UV light (----). Irradiation with x-rays (1 kR) occurred during the preceding mitosis (M-1); UV light (2600 ergs/mm^2) was applied 5 h prior to mitosis (M). (Sachsenmaier and Hansen, 1973)

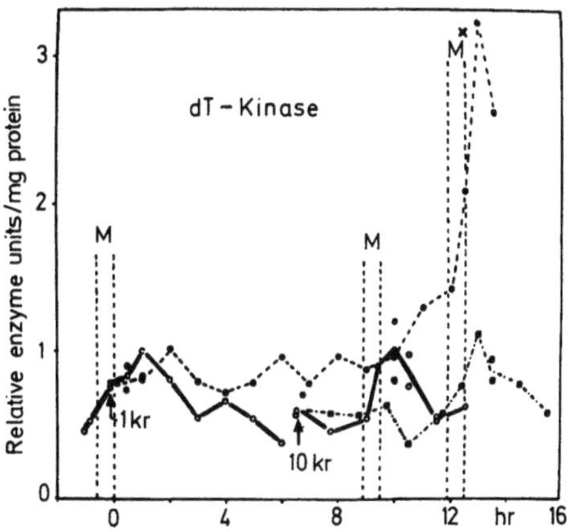

Fig. 7. Effect of x-irradiation on cycle-dependent fluctuation of thymidine kinase. 6 replicate macroplasmodia were divided into halves at time of mitosis (III) and one moiety was irradiated immediately with 1 kR. A different set of cultures was treated similarly with 10 kR 2 h prior to expected onset of mitosis. Quarter segments (~ 5 cm^2) were harvested at various time intervals, homogenized, and assayed for thymidine kinase activity in high-speed supernatant. Abscissa: time of harvesting. Ordinate: relative specific enzyme activity. Controls (——) irradiated with 1 kR (----) and 10 kR (-.-.-), respectively; mitosis in controls (M) and irradiated (M^x) plasmodia (Sachsenmaier et al., 1970a)

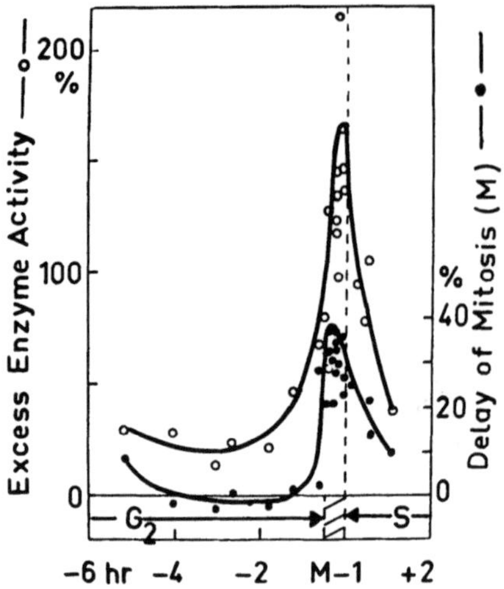

Fig. 8. Correlation of x-ray-induced delay of mitosis and stimulation of thymidine kinase production. Individual plasmodia from series of replicate synchronous cultures were irradiated at various points of nuclear division cycle indicated on abscissa. Delay of mitosis (M), which followed mitosis (M-1), was observed compared to unirradiated controls, and thymidine kinase activity was measured at expected peak within 30 min after telophase. Right ordinate indicates delay of mitosis (M) in % of normal cycle length; left ordinate indicates excess enzyme activity in % over control peak (Sachsenmaier et al., 1973)

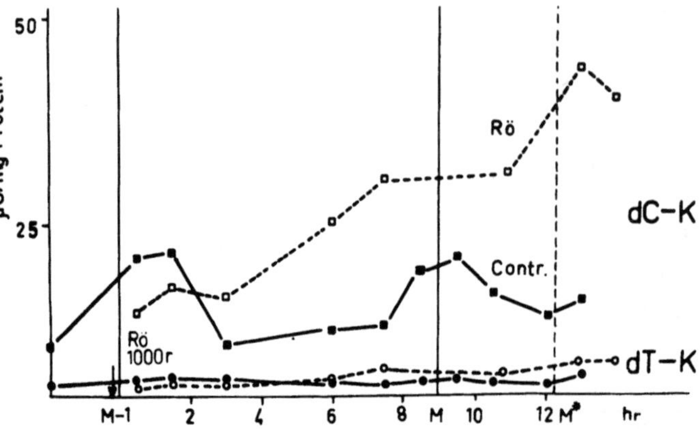

Fig. 9. Effect of x-rays on cycle dependent variation of deoxycytidine kinase activity in BUdR-resistant mutant. Experimental design was similar to that in Fig. 7. Deoxycytidine kinase (dC-K) is largely stimulated by 1 kR x-irradiation whereas thymidine kinase (dT-K), which is almost completely absent in mutant, cannot be induced by treatment. BUdR-resistant mutant (#84) was kindly provided by F. Haugli, Tromsö

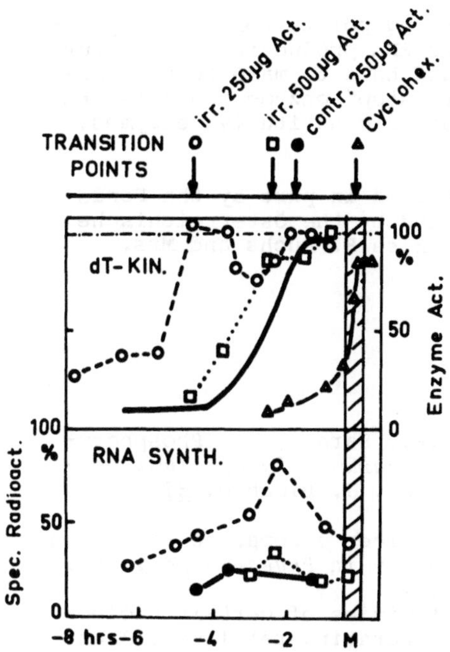

Fig. 10. Effect of actinomycin C on RNA synthesis, thymidine kinase production, and nuclear mitosis following x-irradiation. Synchronous macroplasmodia were irradiated with 1 kR during metaphase (M-1), i.e., 10.5 h prior to control mitosis (M). Plasmodial segments (~ 2 cm^2) were dissected from normal and irradiated cultures and transferred onto medium containing 250 or 500 µg/ml actinomycin C at various times indicated on abscissa. One set of subcultures was pulse-labeled with ^{14}C-uridine as described in Methods. Remaining subcultures were harvested 30 min after mitosis of corresponding normal and irradiated controls and analyzed for thymidine kinase activity. Ordinate of lower frame indicates specific radioactivity of RNA of inhibitor-treated cultures relative to controls (= 100%), which were not treated with actinomycin C. Ordinate on upper frame indicates relative specific enzyme activity of inhibitor-treated plasmodia compared to untreated controls (= 100%). Mitosis of irradiated cultures was delayed by 2.5 to 3 h. Unirradiated plasmodia treated with 250 µg/ml actinomycin C (——), irradiated plasmodia treated with 250 µg/ml (---), or 500 µg/ml (....) actinomycin C. For comparison, one set of unirradiated plasmodial segments was treated with 20 µg/ml cycloheximide in order to compare effect of this protein inhibitor on enzyme production (Δ). Arrows on top indicate transition points in cycle for inhibitory action on mitosis of actinomycin C and cycloheximide in normal and irradiated plasmodia. Results indicate that pretreatment with ionizing radiation diminishes the inhibitory action of 250 µg/ml actinomycin C on RNA synthesis, enzyme production, as well as on mitosis; increasing drug concentration to 500 µg/ml overcomes radiation-induced resistance to actinomycin C (Linser, 1975)

"fixation" during DNA synthesis are involved. These unrepaired lesions may interfere with the formation of a repressor that normally restricts the amount of enzyme formed during the induction period. Such a repressor could function at the transcription or translation level. Damaging the template of this repressor would remove the partial block and permit unrestricted enzyme production. Alternatively, unrepaired

DNA lesions could provide additional starting points for RNA polymerase, which increases the rate of messenger RNA production and in turn of enzyme synthesis. In this case one would have to make further assumptions to explain why transcription of "step" enzyme genes is preferentially stimulated as compared to genes of constitutive enzymes.

Acknowledgements. These studies were supported in part by the Fonds Österreichische Krebsforschungsinstitute and by the Österreichische Nationalbank. The technical assistance of Miss E. Fuchs and Mrs. E. Höllwarth is greatly appreciated.

References

Bradbury, E.M., Inglis, R.J., Matthews, H.R., Sarner, N.: Phosphorylation of very lysine rich histone in *Physarum polycephalum*: correlation with chromosome condensation. Eur. J. Biochem. 33, 131-139 (1973)

Brewer, E.N., Nygaard, O.F.: Correlation between unrepaired radiation induced DNA strand breaks and mitotic delay in *Physarum polycephalum*. Nature New Biol. 239, 108-110 (1972)

Chiga, M., Oda, A., Holtzer, R.L.: The activities of certain nucleoside diphosphokinases of normal and regenerating rat liver. Arch. Biochem. Biophys. 103, 366 (1963)

Coutsogeorgopoulos, C., Hacker, B., Mantsavino, R.: On the formation of thymidine-3'-triphosphate by a mammalian cell-free system. Biochem. Biophys. Res. Comm. 20, 129 (1965)

Daniel, J.W., Baldwin, H.H.: Methods of culture for plasmodial myxomycetes. In: Methods in Cell Physiology (Prescott, D.M., ed.). New York: Academic Press, 1964, Vol. 1, pp. 9-41

Devi, V.R., Guttes, E.: Macromolecular syntheses and mitoses in UV-irradiated plasmodia of *Physarum polycephalum*. Radiat. Res. 51, 410-430 (1972)

Guttes, E., Guttes, S., Rusch, H.P.: The morphology and development of *Physarum polycephalum* grown in pure culture. Develop. Biol. 3, 588-614 (1961)

Howard, F.L.: Nuclear division in plasmodia of *Physarum*. Ann. Bot. 46, 461-478 (1932)

Linser, W.: Untersuchungen zur Regulation Pyrimidindesoxyribonucleosid-phosphorylierender Enzyme im synchronen Mitosecyclus von *Physarum polycephalum*. Ph.D. Thesis, University of Innsbruck, 1975

Mitchison, J.M.: The Biology of the Cell Cycle. Cambridge University Press, 1971

Rupff, H.: Untersuchungen über die mitosehemmende Wirkung von UV-Strahlung an natürlich-synchronen Plasmodien von *Physarum polycephalum*. M.D. Thesis, University of Heidelberg, 1967

Sachsenmaier, W.: Analyse des Zellcyclus durch Eingriffe in die Makromolekül - Biosynthese. In: Probleme der biologischen Reduplikation (Sitte, P., ed.). Berlin-Heidelberg-New York: Springer, 1966, pp. 139-160

Sachsenmaier, W., Becker, J.E.: Wirkung von Actinomycin D auf die RNS-Synthese und die synchrone Mitosetätigkeit in *Physarum polycephalum*. Monatshefte Chem. 96, 754-765 (1965)

Sachsenmaier, W., Bohnert, E., Clausnizer, B., Nygaard, O.F.: Cycle dependent variation of x-ray effects on synchronous mitosis and thymidine kinase induction in *Physarum polycephalum*. FEBS Letters 10, 185-189 (1970a)

Sachsenmaier, W., Dönges, K.H., Rupff, H., Czihak, G.: Advanced initiation of synchronous mitoses in *Physarum polycephalum* following UV-irradiation. Z. Naturforsch. 25b, 866-871 (1970b)

Sachsenmaier, W., Finkenstedt, G., Linser, W., Madreiter, H., Woertz, G., Wolf, H.: Regulation of thymidine phosphorylating enzymes in the synchronous nuclear division cycle of *Physarum polycephalum*. In: Proc. 2nd Eur. Symp. on the Cell Cycle. Veröffentlichungen der Universität Innsbruck, Vol. 77, 36-43 (1973)

Sachsenmaier, W., Fournier, D., von, Gürtler, K.F.: Periodic thymidine kinase production in synchronous plasmodia of *Physarum polycephalum*: inhibition by actinomycin and actidion. Biochem. Biophys. Res. Comm. 27, 655-660 (1967)

Sachsenmaier, W., Hansen, K.: Long- and short-period oscillations in a myxomycete with synchronous nuclear divisions. In: Biological and Biochemical Oscillators (Chance, B., Ghosh, A.K., Pye, K.E., Hess, B., eds.). New York-London: Academic Press, 1973, pp. 429-447

Sachsenmaier, W., Ives, D.H.: Periodische Änderungen der Thymidinkinase-Aktivität im synchronen Mitosecyclus von *Physarum polycephalum*. Biochem. Z. 343, 399-406 (1965)

Sachsenmaier, W., Remy, U., Plattner-Schobel, R.: Initiation of synchronous mitosis in *Physarum polycephalum*; a model of the control of cell division in eukaryots. Exp. Cell Res. 73, 41-48 (1972)

Sachsenmaier, W., Rusch, H.P.: The effect of 5-fluoro-2'-deoxyuridine on synchronous mitosis in *Physarum polycephalum*. Exp. Cell Res. 36, 124-133 (1964)

Scholze, H.U.: Untersuchungen über den molekularen Mechanismus der synchronen Kernteilung bei *Physarum polycephalum*. M.D. Thesis, University of Heidelberg, 1965

Zeuthen, E., Rasmussen, L.: Synchronized cell division in protozoa. In: Research in Protozoology (Chen, T.T., ed.). Oxford: Pergamon Press, 1972, Vol. 4, pp. 9-145

2. The Molecular Basis for the Control of Cell Division

R. J. INGLIS, H. R. MATTHEWS, and E. M. BRADBURY

A. Introduction

The plasmodial stage of the life cycle of the slime mould *Physarum polycephalum* has a unique feature that makes it a model system for the study of the processes involved in the initiation and control of mitosis. The 10^8 diploid nuclei in a 7-cm diameter plasmodium divide within a few minutes of each other. This natural synchrony is persistent, with an intermitotic time of about 9 h in our laboratory. The mitotic stages are easily seen in the phase contrast microscope and take about 30 min from early prophase to the end of telophase. Using this natural mitotic synchrony it has been possible to investigate changes in molecular structure and concentration through the cycle with a precision that is lacking in other artificially synchronized eukaryotic systems (see review by Mohberg, 1974).

B. Proposed Scheme of Events Involved in the Initiation and Control of Mitosis

We had paid particular attention to changes in histones through the cycle in order to correlate these changes with changes in chromosome structure and function. The following proposal has recently been made (Bradbury et al., 1974a) on the molecular events involved in the initiation and control of mitosis: that growth-associated histone H1 phosphokinase activity increases, leading to an increase in histone H1 phosphate content, which initiates chromosome condensation and mitosis.

C. Summary of Some of the Evidence for the Proposal

Measurement of the històne H1 phosphokinase activity of nuclei isolated at defined stages of the mitotic cycle has shown that the activity increases steadily through the cycle from a minimum near metaphase to a maximum in late G2 phase, 1.75 h before the next metaphase. After this time it falls rapidly to a minimum near metaphase (Bradbury et al., 1974a). The peak in H1 phosphokinase activity precedes the maximum level of H1 phosphate content in *P. polycephalum* as might be expected if the change in the phosphate content was controlled by the change in enzyme activity (Fig. 1). This peak of phosphate content of H1 coincides exactly with the onset of visible chromosome condensation seen by phase contrast microscopy (Bradbury et al., 1973b). These data have gained support from work on mammalian systems, which have shown, for example, that growth-associated histone H1 phosphokinase (HKG) activity is greatly enhanced in cells arrested in metaphase (Lake, 1973; Lake and Salzman, 1972), H1 phosphate content is high

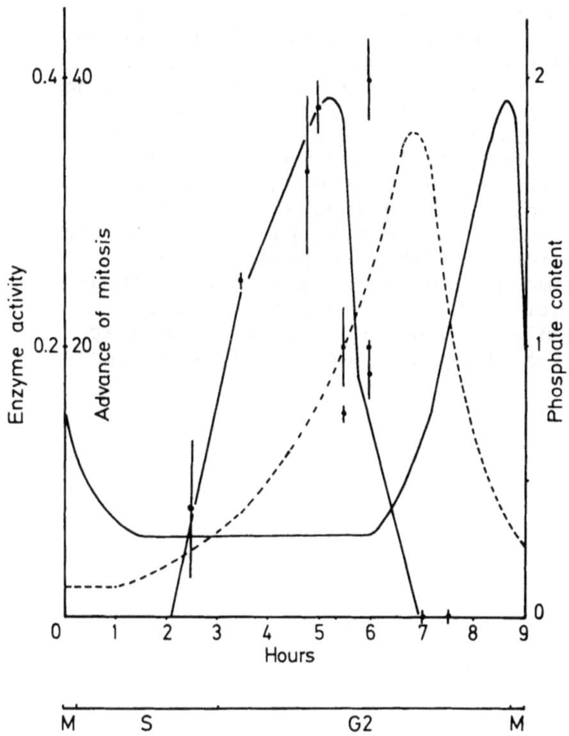

Fig. 1. Advance of mitosis —•—, endogenous H1 phosphokinase activity ----, and H1 phosphate content ——, in the mitotic cycle of *P. Polycephalum* (reproduced with the kind permission of the editor of Nature)

in metaphase chromosomes (Lake and Salzman, 1972) and in late G2 and mitotic cells (Gurley et al., 1973), and by the correlation between H1 phosphate content and growth rate in a variety of tumor cells (Balhorn et al., 1972). It has also been shown that addition of an HKG preparation from Ehrlich ascites cells to growing plasmodia of *P. polycephalum* causes mitosis to occur sooner (Bradbury et al., 1974b). Providing the enzyme was added at the correct time in the cycle, the enzyme-treated plasmodia entered mitosis up to 40 min before a control. This observed behavior agrees well with the time course of endogenous enzyme activity and phosphorylation of H1 (see Fig. 1) and provides strong support for the contention that the observed effect is due to a specific action of H1 phosphokinase.

However, with the availability of a more purified enzyme preparation, additional studies were carried out to confirm the earlier conclusions. The results of these and other experiments are described below.

D. Specificity of Histone H1 Phosphokinase (HKG)

An additional purification step was used during the preparation of the HKG used in the following experiments. HKG was prepared as described previously (Bradbury et al., 1974b) and dissolved in 0.025 *M* phosphate buffer, pH 7.5, and adsorbed onto calcium phosphate gel. The gel was washed with 0.1 *M* phosphate buffer, and the enzyme activity was then eluted with 0.2 *M* phosphate buffer. This eluent is called 'step 2 HKG'.

The specificity of HKG was measured by incubating freshly prepared calf thymus chromatin, γ-^{32}P-ATP, and the step 2 enzyme in the presence of Mg^{++}. The histones were extracted and the distribution of radioactivity monitored by analysis on 15% acrylamide gels containing urea, as described previously (Bradbury et al., 1973b). (^{32}P) was found to be associated only with H1 histone (Fig. 2). A control experiment contained chromatin, γ-^{32}P-ATP, Mg^{++}, and no enzyme. The presence of Mg^{++} is essential for the action of the enzyme but results in precipitation of the chromatin in the incubation mixture. The inaccessibility of the H1 molecules in the precipitated chromatin is reflected by the low incorporation of (^{32}P) (one phosphate per 5000 molecules), whereas at the same enzyme and substrate concentrations using free H1, the incorporation of (^{32}P) is substantially higher (one phosphate per 55 molecules). Under conditions of higher enzyme/substrate ratio up to 2 moles phosphate per mole free H1 can be incorporated.

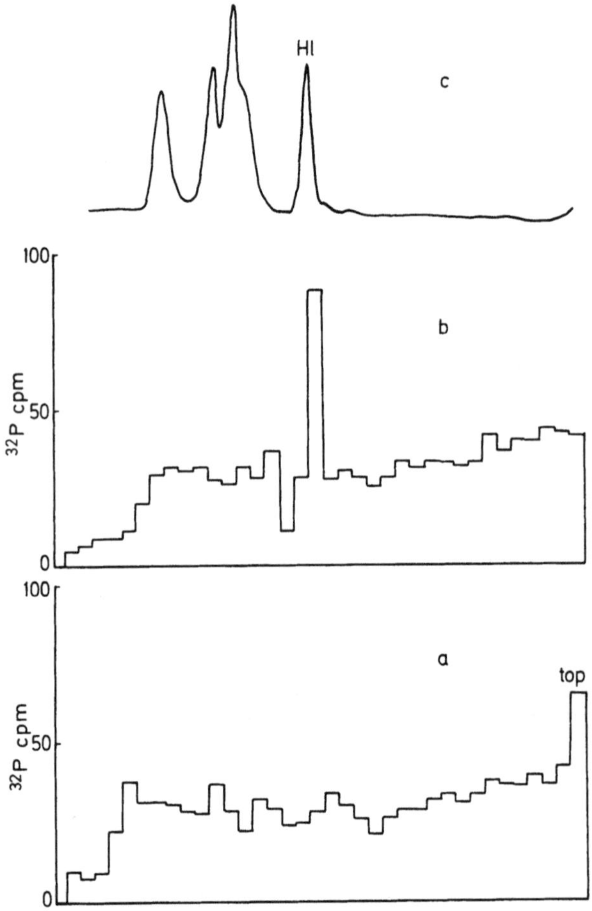

Fig. 2 a-c. Acrylamide gel electrophoresis of histones extracted from calf thymus chromatin incubated with γ-^{32}P-ATP (a) incubated without step 2 HKG; (b) incubated with HKG; (c) stained gel, same with or without HKG (reproduced with the kind permission of the editor of Exptl.Cell Res.)

E. Advancement of Mitosis

The effect of the step 2 HKG enzyme on advancing mitosis was measured as described previously (Bradbury et al., 1974b). Synchronous plasmodia were prepared as before by fusion of microplasmodia on filter paper in petri dishes (Bradbury et al., 1973b) and a plasmodium 3 h before the third mitosis after fusion was cut into quarters and each quarter treated with 0.1 ml, 0.02 *M* Tris buffer containing 0, 25, 50, and 100% of the total HKG activity available. The time of the subsequent (third) metaphase was monitored by phase contrast microscopy. Metaphases occurred in order, the quarter plasmodium treated with 100% enzyme dividing first and the quarter treated with 0% enzyme dividing last (Fig. 3).

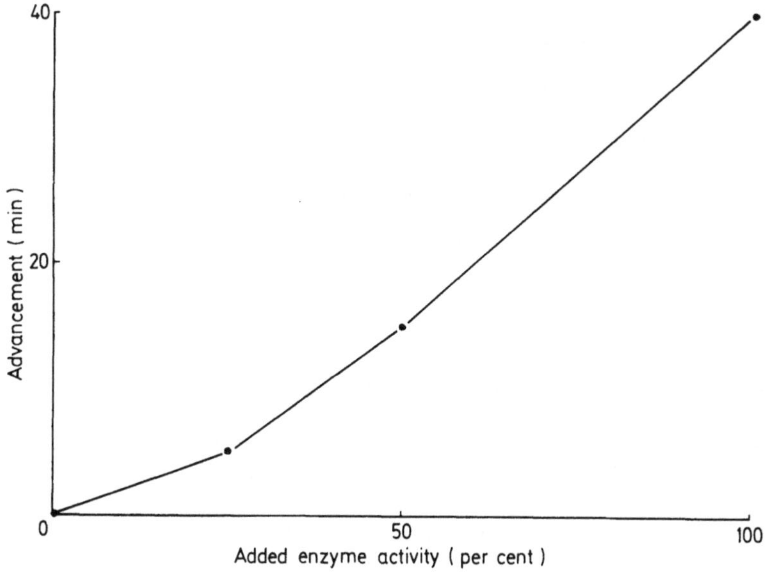

Fig. 3. Advancement of mitosis as function of concentration of HKG added (reproduced with the kind permission of the editor of Exptl.Cell Res.)

In the remaining experiments at 3 h before the third metaphase after fusion as before, the plasmodium was cut into quarters, two quarters were treated with enzyme, either step 1 or step 2 HKG, and the remaining two quarters were control solutions (treated with 0.02 *M* Tris buffer, pH 7.2). In previous experiments control solutions included albumin, creatine phosphokinase, step 1 HKG inactivated by freezing and thawing, or highly diluted (Bradbury et al., 1974b). The time difference between identical controls was normally not more than 5 min. These experiments are summarized by histograms in Figure 4. This shows the effect of step 1 and step 2 HKG on advancing mitosis. Each experiment was done in duplicate. The results show that both HKG enzyme extracts have a positive effect on advancing mitosis, but the degree of advancement was variable. The reason for this variation in the values observed is not clear, but may reflect either a variable uptake of the enzyme by the plasmodium since the method of application was by pipetting onto the plasmodial surface, or addition of different amounts of enzyme activity since a variable amount

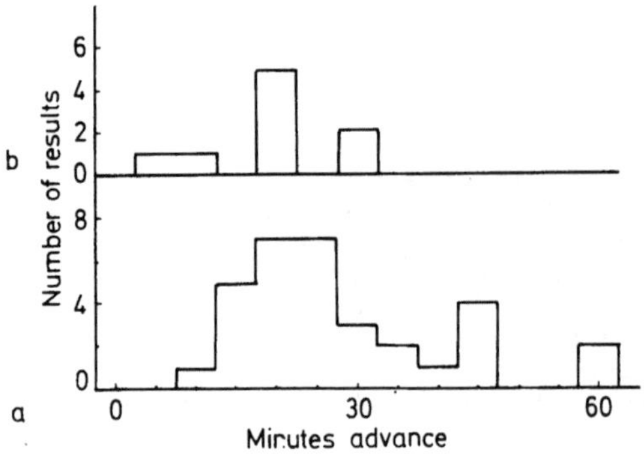

Fig. 4 a and b. Histograms showing number of times any particular change in time of mitosis occurred due to adding HKG. (a) observations using step 1 HKG; (b) observations using step 2 HKG

of inactivation occurs in the dialysis step that precedes addition of the enzyme. It should be added that the amount of HKG added was many times that of the level of endogenous HKG.

The mean advance for HKG step 1 was 28 ± 3.1 min, 32 measurements, and for HKG step 2 was 19.4 ± 2.4 min, 9 measurements. The probability of the results occurring by chance was calculated by using the paired t test and was found to be less than 1 in 1000 for both HKG step 1 and step 2 enzymes. The advanced mitosis was not affected in any way in terms of morphology of the mitotic figures, or duration of the mitotic stages. These results confirm our earlier work (Bradbury et al., 1974b) that the only effect of these extracts appears to be a stimulation of mitosis, but until a pure HKG preparation is available, there is no proof that the active agent is HKG.

No advance of mitosis occurred when the enzyme was deliberately inactivated or diluted, or when the time of enzyme addition was later in the cycle than the peak of endogenous enzyme activity. On one occasion when both HKG step 1 and step 2 enzymes were used in simultaneous experiments no advance occurred under conditions that produced an advance on other occasions. These experiments have been omitted from the statistical calculations but even when included, the probability of the results occurring by chance was less than 1 in 1000.

F. DNA Synthesis in Enzyme-Treated Plasmodia

In *P. polycephalum* there is no G1 phase, and DNA synthesis starts immediately after mitosis. We have asked the following question: does DNA synthesis start immediately after a mitosis that has been advanced by the addition of HKG? A double label experiment was carried out in which a plasmodium was labelled with methyl-^{14}C-thymidine continuously from fusion. At 3 h before the third mitosis after fusion, the plasmodium was cut in half, one half being treated with enzyme and the other with 0.02 *M* Tris buffer, pH 7.2. Each half was then cut into 3 equal segments. One enzyme and one control segment was pulse-labelled with methyl-^{3}H-thymidine for a 20-min period beginning 20 min before metaphase. The plasmodium was harvested and the DNA extracted and purified by hydroxylapatite (HAP) binding. The remaining segments were pulse-labelled for 20-min periods beginning 10 min after metaphase, and 60 min after metaphase, and the DNA isolated by HAP binding. The HKG-treated segments entered mitosis 20 min before control segments. The results, given in Table 1, show a small (^{3}H) incorporation before metaphase but a large incorporation after metaphase in both treated and control plasmodia, consistent with the occurrence of DNA synthesis at this time.

Table 1. DNA synthesis in enzyme-treated plasmodia

	Radioactivity in DNA			Time of pulse	
	(^{14}C) dpm (continuous)	(^{3}H) dpm (pulse)	(^{3}H)/(^{14}C)	min after metaphase begin	end
Control	1019	1396	1.4	-20	0
Treated	1426	2561	1.8	-20	0
Control	2374	34294	14	10	30
Treated	3468	42984	12	10	30
Control	3125	49282	16	60	80
Treated	3030	52559	17	60	80

G. DNA/Histone Interaction

H1 has been implicated in chromosome condensation for a variety of reasons. When a chromatin gel is dialyzed against increasing salt molarities, the gel contracts to about 10% of its original volume, the maximum contraction occurring at 0.15 *M* NaCl. This phenomenon is not observed with H1-depleted chromatin. Nuclear magnetic resonance studies (Bradbury et al., 1973a) have shown that in H_2O, H1 molecules are tightly bound to chromatin. On increasing the salt concentration to 0.15 *M*, certain apolar regions of the H1 molecule become free, while the regions rich in lysine residues remain firmly bound. At higher salt concentrations H1 molecules are progressively dissociated from chromatin, with an expansion of the gel to its original volume. Similar behaviour is observed for the contraction of H1/DNA complexes.

Evidence connecting H1 phosphate content and chromosome structure is sparse. The correlation between the increased H1 phosphate content and chromosome condensation is purely temporal in the observations on *P. polycephalum* (Bradbury et al., 1973b). Lake and Salzman (1972), however, found that in metaphase cells, in which H1 phosphate is high, the H1 molecules are less strongly bound than in interphase cells, when the H1 phosphate is low. Recently, by measuring changes in the turbidity of DNA and H1 solutions, it has been possible to follow precisely the interaction of DNA and H1 molecules at different salt concentrations, using phosphorylated and non-phosphorylated H1. H1 from calf thymus was phosphorylated in vitro by using a growth-associated histone phosphokinase (HKG) from rat tumor cells and ATP. The H1 contained an average of 2.2 phosphates per molecule. The enzyme phosphorylates H1 in vitro mainly in the N and C terminal regions at sites other than those phosphorylated by HK1 (cyclic AMP-dependent) and HK2 (cyclic AMP-independent) (Langan, 1971). As can be seen in Figure 5, the salt-induced interaction, which results in increased turbidity, occurs at a lower salt concentration in phosphorylated H1/DNA mixtures; moreover some turbidity is already apparent before adding any NaCl. The maximum turbidity measurement occurs at about 0.15 *M* NaCl. When a solution of H1 is incrementally added to a DNA solution (initial concentration 40 μg/ml) in 0.12 *M* NaCl, the tur-

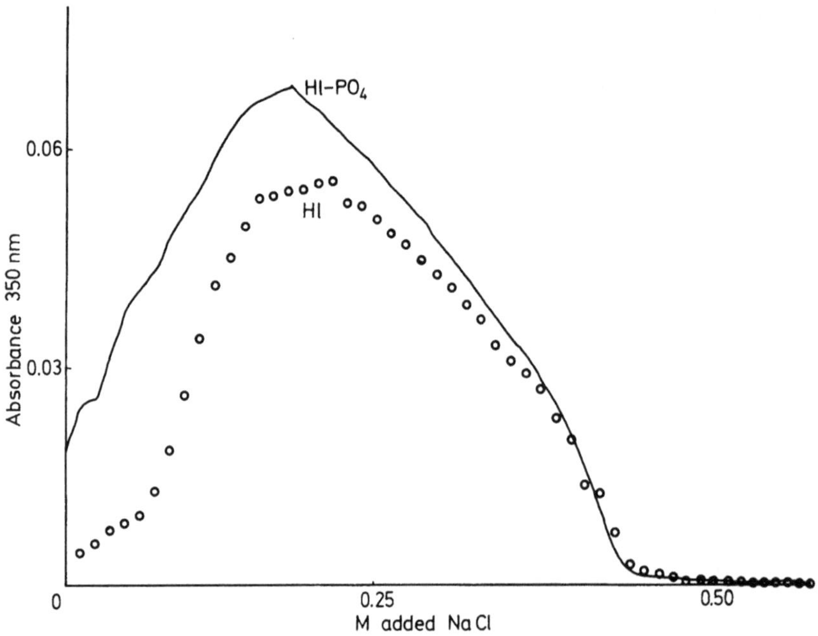

Fig. 5. Optical density at 350 nm as function of NaCl molarity for DNA concentration of 10 µg/ml. DNA histone ratio was 1:1 w/w

bidity increases. The effect of adding phosphorylated versus non-phosphorylated H1 is shown in Figure 6. It can be clearly seen that the aggregation of the DNA molecules is increased with phosphorylated H1 over most of the range of H1 concentration, the difference being most marked at the final concentration (H1 10 µg/ml; DNA 20 µg/ml). This is direct evidence that the interaction of DNA and H1 is substantially altered by the presence of phosphate groups, and furthermore that the in vitro cross-linking of DNA molecules by H1 is enhanced. We assume, from consideration of the physico-chemical forces involved, that this mechanism is dependent on a conformational change in the H1 molecule. It is not known at present whether the turbidity reflects cross-linking by H1:H1 or H1:DNA interactions. Experiments to clarify this situation are currently being conducted. The in vitro condensation of DNA and chromatin cannot, however, account for the coiling of DNA into the compact structure of the metaphase chromosome and is therefore suggested as the initial stage of such a process.

H. Conclusion

There is currently a wide range of evidence from mammalian and *P. polycephalum* (see Oppenheim and Katzir, 1971; Chin et al., 1972) systems consistent with the proposed molecular basis of control of cell division. However, most of the evidence is circumstantial, and direct evidence is not as yet available.

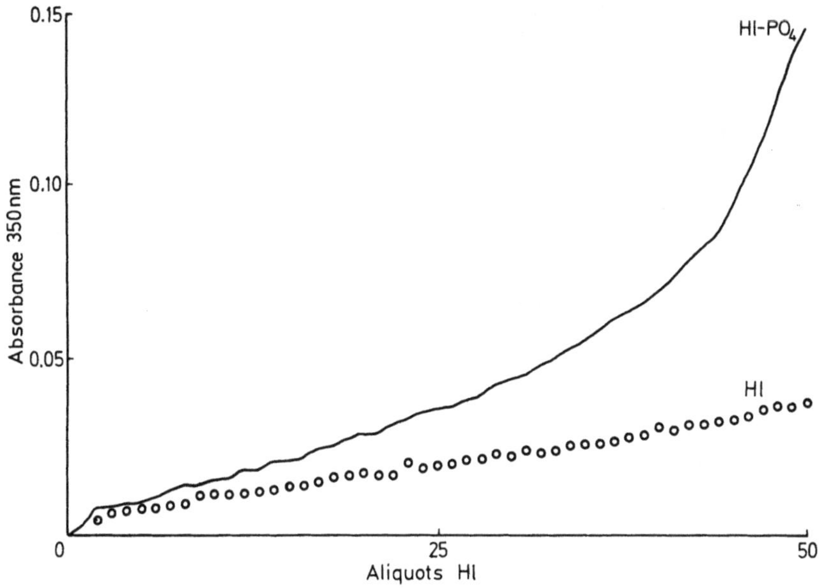

Fig. 6. Optical density at 350 nm as function of H1 concentration. Initial DNA concentration 40 µg/ml in 0.12 *M* NaCl. Final concentration DNA 20 µg/ml, H1 10 µg/ml

Acknowledgements. We are particularly grateful to Professor T.A. Langan, for a gift of the HKG enzymes, and to Mr. K. Simpson for a gift of calf thymus chromatin. The work was supported by research grants from the Cancer Research Campaign, the Science Research Council, and NATO.

References

Balhorn, R., Balhorn, M., Morris, H.P., Chalkley, R.: Comparative high resolution electrophoresis of tumour histones: variation in phosphorylation as a function of cell replication rate. Cancer Res. 32, 1775-1784 (1972)

Bradbury, E.M., Carpenter, B.G., Rattle, H.W.E.: Magnetic resonance studies of deoxyribonucleoprotein. Nature (Lond.) 241, 123-126 (1973a)

Bradbury, E.M., Inglis, R.J., Matthews, H.R.: Control of cell division by very lysine rich histone (F1) phosphorylation. Nature (Lond.) 247, 257-261 (1974a)

Bradbury, E.M., Inglis, R.J., Matthews, H.R., Langan, T.A.: Molecular basis of control of mitotic cell division in eukaryotes. Nature (Lond.) 249, 553-556 (1974b)

Bradbury, E.M., Inglis, R.J., Matthews, H.R., Sarner, N.: Phosphorylation of very lysine rich histone in *Physarum polycephalum*: correlation with chromosome condensation. Eur. J. Biochem. 33, 131-139 (1973b)

Chin, B., Friedrich, P.D., Bernstein, I.A.: Stimulation of mitosis following fusion of plasmodia in the myxomycete *Physarum polycephalum*. J. Gen. Microbiol. 71, 93-101 (1973)

Gurley, L.R., Walters, R.A., Tobey, R.A.: Histone phosphorylation in late interphase and mitosis. Biochem. Biophys. Res. Comm. 50, 744-750 (1973)
Lake, R.S.: Further characterization of the Fl histone phosphokinase of metaphase arrested animal cells. J. Cell. Biol. 58, 317-331 (1973)
Lake, R.S., Salzman, N.P.: Fl-histone modification at metaphase in Chinese hamster cells. Exp. Cell Res. 73, 113-121 (1972)
Langan, T.A.: Cyclic AMP and histone phosphorylation. Anal. N.Y. Acad. Sci. 185, 166-180 (1971)
Mohberg, J.: The nucleus of the plasmodial slime moulds. In: The Cell Nucleus (Busch, H., ed.). New York: Academic Press, 1974, Vol. 1, pp. 187-218
Oppenheim, A., Katzir, N.: Advancing the onset of mitosis by cell free preparations of *Physarum polycephalum*. Exp. Cell Res. 68, 224-226 (1971)

3. Cell Division and the Effects of Cell Age upon Radiation Sensitivity in Yeast

J. M. Parry, E. M. Parry, P. J. Davies, and W. E. Evans

A. Introduction

The sensitivities of both prokaryotic and eukaryotic organisms to chemical and physical mutagens depend to a large extent upon their intrinsic genetic capabilities for the repair of the induced DNA damage. Many repair deficient mutants have been isolated that show differences in their sensitivites to particular mutagens compared to their wild-type parental strains. For example, UV-sensitive strains of *Escherichia coli* (Howard-Flanders et al., 1966) have been shown to be defective in the process of excision repair whilst the x-ray-sensitive recombination-deficient mutants originally isolated by Clarke and Margulies (1965) have proved to be incapable of post-replication repair.

However, the processes leading to DNA repair can also be modified by changes in the cellular environment in cultures at different stages of the growth cycle, i.e., the physiological state of the cell may influence the expression of the different repair processes and thus affect the sensitivity of the cells to mutagens. For example, Hanawalt (1966) demonstrated that an *E. coli* strain TAU, which had completed the DNA replication cycle, showed increased UV resistance compared to UV-induced inactivation in an exponentially growing culture.

The term "cell age" has been used to describe two separate phenomena. Firstly, in synchronously dividing cultures cell age may refer to the position of a culture in the cell cycle. A second usage refers to the stage of a culture that has been allowed to grow from initial inoculation into fresh medium until all the nutrients have been exhausted and waste products accumulate. It is in this second context that the term cell age is used in this paper.

Tyrrell et al. (1972) examined the variations in UV sensitivity of various strains of *E. coli* as a function of their stage of growth. They found that the wild-type and an excision deficient uvr^- rec^+ strain showed high UV resistance in the early exponential phase of growth with a large increase in sensitivity during late exponential and stationary phase. In contrast the uvr^+ rec^- strain, deficient in recombination, was sensitive during early exponential phase with increased resistance as the culture progressed into late exponential and early stationary phase.

A similar UV response to the influence of growth phase was demonstrated in the fission yeast, *Schizosaccharomyces pombe*, by Fabre (1970). In this eukaryote the wild-type cultures showed maximum UV resistance during the exponential phase with increased sensitivity during the stationary phase. The response of UV-sensitive mutants varied depending upon their degree of sensitivity to UV, the more sensitive mutants showing the opposite effect to wild type with exponential cells being more sensitive than stationary cells.

The increased resistance of exponential phase cells of yeast was also shown after x-irradiation of *Saccharomyces cerevisiae* by Beam et al. (1954), who correlated the radioresistance with the number of budding cells in the population. A considerable literature (see Pepper, 1975) now exists, which suggests that budding cells of *S. cerevisiae* show resistance to a range of x-ray-induced events such as cell lethality; mitotic recombination, respiratory deficiency, division delay and mutation compared with the more sensitive non-budding cells.

In contrast to the effects of UV and x-ray inactivation, Schoenberg-Frasino and Moustacchi (1972) demonstrated that exponential phase cells of yeast were more sensitive than stationary phase cells to heat shock at 52°C. The increased sensitivity of exponential phase cells of yeast was also shown after treatment with the chemical mutagens ethylmethane sulphonate (Evans, 1974), nitrous acid, and mitomycin C (Davies, 1975).

In this paper we describe studies of the changes in UV-induced cell death, mitotic recombination and the rates of macromolecular synthesis in populations of cells of various ages sampled during the transition period between the exponential and stationary phases of cell growth in the yeast *S. cerevisiae*.

B. Materials and Methods

I. Strains

The wild-type diploid cultures of *S. cerevisiae* were:

2NJ. genotype a/α, ade-2/+

JD1. genotype a/α, ade-2/+, ser-1/+, his 4C/his 4ABC, trpS-u9/trp 5-u6; the properties of the strain have been described in Davies et al. (1975).

The repair deficient mutant cultures were:

rad1/rad1. genotype a/α, *rad1-e5/rad1-e5* (Cox and Parry, 1968),

rad2/rad2. genotype a/α, *rad2-e9/rad2-e9* (Cox and Parry, 1968),

both *rad1* and *rad2* are mutant genes that result in defects in the excision repair of UV damage (Parry and Parry, 1969; Unrau et al., 1971).

rad50/rad50. genotype a/α *rad50-e43/rad50-e43*, the allele$_{e43}$ results in increased sensitivity to x-irradiation and results in the abolition of liquid holding recovery after UV (Parry and Parry, 1969).

II. Growth Conditions

Cultures were grown in the liquid yeast complete (YC) medium of Cox and Bevan (1962) in an orbital incubator at 28°C. In general, cultures were inoculated at a concentration of 1×10^3 cells/ml from a single colony on a YC plate, grown overnight and sampled the following day at concentrations between 5×10^6 and 5×10^8 cells/ml.

III. Cell Counts

All cell counts were performed with a haemocytometer. None of the strains used showed any appreciable clumping under the growth conditions described. Buds were scored separately but included in the total cell counts.

IV. UV Treatment

Samples containing 10^8 cells were taken at 1 h intervals and washed three times by centrifugation and resuspended at 10^7 cells/ml in saline. Five-ml portions of the culture were irradiated in agitated open petri dishes. The source of UV light was a Hanovia 11a low-pressure mercury discharge tube generating almost monochromatic radiation at 254 nm. Dose rate was determined with a calibrated photocell. All manipulations were performed in red light to avoid photoreactivation.

V. Determination of DNA, RNA, and Protein Content

Samples containing 10^8 cells were harvested by centrifugation, washed twice with distilled water and the yeast pellet was frozen at -20°C. The pellets were later thawed and extracted by the following modification of the procedure of Parry and Cox (1972). The cells were extracted twice with 2.5 ml 10% (W/V) trichloracetic acid (TCA) for 20 min at 0°C, washed twice with 95% ethanol, extracted three times with 1.0 ml 5% (V/V) perchloric acid (PCA) for 20 min at 70°C, washed twice with 95% ethanol and extracted twice with 1.5 ml 1N·NaOH for 30 min at 60°C. The PCA extracts were pooled and assayed for DNA by the diphenylamine method of Burton (1956) and for RNA by the orcinol method (Schneider, 1957), with calf thymus DNA and yeast RNA as the respective standards. The NaOH extracts were combined and assayed for protein by the method of Lowry et al. (1951) with bovine serum albumin as the standard.

VI. Incubation and Scoring

Treated cells were incubated on agar plates at 28°C in the dark. Mitotic gene conversion was scored on plates of minimal medium supplemented with histidine or tryptophan to detect tryptophan and histidine prototrophs, respectively. Mitotic crossing-over was measured after 6 days by scoring whole red and half-sectored red and white colonies on YC medium. At high survival levels the majority of red colonies result from reciprocal crossing-over in yeast strains heterozygous at the adenine-2 locus.

All the experiments described were performed at least twice and representative results are presented here.

C. Results

I. Cell Viability

The transition of yeast cells from exponential to the stationary phase of growth has been studied in detail by Deutch and Parry (1974) and a theoretical graph derived from this data is shown in Figure 1. This

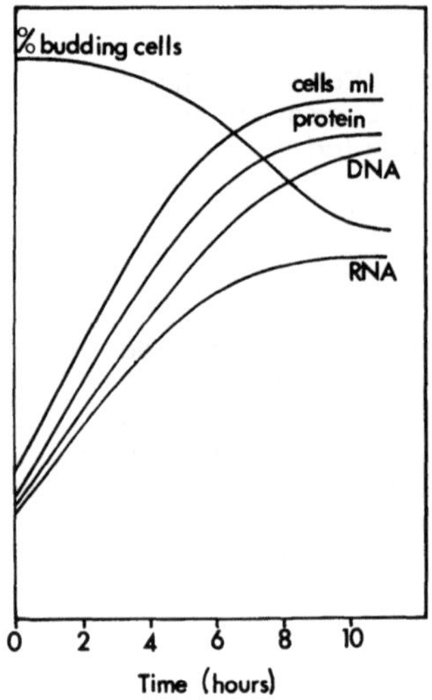

Fig. 1. Transition of yeast cells from exponential to stationary phase of growth. (Data from Deutch and Parry, 1974)

figure summarizes the macromolecular changes that occur during this period of culture growth. Our experiments were initiated when cell numbers reach a density of ~ 10^7 cells/ml (approximately 18 h after initial inoculation) and samples were taken over a time period of 12 h during which the culture "ages", i.e., it undergoes the transition from exponential to the stationary phase of growth.

At time 0, the cultures are in a period of balanced exponential growth during which total cell counts, DNA, RNA, and protein content increase at the same rate (doubling time ~ 7 h). At approximately 4 h the culture enters a transition period that lasts 8 to 9 h. During this period RNA synthesis decreases first, followed by parallel decreases in DNA synthesis, protein synthesis and total cell counts. These decreases were accompanied by gradual decreases in the percentage of budding cells in the population. It is during this transition period that we have investigated the effects of UV irradiation upon cell viability, mitotic recombination and macromolecular syntheses.

The effects of cell age upon cell viability after UV irradiation of a wild-type diploid culture of yeast are shown in Figure 2. The results demonstrate that in the wild-type culture during exponential growth a UV dose of 132 J/m^2 reduces cell viability to 50% compared with 21% in stationary phase cells. The increase in cell sensitivity occurred between 3 to 5 h approximately one hour after a significant reduction was observed in the frequency of budding cells. Similar increases in cell sensitivity as culture enters the stationary phase are also shown in Figure 2 after a UV dose of 264 J/m^2.

The sensitivity of yeast cells during the transition period was further investigated using two excision-deficient diploid cultures of yeast, i.e., *rad1/rad1* and *rad2/rad2*. Because of the reduced repair capacities of these mutant cultures the experiments were performed at UV doses giving comparable survival levels in exponential phase cells, i.e., 28 J/m^2 for *rad1/rad1* and 21 J/m^2 for *rad2/rad2*.

The effects of UV irradiation during the transition period upon *rad1/rad1* and *rad2/rad2* cultures are shown in Figures 3 and 4, respectively. In both mutant cultures cell sensitivity increased as the cultures entered the stationary phase of growth but with both cultures the differences in cell sensitivity between exponential and stationary phase

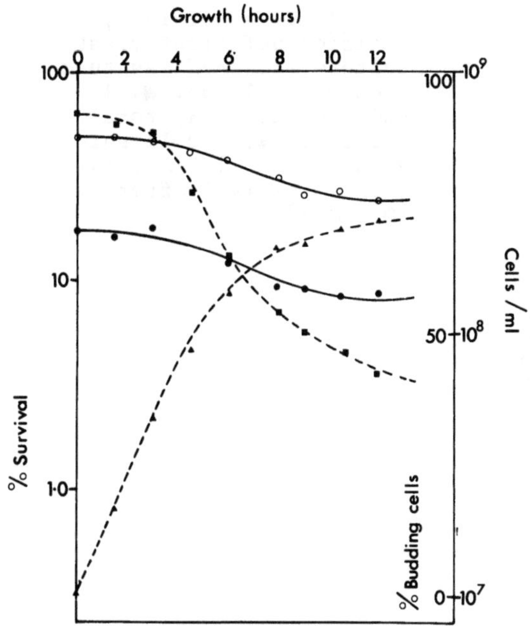

Fig. 2. UV sensitivity of wild-type yeast culture *JD1* as function of cell age. ■, % budding cells; ▲, total cells/ml; ●, % cell survival after 264 J/m^2; o, % cell survival after 132 J/m^2

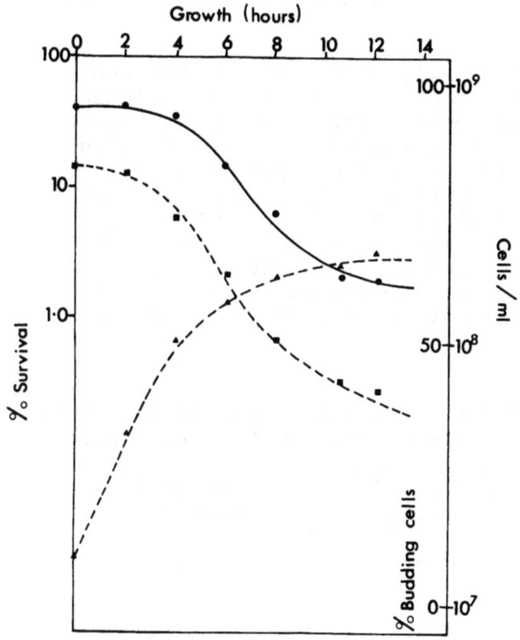

Fig. 3. UV sensitivity of excision-deficient yeast culture *rad1/rad1* as function of cell age. ■, % budding cells; ▲, total cells/ml; ●, % cell survival after 28 J/m^2

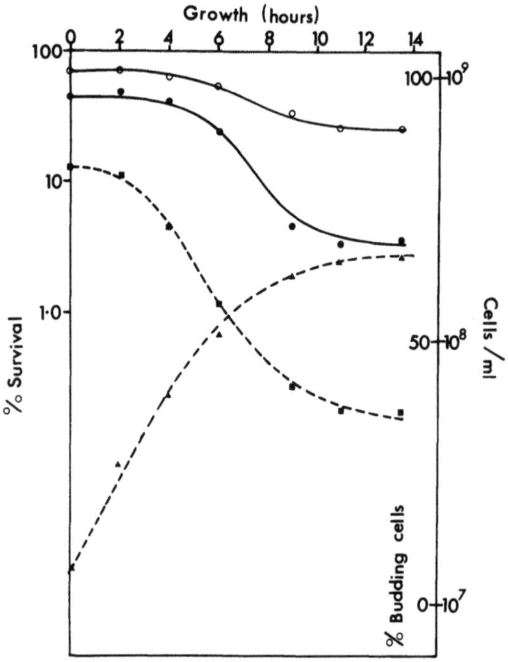

Fig. 4. UV sensitivity of excision-deficient yeast culture *rad2/rad2* as function of cell age. ■, % budding cells; ▲, total cells/ml; ●, % cell survival after 21 J/m^2; o, % cell survival after 10.5 J/m^2

cells were accentuated compared to wild-type cells. In the case of the *rad1/rad1* cultures, survival after 28 J/m^2 was reduced from 40% during exponential growth to 3.5% during stationary whereas in *rad2/rad2* the reduction was from 45% to 3.1%.

Comparisons of the changes in UV sensitivity with cell age between wild type and the excision-deficient mutants may be made by calculation of the dose-modifying effects of these culture changes. These calculations were made by comparing the UV survival curves of cultures of wild-type and *rad2/rad2* cells derived from the exponential phase (time 0 on Fig. 1) and stationary phase (time 10 h on Fig. 1). The dose modification factors (DMF) for cell age are shown in Table 1, calculated at 37% and 10% survival. For wild-type and *rad2/rad2* cultures the DMF's with cell age are 0.82 and 0.37, respectively, at 37% survival. These calculations thus confirm the increased cell age effect found in excision-deficient yeast mutants during the transition period.

Table 1. Dose modification factors (DMF) of cell age upon UV sensitivity in wild-type diploid *JD1* and excision-deficient mutant *rad2/rad2*

Strain	Growth phase comparison	DMF at 37% survival	DMF at 10% survival
JD1	exponential stationary	0.82	0.77
rad2/rad2	exponential stationary	0.38	0.48

The results obtained with the x-ray-sensitivie mutant *rad50/rad50* shown in Figure 5 were in contrast to the changes in UV sensitivity of wild type and excision-deficient mutants with changes in cell age. The results presented in Figure 5 demonstrate that in this culture

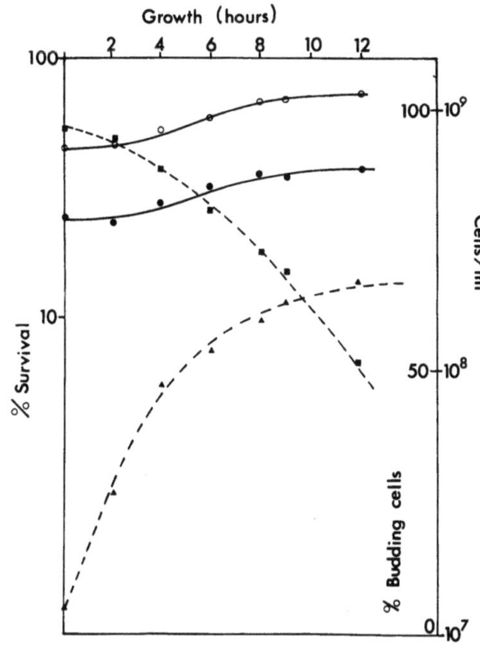

Fig. 5. UV sensitivity of recombination-deficient yeast culture *rad50/rad50* as function of cell age. ■, % budding cells; ▲, total cells/ml; ●, % survival after 132 J/m^2; o, % survival after 66 J/m^2

no increases in UV sensitivity were detectable as the culture progresses from exponential to stationary phase. In fact at the two UV doses of 66 J/m^2 and 132 J/m^2 utilized, small increases in cell resistance were detectable over a period from 4 to 9 h growth.

II. Macromolecular Synthesis

The effects of UV irradiation during the transition from exponential to stationary growth has been further investigated by studying the changes produced in macromolecular synthesis (cell division, DNA, RNA, protein) after UV growth of cells derived from time 0 and time 10 h of the transition period.

The effects in a wild-type culture of UV irradiation upon cell division, DNA, RNA and protein synthesis are shown in Figures 6 and 7 for exponential and stationary phase cells, respectively. The results in both figures demonstrate that the reinoculation of unirradiated cells from both exponential and stationary phase cultures into fresh medium resulted in little or no growth delay with the culture immediately entering a period of exponential growth. However, UV irradiation of exponential and stationary cultures gave significantly different results. As shown in Figure 6, in exponential phase cultures cell growth and macromolecular synthesis commenced immediately and the cells entered a period of rapid increase in cell division, DNA, RNA, and protein synthesis for the first 2 to 3 h of incubation. After this time period the growth rate was reduced until it reached a dose-dependent plateau at 7 to 8 h. In contrast UV-irradiated stationary phase cultures in Figure 7 show a growth delay of 1 1/2 to 2 h after inoculation into fresh medium before growth commenced.

The absence of any UV-induced growth delay in exponential phase cells was also found to be characteristic of excision-deficient yeast cultures. Figure 8 demonstrates the effect of UV irradiation at 40 J/m^2

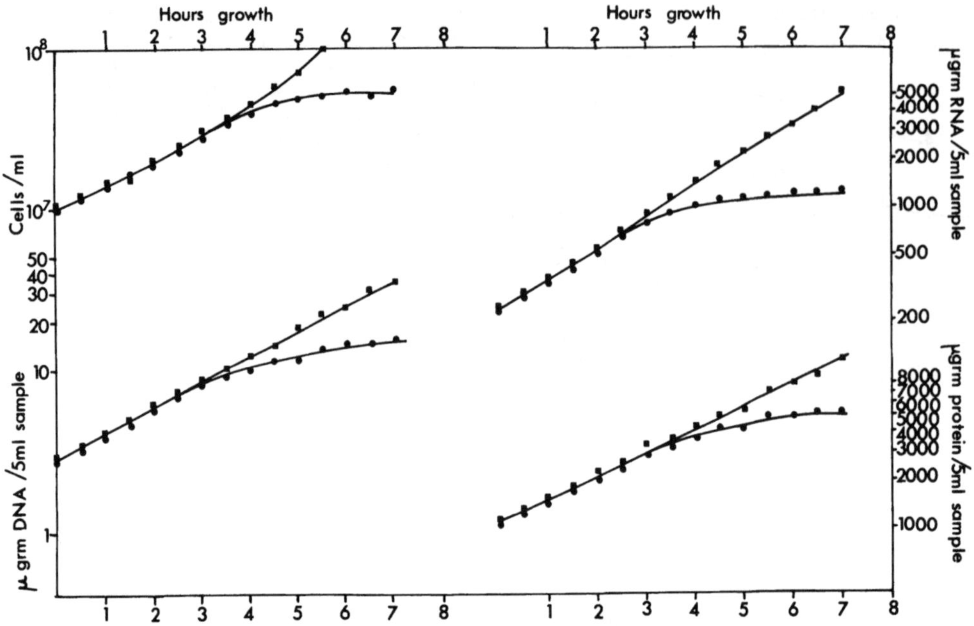

Fig. 6. Effects of UV irradiation (400 J/m^2) upon cell division, DNA, RNA, and protein synthesis in cultures of yeast (strain *JD1*) derived from exponential phase of growth. Cells reinoculated into fresh medium after UV treatment. ■, unirradiated control culture; ●, UV-irradiated culture (400 J/m^2)

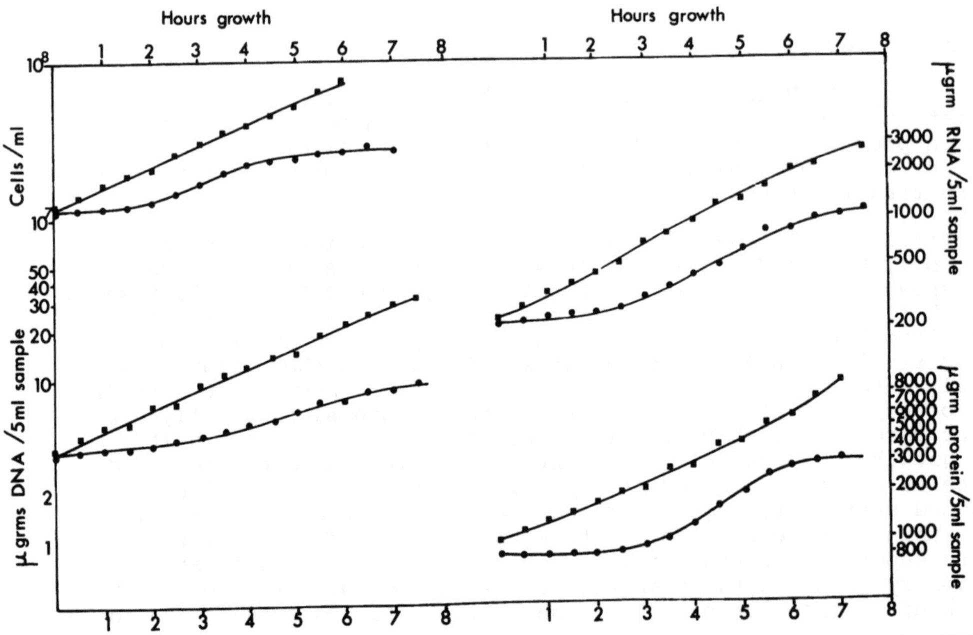

Fig. 7. Effects of UV irradiation (J/m^2) upon cell division, DNA, RNA, and protein synthesis in culture of yeast (strain *JD1*) derived from stationary phase of growth. Cells reinoculated in fresh medium after UV treatment. ■, unirradiated control culture; ●, UV-irradiated culture (400 J/m^2)

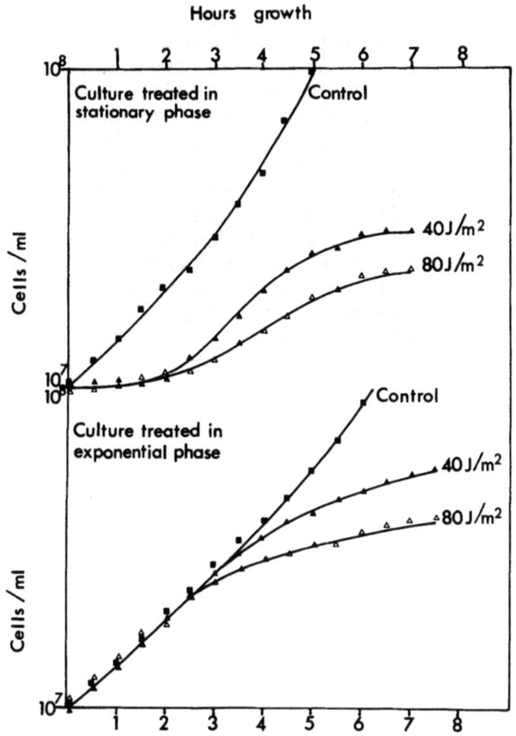

Fig. 8. Effects of UV irradiation (40 and 80 J/m^2) upon cell division in culture of *rad2/rad2* derived from exponential and stationary phases of growth. Cells reinoculated into fresh medium after UV treatment. ■, unirradiated control culture; ▼, UV-irradiated culture (40 J/m^2); ▽, UV-irradiated culture (80 J/m^2)

and 80 J/m^2 upon cell division in *rad2/rad2* cultures. The results show the presence of a UV-induced growth delay in stationary phase cells that is absent from exponential cultures.

In contrast to the results with wild-type and excision-deficient yeast strains are those of the *rad50/rad50* culture shown in Figure 9. This figure demonstrates that after UV irradiation of *rad50/rad50*, both exponential and stationary phase cells showed a UV-induced growth delay. Both exponential and stationary phase cells showed a delay in cell division of 1 1/2 to 2 h before significant increase in cell numbers could be detected.

The results obtained thus implicate a product of the RAD50 gene in the abolition of the growth delay found after UV irradiation of stationary phase cells of wild-type and excision-deficient mutants of yeast.

III. Mitotic Recombination

Mitotic recombination in yeast may be produced within a gene by gene conversion or between genes by crossing-over (see Davies et al., 1975). Both events may be readily detected in yeast strains of the appropriate genotypes. We have examined the effects of UV irradiation of exponential (time 0) and stationary phase (time 10 h) cells upon the production of recombinant cells by both mitotic gene conversion and crossing-over.

Figure 10 demonstrates the production of histidine and tryptophan prototrophs by mitotic gene conversion and cell viability after UV irradiation of exponential and stationary phase cells of the wild-type

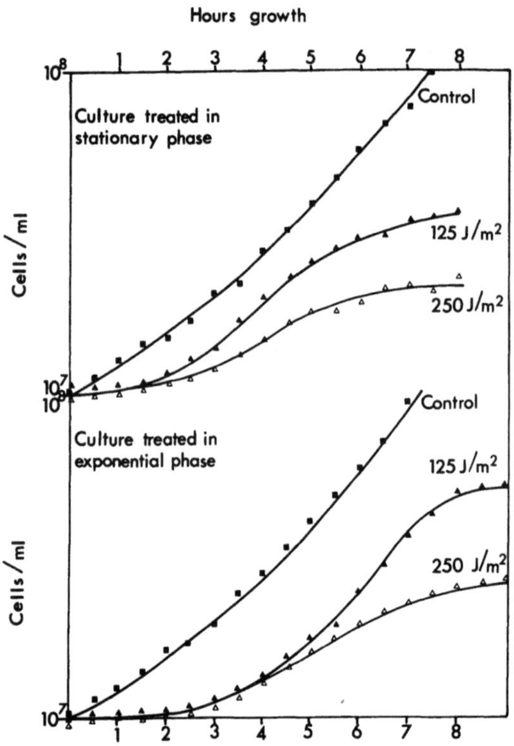

Fig. 9. Effects of UV irradiation (125 and 250 J/m^2) upon cell division in cultures of *rad50/rad50* derived from the exponential and stationary phases of growth. Cells reinoculated into fresh medium after UV treatment. ■, unirradiated control culture; ▼, UV-irradiated culture (125 J/m^2); ▽, UV-irradiated culture (250 J/m^2)

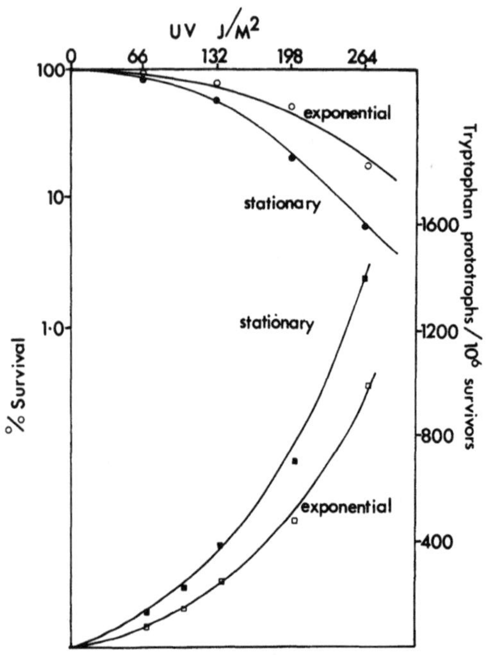

Fig. 10. Induction of tryptophan prototrophs produced by mitotic gene conversion and cell viability after UV irradiation of exponential and stationary phase cells of strain *JD1*. ●, cell viability of stationary phase cells; o, cell viability of exponential phase cells; ■, increase in frequency of trp^+ colonies/10^6 survivors in stationary phase cells; □, increase in frequency of trp^+ colonies/10^6 survivors in exponential phase cells

yeast diploid *JD1*. As shown previously, stationary phase cells of this culture are more sensitive than exponential phase cells to the lethal effects of UV irradiation. In contrast, the induction of both histidine and tryptophan prototrophs was higher at each UV dose in stationary phase cells as compared to exponential cells. The results thus indicate that the cell age UV sensitivity to the induction of mitotic gene conversion is opposite to the lethal effect.

The results of the gene conversion experiments may also be expressed as the frequency of UV-induced prototrophs per surviving cell. The data calculated in this manner are shown in Figure 11. The figure demonstrates that at both the histidine 4 and tryptophan 5 loci the frequencies of UV-induced prototrophs were the same at each survival level, irrespective of whether the treated cells were derived from exponential or stationary phase cultures.

The effects of cell age upon the production of UV-induced recessive homozygosis at the adenine-2 locus produced by mitotic crossing-over was determined in both wild-type and excision-deficient cultures of yeast. The cultures used are heterozygous at the adenine-2 locus and produced red adenine-requiring colonies by mitotic crossing-over between the adenine-2 gene and the centromere of chromosome XV.

The frequencies of the red adenine-requiring colonies produced after UV irradiation of exponential and stationary phase cells of a wild-type and an excision-deficient strain *rad2/rad2* are shown in Figure 12. The results demonstrate that UV irradiation produced increases in mitotic crossing-over in exponential and stationary phase cells of both wild-type and *rad2/rad2* cultures although much lower doses of UV were used in the latter culture. In both exponential and stationary cultures UV irradiation produced a maximum frequency of homozygosis of approximately 10% with a reduction at doses giving high level of cell death. The most important observation in Figure 12 concerns the relative sensitivities of exponential and stationary cells to the induction of recessive homozygosis. The results demonstrate that per unit dose of UV light a higher frequency of homozygotes was induced in the more UV-sensitive stationary phase cells of both wild-type and *rad2/rad2* cultures.

The results presented thus demonstrate that the induction of both mitotic gene conversion and mitotic crossing-over are at a maximum during the stationary phase of growth during which the cells show their greatest sensitivity to the lethal effects of UV irradiation.

D. Discussion

The transition experiments described here demonstrate that cultures of diploid cells of the yeast *S. cerevisiae* progressing from exponential to the stationary phase of growth showed increasing sensitivity to UV irradiation. The increase in UV sensitivity correlates closely with the period of decreased cell growth and the reduction in the frequency of budding cells. The increased UV sensitivity of stationary phase cells was accentuated in the cultures *rad1/rad1* and *rad2/rad2*, both defective in excision repair, and abolished in the culture *rad50/rad50*, which is x-ray-sensitive. The results presented suggest that the increased UV resistance of the wild-type and excision-deficient cultures during the exponential phase of growth is conferred by a product coded for by the gene RAD50.

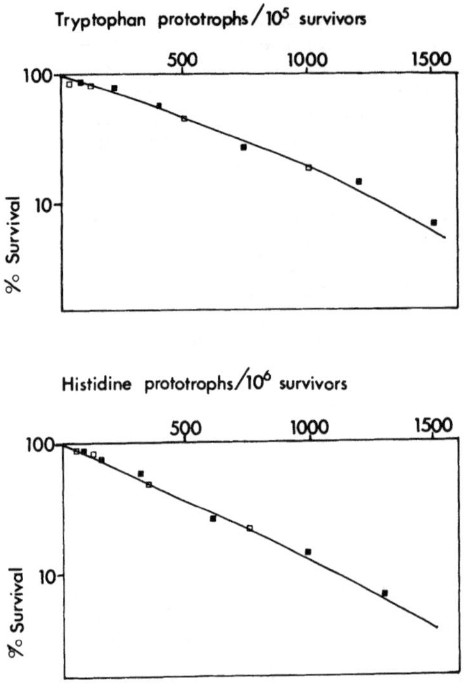

Fig. 11. Induction of tryptophan and histidine prototrophs produced by mitotic gene conversion after UV irradiation of exponential and stationary phase cells of strain *JD1*. Prototroph frequency plotted against cell survival. ■, prototroph frequency produced after UV irradiation of exponential phase culture; □, prototroph frequency produced after UV irradiation of stationary phase culture

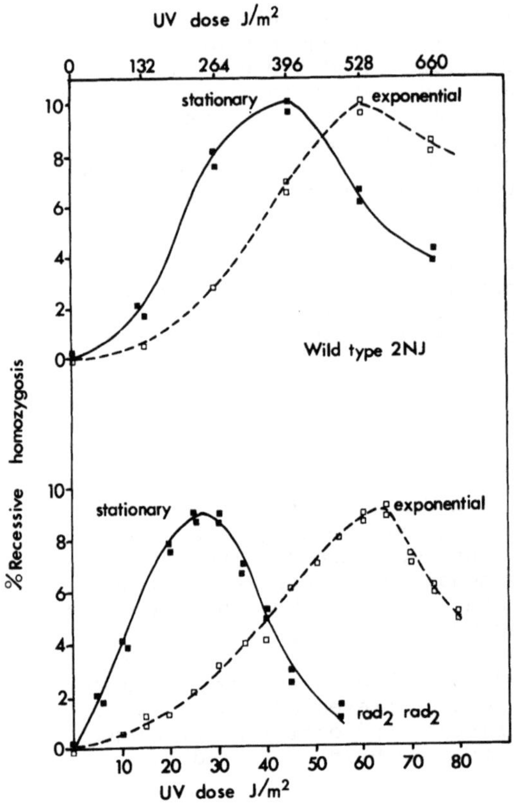

Fig. 12. Induction of red adenine-requiring recessive homozygous colonies by mitotic crossing-over after UV irradiation of exponential and stationary phase cells of strains *JD1* and *rad2/rad2*. ■, % recessive homozygosis in stationary phase cells; □, % recessive homozygosis in exponential phase cells

The measurements of macromolecular synthesis after UV irradiation demonstrate a clear difference in response between exponential and stationary phase cultures. In both wild-type and excision-deficient cultures UV irradiation of stationary phase cells resulted in a growth delay of 1 1/2 to 2 h, whereas no detectable delay was observed in exponentially growing cultures. In contrast, cultures of *rad50/rad50* showed a division delay when both exponential and stationary phase cells were UV-irradiated. These results indicate that the product of the RAD50 gene is involved in the elimination of the UV-induced division delay in exponential cells and that the elimination of such a delay leads to increased cell viability.

Some further information concerning the nature of the UV sensitivity of stationary phase cells has been obtained by the studies of UV-induced mitotic recombination. Our results demonstrate that in both the wild-type and the excision-deficient yeast cultures, the increased UV-resistance of exponential phase cells was correlated with increased levels of mitotic gene conversion and mitotic crossing-over. Some recent work using synchronized yeast cultures (Chanet et al., 1975) indicate differences in the levels of pyrimidine dimers induced by UV light during the cell cycle with lower frequencies of dimers induced per unit dose of UV in budding cells. Such variation in dimer induction may account for a fraction of the differences observed here in UV sensitivity in yeast cultures of differing cell ages. However, the responses of the excision-deficient and x-ray-sensitive cultures indicate that differences in repair capacity make the major contribution to the cell age response.

We previously presented (Davies et al., 1975) a model of UV repair in yeast in which it is postulated that induced mitotic recombination results from the postreplication repair of UV-induced pyrimidine dimers. The model postulates that although the initial lesions may be the same, the steps leading to gene conversion and crossing-over are different. The present results indicate that for a given UV dose, exponential cells show higher cell viability and lower levels of both mitotic gene conversion and crossing-over. Such evidence suggests that UV-induced DNA lesions in exponential phase cells are repaired before they may act as substrates for the production of genetic recombinational events.

In yeast the repair processes described appear to depend upon the functional integrity of the RAD50 gene. We have previously shown that yeast mutants carrying defective alleles of the RAD50 gene are impaired in liquid holding recovery, a process which in yeast appears to involve excision repair (Parry and Parry, 1969). In such cultures dimer removal takes place without the completion of the steps which in liquid holding lead to cell recovery. It would therefore be attractive to postulate that the RAD50 gene is defective in a gap-refilling process necessary in yeast for the repair of both UV and x-ray damage.

We would thus conclude that the differences in response of exponential and stationary phase cells of yeast to UV-induced cell death, division delay, and mitotic recombination derive from a more rapid process of gap-refilling found in exponential phase cells and that this process, which is coded for by the RAD50 gene, slows down as cells enter stationary phase.

E. Summary

During the transition period from the exponential to the stationary phase of growth, yeast cells showed variation in sensitivity to the lethal and non-lethal effects of ultraviolet (UV) irradiation. Exponential phase cells showed maximum resistance to the lethal effects of UV, with an increase in sensitivity as the cells entered stationary phase. In contrast, UV-induced mitotic gene conversion and crossing-over was at a minimum during exponential phase and increased as the cells in the culture aged and entered stationary phase.

Yeast cells irradiated in stationary phase showed a UV-induced growth delay in cell division, DNA, RNA, and protein synthesis, whereas the more resistant exponential phase lacked the growth delay at equivalent UV doses.

The resistance of yeast cells to UV-induced cell death and growth delay during exponential phase was abolished in cultures homozygous for the gene RAD50. A model is presented that suggests that the cell age response of yeast cultures is dependent upon differences in the rates of a "gap-filling" repair process coded for by the RAD50 gene.

Acknowledgements. This work was supported by Euratom grant No. BlO E 1199-72-1. We are grateful to Mrs. Margaret Clatworthy for the preparation of the figures.

References

Beam, C.A., Mortimer, R.K., Wolfe, R.G., Tobias, C.A.: The relation of radioresistance to budding in *Saccharomyces cerevisiae*. Arch. Biochem. Biophys. 49, 110-122 (1954)

Burton, K.: A study of the conditions and mechanisms of the diplenylamine reaction for the colorimetric estimation of deoxyribonucleic acid. Biochem. J. 62, 315-323 (1956)

Chanet, R., Waters, R., Moustacchi, E.: The induction of pyrimidine dimers in nuclear DNA after UV-irradiation during the synchronous cycle of *Saccharomyces cerevisiae*. Int. J. Radiat. Biol. 27, 481-485 (1975)

Clarke, A.J., Margulies, A.D.: Isolation and characterization of recombination deficient mutants of *E. coli* K12. Proc. Nat. Acad. Sci. USA 53, 451-459 (1965)

Cox, B.S., Bevan, E.A.: Anueploidy in yeast. New Phytologist 61, 342-355 (1962)

Cox, B.S., Parry, J.M.: The isolation, genetics and survival characteristics of ultraviolet light sensitive mutants in yeast. Mutation Res. 6, 37-55 (1968)

Davies, P.J.: The chemical induction of genetic damage in eukaryotes. Ph.D. Thesis, University of Wales, 1975

Davies, P.J., Evans, W.E., Parry, J.M.: Mitotic recombination induced by chemical and physical agents in the yeast *Saccharomyces cerevisiae*. Mutation Res. 29, 301-314 (1975)

Deutch, C.E., Parry, J.M.: Sphaeroplast formation in yeast during the transition from exponential phase to stationary phase. J. Gen. Microbiol. 80, 259-268 (1974)

Evans, W.E.: Studies upon the inactivation and repair of cellular damage induced by chemical and physical agents. Ph.D. Thesis, University of Wales, 1974

Fabre, F.: UV sensitivity of the wild-type and different UVS mutants of *Schizosaccharomyces pombe*: influence of growth stages and DNA content of the cells. Mutation Res. 10, 415-426 (1970)

Hanawalt, P.C.: The UV sensitivity of bacteria: its relation to the DNA replication cycle. Photochem. Photobiol. 5, 1-12 (1966)

Howard-Flanders, P., Boyce, R.P., Theriot, L.: Three loci in *E. coli* K12 that control the excision of pyrimidine dimers and certain mutagen products from DNA. Genetics 53, 1119-1136 (1966)

Lowry, O.H., Rosebrough, N.J., Farr, A.L., Randall, R.J.: Protein measurement with the Folin phenol reagent. J. Biol. Chem. 193, 265-275 (1951)

Parry, E.M., Cox, B.S.: The general properties of a strain of yeast with less DNA than normal. J. Gen. Microbiol. 70, 129-149 (1972)

Parry, J.M., Parry, E.M.: The effects of UV-light posttreatment on the survival characteristics of 21 UV-sensitive mutants of the yeast *Saccharomyces cerevisiae*. Mutation Res. 8, 545-556 (1969)

Pepper, T.: Cyclic x-ray resistance to lethal and non-lethal damage in *Saccharomyces cerevisiae*. Radiation Res. 63, 97-118 (1975)

Schneider, W.C.: Determination of nucleic acids in tissues by pentose analysis. In: Methods in Enzymology (Colowick, S.P., Kaplan, N.O., eds.). New York: Academic Press, 1957, Vol. 3, pp. 680-684

Schoenberg-Frascino, A., Moustacchi, E.: Lethal and mutagenic effects of elevated temperature on haploid yeast. I. Variations in sensitivity during the cell cycle. Molec. Gen. Genetics 115, 243-257 (1972)

Tyrrell, R.M., Moss, S.H., Davies, D.J.G.: The variation in UV sensitivity of four K-12 strains of *E. coli* as a function of their stage of growth. Mutation Res. 16, 1-12 (1972)

Unrau, P., Wheatcroft, R., Cox, B.S.: The excision of pyrimidine dimers from DNA of UV irradiated yeast. Molec. Gen. Genetics 113, 359-362 (1971)

4. Budding and Division Delay in Diploid Yeast after Irradiation

G. FABER and J. KIEFER

A. Introduction

It has been shown that irradiation of eukaryotes with low doses of ultraviolet (UV) light, which lead to high survival fractions, cause progression and division delay (Bootsma and Humphrey, 1967; Domon and Rauth, 1968, 1969; Djordjevic and Tolmach, 1967; Goodin et al., 1971; Johnson et al., 1973; Phillips and Tolmach, 1966; Trosko and Kaschau, 1967). Most of these experiments were performed with mammalian cells (Bootsma and Humphrey, 1969; Domon and Rauth, 1968, 1969; Djordjevic and Tolmach, 1967; Phillips and Tolmach, 1966; Trosko and Kaschau, 1967). With regard to the progression through the cell cycle the S phase was found to be most sensitive (Bootsma and Humphrey, 1969; Domon and Rauth, 1968, 1969; Djordjevic and Tolmach 1967).

Phillips and Tolmach (1966) hypothesized that progression delay is dependent upon the time necessary to repair the radiation-induced damage. This causes division delays that vary in length of time, depending on dose.
For us, it was of interest to investigate whether and to what extent budding and division delay would occur in yeast after irradiation with low doses of UV light.

We hoped to answer questions concerning radiation-induced damage and repair, and the relationship between budding and division delay and between colony forming ability and progression through the cell cycle.

B. Material and Methods

Experiments were performed with two differently sensitive strains of *Saccharomyces cerevisiae*, the wild type 211 and the excision negative mutant No. 2094 (*rad2*). Strains were obtained from Laskowski and Averbeck (Berlin).
Experimental procedure was:
Stationary phase cells (1.2×10^8 cells/ml) from nutrient agar plates were irradiated in distilled water with a mercury low-pressure lamp (Osram HNS 12 ofr).
The cell suspensions were then incubated up to 8 h in liquid nutrient at 30°C. They were kept in the dark to avoid photoreactivation. During incubation period samples were taken every half hour and cell number/ml as well as number of budding cells of 300 cells was determined microscopically.

Two different types of curves were obtained:
budding curves and growth curves.

C. Results

I. Strain No. 211

Figure 1 shows the budding curves of the wild type. Per cent budding cells are plotted against incubation time. Cells were irradiated with

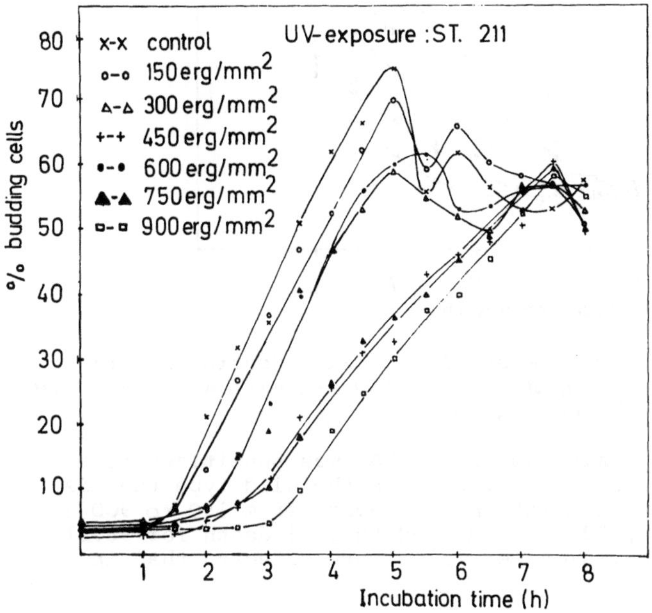

Fig. 1. Budding curves for wild-type strain No. 211: control and irradiated samples

doses that garanteed nearly 100% survival of cells.
After a lag phase in which only 2-3% of the cells formed buds, the index increased more or less rapidly for several hours until it had reached a maximum value. Whereas the control curve shows several maxima and minima, the curves for the irradiated cultures show a decreased number of maxima and minima with increasing UV exposure. This is caused by a prolongation of the lag phases and a slow increase of the budding index. Minimum values in a budding curve indicate cell division. The growth curves (Fig. 2) confirm this. In analogy to the budding curves, the relative cell number N/N_0 is plotted against the incubation time.

Because of the relatively short incubation period the growth curves only show the lag phase of the budding curves and the beginning of the exponential phase. The first generation cycle of the unirradiated cells extended over about 5-5 1/2 hrs. At this point of time the relative cell number had the value 2 and/or in the budding curves (Fig. 1) the first minimum was reached. With increased UV exposure, however, the lag times prolonged rapidly up to 6 1/2 hrs for the highest doses. Changes in the budding curves of irradiated cultures indicate that budding as well as division is affected by low doses of UV light.

II. Strain No. 2094

This strain was found to be extremely sensitive to UV light. Though we reduced the UV doses by a factor of 6.8, compared with the doses

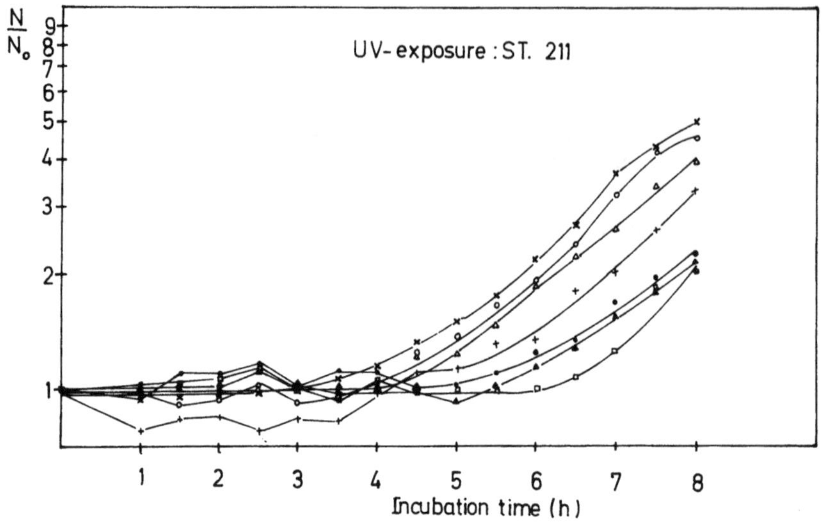

Fig. 2. Growth curves for wild-type strain No. 211: Control x - x; o - o 150 erg/mm^2; Δ - Δ 300 erg/mm^2; + - + 450 erg/mm^2; ● - ● 600 erg/mm^2; ▲ - ▲ 750 erg/mm^2; □ - □ 900 erg/mm^2

for the wild type, colony forming ability (CFA) was sensitized to a much greater degree than in strain No. 211. In the wild type the CFA was virtually unaffected by irradiation with doses from 150 to 900 erg/mm^2; in the mutant, however, 132 erg/mm^2 led to cell death for 98.5% of the cells. Budding and division were also more affected than in the wild type.

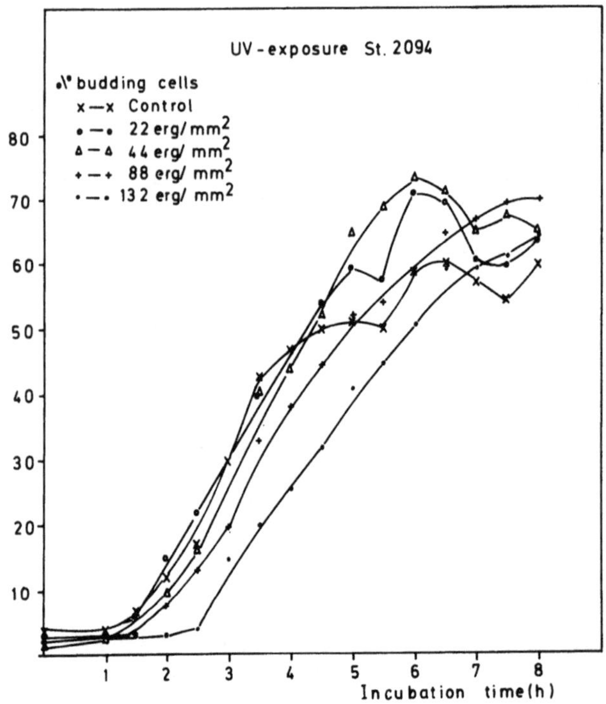

Fig. 3. Budding curves for excision negative mutant No. 2094: control and irradiated samples

Wild-type cells irradiated with 150 erg/mm^2 behaved exactly like the control. 132 erg/mm^2 in the mutant caused extreme prolongations of the first generation cycle. Budding curves (Fig. 3) show that compared to the control culture the lag times of all irradiated cultures were relatively short, but the time till the budding index had reached a maximum value was rapidly prolonged with increased UV exposure. Whereas cells irradiated with 22 and 44 erg/mm^2 divided during incubation time, cells irradiated with 88 and 132 erg/mm^2 were still in the ascending phase of budding (Fig. 3) and/or in the lag phase of growth (Fig. 4).

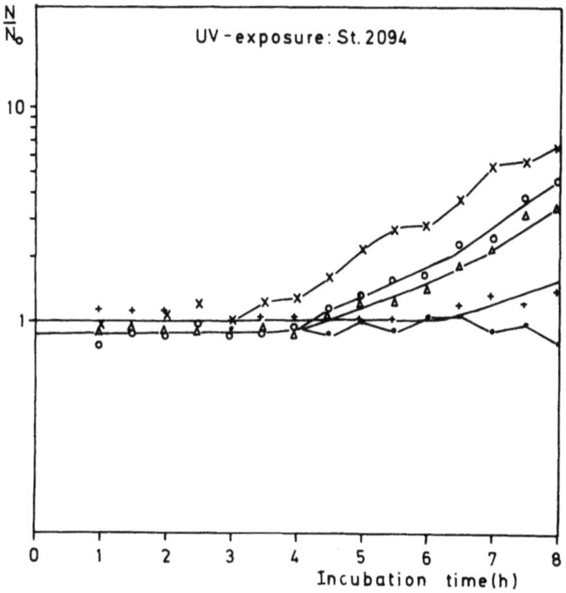

Fig. 4. Growth curves for excision negative mutant No. 2094: control x - x; o - o 22 erg/mm^2; Δ - Δ 44 erg/mm^2; + - + 88 erg/mm^2; ● - ● 132 erg/mm^2

D. Calculation of Budding and Division Delays

Budding and division delays were calculated in the same way for both strains.
To calculate *budding delay*, that point of time when 15% of the irradiated cells had formed a bud was compared with that of the control.

In analogy *division delay* was determined by comparing that point of time when the relative cell number of the irradiated cells had reached the value 1.5 with the corresponding value of the control. The value 1.5 was chosen for calculation because with increased UV exposure the lag times became so long that not all cells divided during the incubation time.

E. Comparison of Budding and Division Delays between the Wild Type and the Mutant

Figure 5 shows budding delay for both strains, dependent on dose. In both cases budding delay increased linearly with dose. In contrast to the wild type, budding in the mutant was much more affected at the same doses.

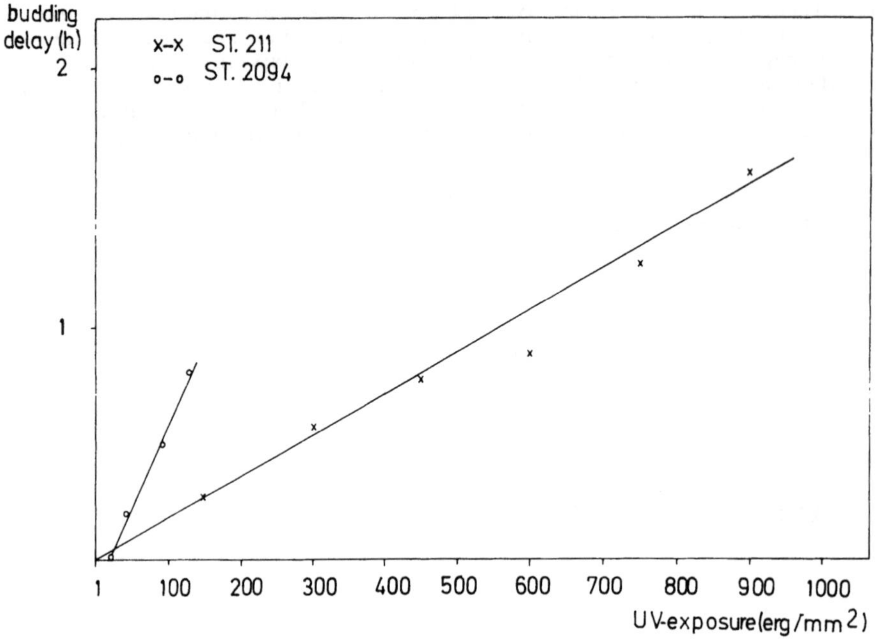

Fig. 5. Budding delay of strain No. 211 (x - x) and strain No. 2094 (o - o)

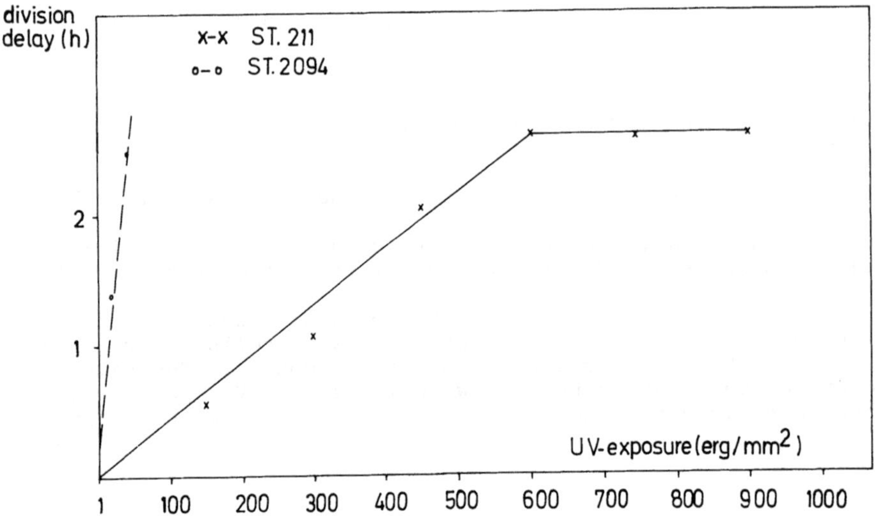

Fig. 6. Division delay of strain No. 211 (x - x) and strain No. 2094 (o - o)

If one compares the curves for budding delay with those for division delay (Fig. 6), one can see that in both strains division delay was greater than budding delay.

Maximum budding delay in the wild type was 1.6 h, in the mutant 0.82 h. Maximum division delay in the wild type was 2.63 h. For strain No. 2094 division delay could only be determined for cells irradiated with 22 and 44 erg/mm^2 because cells irradiated with higher doses did not reach the relative cell number 1.5. The fact that division delay of the mutant is extremely delayed indicates the high sensitivity of this strain.

F. Discussion

The question arises as to whether budding and division delays in both strains were caused by the same processes or not. As mentioned already, the wild type was irradiated with doses that virtually did not affect its CFA. Budding as well as division was considerably more sensitive. But both processes were only delayed during the first cell cycle (Fig. 1). After the first division the irradiated cultures behaved like the control culture. From that, we concluded that the cells possess repair systems that allow them to eliminate the radiation-induced damage in the DNA during the first cell cycle. Therefore we can assume that budding and division delays represent the time necessary to repair the radiation-induced damage. From the fact that budding delay was less than division delay, we conclude that only part of the damage must be repaired for the initiation of budding; for division, however, it seems necessary that nearly all lesions are repaired.

In contrast to the wild type, the delay times in the mutant were probably not governed by repair processes, because the CFA was sensitized to a much greater degree than in the wild type.
To explain budding delay times, we analyzed the budding curves by assuming that the apparent delay is caused by the permanent loss of budding ability in a fraction of the cell population. We normalized the values of the irradiated cultures for the increasing phase of budding cells, till the first maximum value of the control.

Figure 7 shows the relative index given in per cent, in relation to dose and in comparison with the experimental survival curve. At the lowest dose budding as well as CFA was unaffected. At higher doses, however, the percentage of cells that had lost the ability to form a bud decreased. But this did not occur to the same extent as the CFA was sensitized. From that we conclude that some of the cells that were capable of budding died at a later point of time and that this is probably a reason for the enormous division delay.

G. Summary

In summary, we can say that in the wild type the extent of the delay periods is governed mainly by repair times; in the sensitive mutant, however, it is largely due to a reduced number of cells capable of forming buds and/or of division.

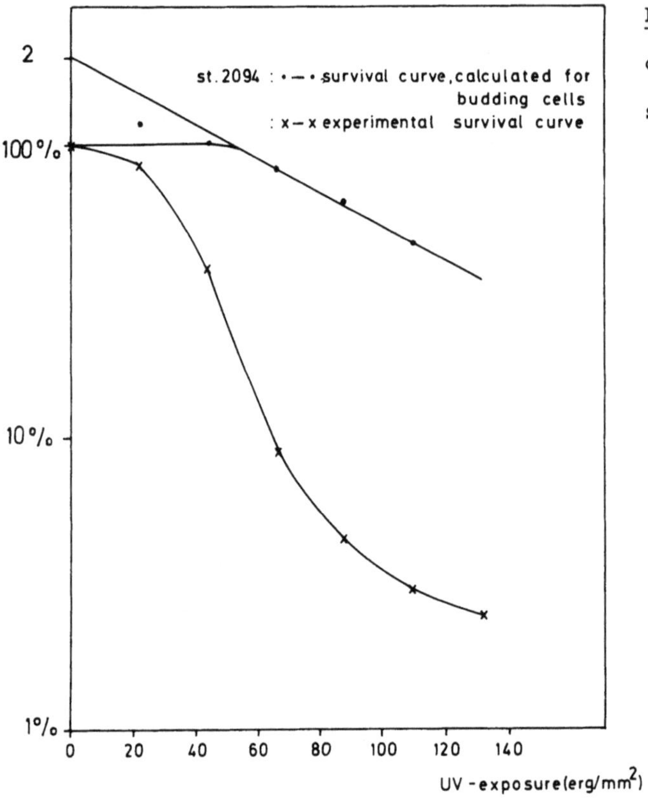

Fig. 7. Strain No. 2094: (a) survival curve, calculated for budding cells (o - o); (b) experimental survival curve (x - x)

References

Bootsma, D., Humphrey, R.M.: The progession of mammalian cells through the division cycle following ultraviolet irradiation. Mutation Res. 5, 289-298 (1967)

Djordjevic, B., Tolmach, L.J.: Response of synchonous populations of HeLa cells to ultraviolet irradiation at selected stages of the generations cycle. Radiat. Res. 32, 327-346 (1967)

Domon, M., Rauth, H.M.: Ultraviolet irradiation of mouse L cells: Effects on DNA synthesis and progression through the cell cycle. Radiat. Res. 35, 350-368 (1968)

Domon, M., Rauth, H.M.: Ultraviolet irradiation of mouse L cells: Effects on cells in the DNA synthesis phase. Radiat. Res. 40, 414-429 (1969)

Goodin, L.J., Johnson, R.G., Morrison, H.G.: Photoreversal of ultraviolet-induced division delay in Saccharomyces cerevisiae. Radiat. Res. 45, 440-448 (1971)

Johnson, R.G., Wahnsley, J.G., Goodin, L.J., Morrison, H.G.: Ultraviolet-induced delay of budding in synchronized yeast cells. Radiat. Res. 53, 134-142 (1973)

Phillips, R.A., Tolmach, L.J.: Repair of potentially lethal damage in X-irradiated HeLa cells. Radiat. Res. 29, 413-432 (1966)

Trosko, J.E., Kaschau, M.R.: Study of pyrimidine dimers in mammalian cells surving low doses of ultraviolet radiation. Photochem. Photobiol. 6, 215-219 (1966)

5. Synthesis of Histones in the Cell Cycle of *Saccharomyces cerevisiae*

E. WINTERSBERGER and R. MOLL

A. Introduction

Replication of the genetic material in higher organisms involves not only the duplication of DNA but also the synthesis of nuclear proteins, especially the histones, as well as the assembly of these constituents into chromatin. The production of the structural components of chromatin is in general highly coordinated. In most higher cells, DNA and histones are synthesized exclusively in the S phase of the cell cycle. Whereas inhibition of protein synthesis by cycloheximide or puromycin at any time in the S phase impairs DNA replication, inhibition of DNA synthesis with hydroxyurea or cytosine arabinoside blocks histone formation (see for instance Seale and Simpson, 1975 and papers cited therein). DNA replication in higher cells thus depends on concomitant protein synthesis throughout the S phase. This is in contrast to the replication mechanism of bacterial DNA, which requires protein synthesis only for its initiation (Lark, 1969).

The yeast *Saccharomyces cerevisiae* holds an intermediate position between higher cells and prokaryotes: In this primitive eukaryote, DNA replication is dependent on protein synthesis only in its initial phase; ongoing DNA synthesis is not influenced by addition of cycloheximide (Williamson, 1973; Hereford and Hartwell, 1973; Golombek et al., 1974). More specifically, protein synthesis is required only during the first 10 min of a 30-min S phase of *S. cerevisiae*. This could mean that initiation of the numerous replicons of this organism occurs asynchronously during the first third of the S phase. Since this step is known to depend on protein synthesis, addition of cycloheximide at any time during this period would inhibit the initiation of at least part of the replicons, resulting in an incomplete doubling of the genome.

The completion of initiated replicons, independent of simultaneous protein, and therewith, histone synthesis, must either mean that there exists a pool of histones from which these proteins are taken for chromatin assembly, or that dublication of the DNA in *S. cerevisiae* can be completed in the absence of a full set of histones. To solve this question we have studied the synthesis of histones in synchronously growing yeast cells.

B. Further Characterization and Classification of Yeast Histones

Extraction of isolated yeast nuclei or of yeast chromatin with acid solutions solubilizes proteins that upon electrophoresis in polyacrylamide gels in the presence of urea migrate similarly to mammalian histones (Wintersberger et al., 1973). Three main bands and a shoulder could clearly be distinguished in the previous work, indicating that *S. cerevisiae* contains four different histones, which correspond in mobility to the histones H2a, H2b, H3, and H4 of mammalian cells. The incomplete separation of one pair of these histones is a technical problem that can be overcome by slightly altering the conditions of electrophoresis. Figure 1 shows a typical electropherogram in which all four of these proteins are separated well enough to definitively

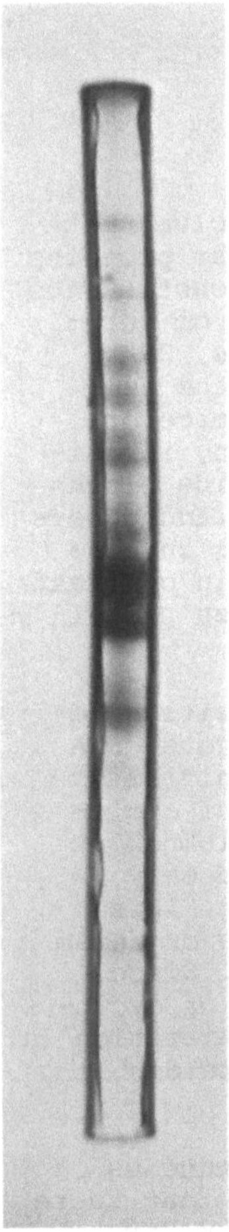

Fig. 1. Electrophoresis of yeast histones in polyacrylamide gels containing 2.5 *M* urea at pH 2.9

demonstrate the existence of four histones in yeast. This point is of considerable importance with regard to the basic unit of chromatin structure in yeast (see below). Histone H1 has never been found in yeast. This protein shows the greatest variation of all the histones among different organisms. Most obvious are the differences in the electrophoretic mobility. For instance, in another lower eukaryote, the slime mold *Physarum polycephalum*, histone H1 migrates in acrylamide gels significantly slower than H1 from mammalian tissues (Bradbury et al., 1973). Since acid extracts of yeast nuclei and to a lesser extent those of yeast chromatin contain in addition to the four histones a heterogeneous population of less basic proteins among which histone H1 could eventually be present, we have used an additional test to look for this histone in yeast: It is known that histones from all organisms tested so far lack the amino acid tryptophan (de Lange and Smith, 1971). We have therefore labelled yeast cells with radioactive tryptophan, prepared nuclei, and isolated the acid-soluble proteins. These were separated by electrophoresis in acrylamide gels in the presence of urea. After electrophoresis the gels were sliced, dissolved, and counted. A gel run in parallel was stained with amido black to visualize the protein bands (Fig. 2). Two important results became

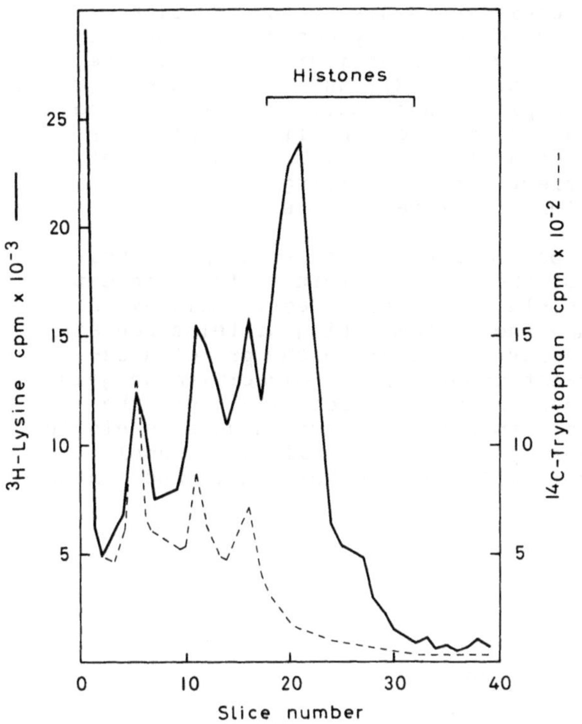

Fig. 2. Electrophoresis of acid-extractable nuclear proteins from yeast cells labelled with either ^{3}H-lysine (——) or ^{14}C-tryptophan (----) respectively

apparent from this experiment: (1) There are no peaks with tryptophan label at positions where the histones migrate in the gel. This proves by another criterion - in addition to the electrophoretic characteristics - that these proteins correspond to the histones 2a, 2b, 3, and 4 of higher organisms. (2) All proteins with a lower mobility than the histones contained tryptophan and therefore are nonhistone proteins. This result serves as an additional argument for the absence of histone H1 from yeast.

According to a recently published hypothesis (Olins and Olins, 1974; Kornberg, 1974; Noll, 1974; Oudet et al., 1975), the above-mentioned four histones together with a piece of DNA of about 200 base pairs, in higher organims, build up the units of chromatin, the nucleosome. Histone H1 is not involved in the formation of this structural element of chromatin. For the idea that the chromatin structure of yeast might be basically similar to that of higher organisms, the presence of the four histones, 2a, 2b, 3, and 4, is of vital importance. Support for the existence of yeast nucleosomes comes from the observation of discrete, nuclease-resistant deoxynucleoprotein particles containing the four histones and about 150 base pairs of DNA (Lohr and van Holde, 1975; Rogall and Wintersberger, in preparation).

C. Synthesis of Histones in the Cell Cycle

Histone synthesis in *S. cerevisiae* was studied by pulse labelling the histones with tritiated lysine at different times in the division cycle of cells that had been synchronized according to a procedure described earlier (Golombek and Wintersberger, 1974). After the pulse cells were blocked in further protein synthesis by the addition of cycloheximide, rapidly cooled, and collected by centrifugation. Nuclei were prepared via spheroplasts, using a slight modification of the published method (Wintersberger et al., 1973), and extracted with 0.4 *M* H_2SO_4. Because the yield of spheroplasts and nuclei varies with the stage of cells in the cycle, DNA was measured in aliquots of cells, spheroplasts, and nuclei during each preparation. These determination, proved necessary in order to correct for the varying recovery of spheroplasts from cells and nuclei from spheroplasts.

The presence of a heterogeneous population of nonhistone proteins, sometimes comprising up to 50% of the total protein in the acid extract of yeast nuclei, makes it impossible to equate directly acid extractable protein with histones. Since the contaminating proteins contain lysine (see Fig. 2), it was necessary to separate the proteins extracted at each time point by electrophoresis in polyacrylamide gels in the presence of urea and to sum up the radioactivity under the peaks of the histones. Only this elaborate procedure have a meaningful measure of the amount of histones synthesized in different samples. Details of the procedure will be published elsewhere (Moll and Wintersberger, in preparation).

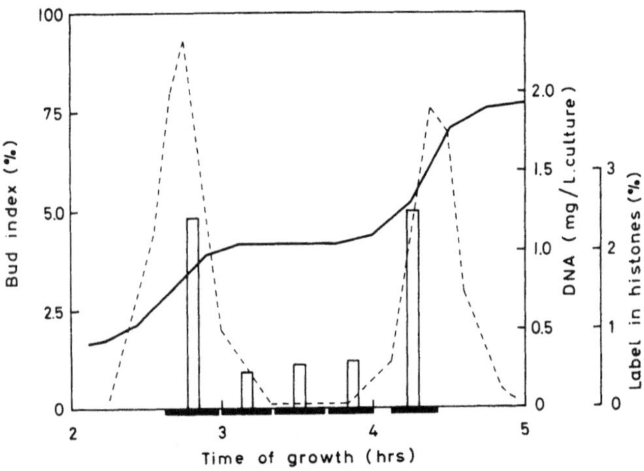

Fig. 3. Synthesis of histones in synchronously growing yeast cells. 500-ml aliquots of culture were pulse-labelled for 20 min with 1.5 μCi/ml of ^{3}H-lysine at time points indicated by heavy lines below abscissa. After the pulse, further protein synthesis was stopped by addition of cycloheximide (200 μg/ml), cells were rapidly cooled, harvested, and nuclei were prepared. Acid-soluble nuclear proteins were separated by gel electrophoresis, and radioactivity under region where histones migrate was determined. From this, percentage of label in histones relative to total radioactivity incorporated into cellular protein was estimated. This is presented as vertical bars. (----) bud index, (——) DNA content

The results of these experiments are summarized in Figure 3: Incorporation of lysine into histones occurs predominantly during the S phase. There is some labelling of histones also at other periods of the cell cycle, but this amounts to only about 20% of that measured in the S phase. This low rate of histone synthesis outside the S period could mean that the coupling of histone and DNA synthesis is less stringent in yeast than in higher organisms. It is more likely, however, that this effect is simply caused by an incomplete synchrony of the cell population used.

From experiments measuring total acid-extractable cellular protein from the yeast *Kluyveromyces fragilis* during the cell cycle, Duffus and Penman (1973) concluded that histones are synthesized mainly in G1 and not in the S phase of this yeast. As it is almost certain that the extracts analyzed by these authors contained contaminating non-histone proteins, this study is not comparable to ours.

In *S. cerevisiae* histone synthesis takes place throughout the entire S phase. This was demonstrated by labelling cells at early, mid-, or late S phase. At each of these time points, lysine incorporation into histones occurred at approximately equal rates.

All our measurements were carried out on histones extracted from isolated nuclei. A slight possibility therefore remains that we actually determined the capacity of cells to transport histones from the cytoplasm into nuclei. This transport could be coupled to active DNA synthesis and thus might occur only in S, while cytoplasmic histone synthesis could take place also at other times in the cell cycle, for instance in G1. In order to exclude this possibility, cells were pulse labelled at a time when histones isolated from nuclei are poorly labelled. A vast excess of cold lysine was then added and cells grown until they were in the S phase. Nuclei were then prepared and histones analyzed. Under these conditions we found only little more radioactivity in the histones than in those isolated immediately following the pulse. The slight increase is probably due to the fact that the addition of even very large amounts of cold lysine did not completely abolish the incorporation of tritiated lysine while cells moved into the S phase. If the assumption of a regulated histone transport into nuclei were correct, we would have expected that the histones isolated in this experiment would be as highly labelled as those pulse-labelled in the S phase. This was clearly not the case. Our experimental procedure therefore gives a fairly accurate picture of the synthesis of histones in the yeast cell cycle. Newly synthesized histones seem to appear rapidly in nuclei; there is no evidence for an accumulation of these proteins in the cytoplasm.

We have to conclude from our results that there is only a small if any pool of histones in yeast cells entering the S phase. Since these cells can complete DNA synthesis in the absence of protein synthesis, free DNA, not complexed with histones, must pile up under these conditions. In animal cells, addition of cycloheximide causes a reduction of the rate of DNA synthesis and the production of DNA fragments of medium size (21 S) that cannot be chased into high-molecular-weight DNA (Seale and Simpson, 1975). Since in these cells clusters of replicons are initiated at different times in the S period, a major effect of cycloheximide could be due to an inhibition of the initiation of those replicons that are turned on late. Chromatin formed in the presence of cycloheximide shows an increased susceptibility to nucleases (Seale and Simpson, 1975), indicating the synthesis of a higher proportion of free DNA. Similar nuclease digestion experiments have not yet been carried out on newly synthesized yeast chromatin, but we suspect that the situation will turn out to be analogous to that in higher cells.

Considering our findings one should expect that DNA syntheis exerts a control over histone formation in yeast. If this were so, inhibition of DNA replication should block histone synthesis. We have studied this question using the inhibitor of DNA replication, hydroxyurea. It was found that histone synthesis is inhibited as soon as DNA replication has come to a complete stop. Rather high concentrations of hydroxyurea (above 0.1 *M*) are necessary to block completely DNA replication in our yeast strain, and the effect of the drug is seen only about two hours after its addition. Control experiments were therefore conducted in which the effect of hydroxyurea on total protein synthesis as measured by the incorporation of ^{3}H-leucine into total cellular proteins was studied. It was found that hydroxyurea under the conditions employed did not interfere with the capacity of cells to synthesize protein. Incorporation of leucine was almost identical in cells held in the presence or absence of hydroxyurea. If the incorporation of ^{3}H-lysine into total cellular protein was determined after incubation of cells in the presence of hydroxyurea, labelling rates was found to be reduced by about 20%. This could at least in part be due to a block of histone synthesis. Thus, despite the somewhat unfavorable conditions that had to be used, our experiments strongly indicate a control of DNA replication over histone synthesis in yeast similar to that known to operate in higher organisms.

Acknowledgements. We thank Waltraud Machnik for her expert technical assistance in some of the experiments. This work was supported by grants from the Deutsche Forschungsgemeinschaft (SFB 105 and Schwerpunktsprogramm "Nucleinsäure- und Proteinbiosynthese").

References

Bradbury, E.M., Inglis, R.J., Matthews, M.R., Sarner, N.: Phosphorylation of very lysine-rich histone in *Physarum polycephalum*. Eur. J. Biochem. 33, 131-139 (1973)

Duffus, J.M., Penman, C.S.: Stepwise accumulation of an acid-extractable protein fraction in the budding yeast, *Kluyveromyces fragilis*, during the cell cycle. J. Gen. Microbiol. 79, 189-194 (1973)

Golombek, J., Wintersberger, E.: Glucose uptake in the cell cycle of *Saccharomyces cerevisiae*. Exp. Cell Res. 86, 199-202 (1974)

Golombek, J., Wolf, W., Wintersberger, E.: DNA synthesis and DNA polymerase activity in synchronized yeast cells. Molec. Gen. Genetics 132, 137-145 (1974)

Hereford, L.M., Hartwell, L.M.: Role of protein synthesis in the replication of yeast DNA. Nature New Biol. 244, 129-131 (1973)

Kornberg, R.D.: Chromatin structure: A repeating unit of histones and DNA. Science 184, 868-871 (1974)

Lange, R.J. de, Smith, E.L.: Histones: Structure and function. Ann. Rev. Biochem. 40, 278-314 (1971)

Lark, K.G.: Initiation and control of DNA synthesis. Ann. Rev. Biochem. 38, 569-604 (1969)

Lohr, D., van Holde, K.E.: Yeast chromatin subunit structure. Science 188, 165-166 (1975)

Noll, M.: Subunit structure of chromatin. Nature 251, 249-251 (1974)

Olins, A.L., Olins, D.E.: Spheroid chromatin units (v bodies). Science 183, 330-332 (1974)

Oudet, P., Gross-Bellard, M., Chambon, P.: Electron microscopic and biochemical evidence that chromatin structure is a repeating unit. Cell 4, 281-300 (1975)

Seale, R.L., Simpson, R.T.: Effects of cycloheximide on chromatin biosynthesis. J. Molec. Biol. 94, 479-501 (1975)

Williamson, D.M.: Replication of the nuclear genome in yeast does not require concomitant protein synthesis. Biochem. Biophys. Res. Comm. 52, 731-740 (1973)

Wintersberger, U., Smith, P., Letnansky, K.: Yeast chromatin. Preparation from isolated nuclei, histone composition and transcription capacity. Eur. J. Biochem. 33, 123-130 (1973)

6. Macromolecular Events Following Nutritional Shift-Up of *Tetrahymena*

R. E. Pearlman

Tetrahymena pyriformis is a unicellular eukaryotic organism that has been much used in biochemical and physiologic studies of the cell cycle and the control of cell growth and division (Hill, 1972; Elliott, 1973). *Tetrahymena* has a short generation time (2 1/2 h under most conditions), can be readily grown in large amounts, is readily labelled by nucleic acid and protein precursors, and both enzymes and nucleic acids from these cells are readily isolated and handled. Also, recent technical advances have made mutant isolation and genetic analysis with this organism very feasible (Gardonio et al., 1975; Orias and Bruns, 1975).

Our interest in recent years has been to study replication and transcription at the molecular level in a eukaryotic organism. In this regard, a nutritional shift-up condition with *Tetrahymena* has recently proved to be very useful. The system as utilized with *Tetrahymena* was first described by Cameron and Jeter (1970). A schematic representation of the experimental design for shift-up experiments is shown in Figure 1. We have been most interested in studying events that occur rapidly following return to nutrient medium.

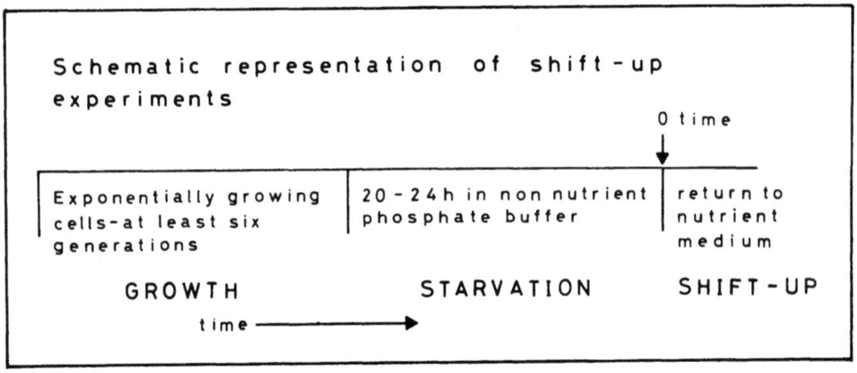

Fig. 1. Schematic representation of the experimental design of nutritional shift-up experiments. Cells are grown either in a complex medium such as 2% proteose peptone (Difco) and 0.1% liver extract (Nutritional Biochemical Co.) or in a completely defined medium (e.g. Rasmussen and Modeweg-Hansen, 1973) to a cell density of between 8 × 10^4 and 1.5 × 10^5 cells/ml. Cells are collected and washed once by centrifugation for 4 minutes at 500 rpm using the 269 rotor of an IEC PR 6000 centrifuge. Cells are washed with starvation buffer (0.6 g KH_2PO_4, 0.15 g K_2HPO_4 and 0.25 g $MgSO_4$ dissolved in 1 liter distilled water and adjusted to pH 6.5 with NaOH) and are supended in this buffer at the appropriate cell concentration for the period of starvation. Shift-up (0 time in most experiments) is accomplished by adding an equal volume of 2 × concentrated growth medium to the cells in starvation buffer. Rudick and Cameron (1972) suggested cells of strain HSM were 75% G1 and 25% G2 at the end of starvation while Mowat et al. (1974) and Wolfe (1973) suggested that cells (both micronucleate and amicronucleate) were exclusively G1 at the end of extensive starvation. Recent data for

The kinetics of uptake of ^{3}H-uridine and ^{3}H-thymidine into RNA and DNA following shift-up are shown in Figure 2. RNA synthesis begins about 30 min after shift-up and rapidly reaches a rate characteristic of that in exponential cells. Further details of transcription studies will be described later. DNA synthesis also begins about 30 min after shift-up. Until about 120 min after shift-up, virtually all of the DNA synthesized is ribosomal DNA (rDNA-nucleolar DNA) and mitochondrial DNA. Macronuclear nucleoplasmic DNA synthesis does not begin until be-

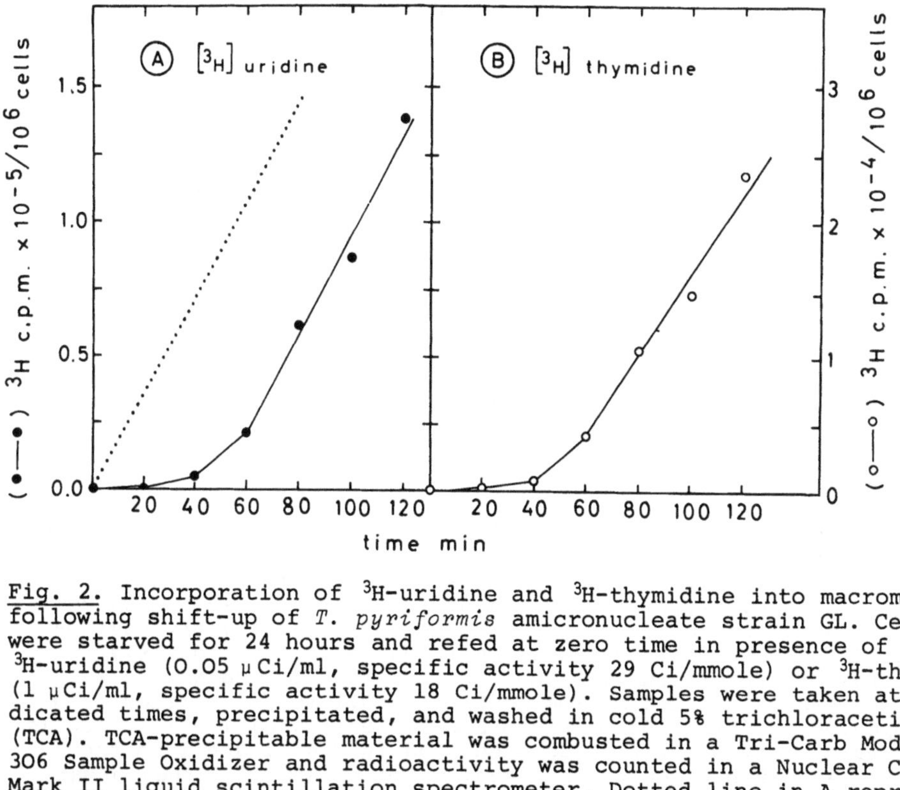

Fig. 2. Incorporation of ^{3}H-uridine and ^{3}H-thymidine into macromolecules following shift-up of *T. pyriformis* amicronucleate strain GL. Cells were starved for 24 hours and refed at zero time in presence of either ^{3}H-uridine (0.05 μCi/ml, specific activity 29 Ci/mmole) or ^{3}H-thymidine (1 μCi/ml, specific activity 18 Ci/mmole). Samples were taken at indicated times, precipitated, and washed in cold 5% trichloracetic acid (TCA). TCA-precipitable material was combusted in a Tri-Carb Model 306 Sample Oxidizer and radioactivity was counted in a Nuclear Chicago Mark II liquid scintillation spectrometer. Dotted line in A represents uptake of ^{3}H-uridine in control exponentially growing culture

amicronucleate strain GL suggest that the population at the end of starvation is heterogeneous and consists of 80-95% cells in G1 and 5 ∿ 20% cells in either late S or G2 (Jeter et al., 1975; Salamone and Pearlman, 1975 and manuscript in preparation). Cameron and Bols (1975 and personal communication) suggest that the fraction of cells in G2 at end of starvation depends on cell density at beginning of starvation. This correlation has not been observed in our laboratory, and we also never see 100% or greater increase in cell number or any given cell dividing more than once during starvation, as has been reported previously (Cameron and Jeter, 1970; Cameron et al., 1971). In our hands, cell number during starvation generally increases between 30% and 75%, depending on strain and medium used during exponential growth (Mowat et al., 1974; Salamone and Pearlman, manuscript in preparation), although early events following shift-up are not affected even if cell number increases only 5% during starvation

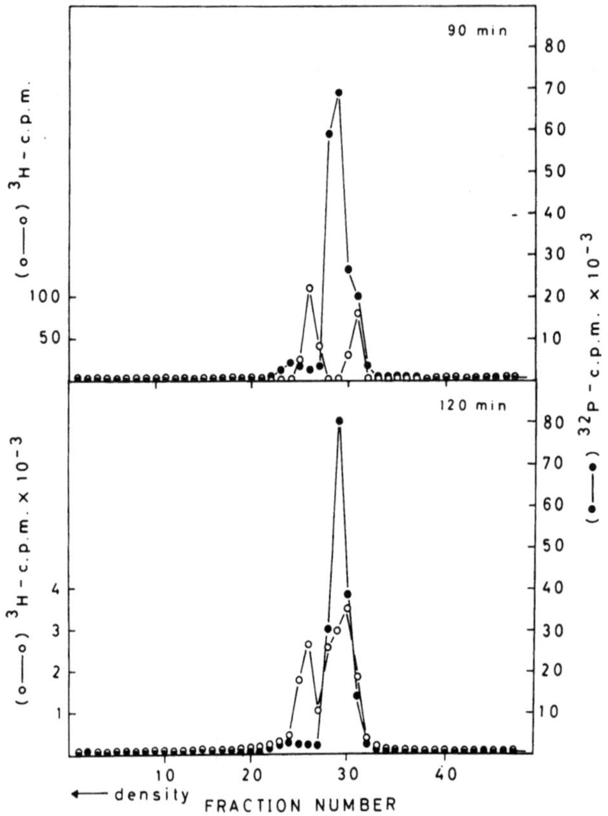

Fig. 3. Preferential replication of rDNA following nutritional shift-up in micronucleate strain of *T. pyriformis*. A strain of syngen 1, inbred family B, mating type IV was used in these experiments. Micronucleus of this strain is homozygous for dominant mutation for resistance to 6-methylpurine but macronucleus and thus phenotype is 6-methylpurine-sensitive. This functional heterokaryon (Bruns and Brussard, 1974) was obtained from Dr. Peter Bruns, Cornell University, Ithaca New York. Cells were grown in defined medium containing 0.1 μCi/ml carrier free ^{32}P-orthophosphate (Risø, Denmark) to a cell density of 1.5 × 10^5 cells/ml. Cells were starved for 20 h as described in Fig. 1 and were then refed with 2 × concentrated defined medium (containing 1/2 the normal concentration of pyrimidine) containing 2 μCi/ml ^{3}H-thymidine (20 Ci/m*M*, New England Nuclear Corp.). Samples were taken at 90 min and at 120 min, harvested by centrifugation, and crude nuclear pellet was prepared as described previously (Engberg and Pearlman, 1972). Nuclei were lysed in 0.5% sodium dodecyl sulfate and DNA was extracted with chloroform-octanol and analyzed on CsCl gradients (7.4 g CsCl, 6 ml DNA) by centrifugation at 37,000 rpm for 60 h in a 40 fixed angle rotor in a Beckman L-50 ultracentrifuge. Fractions were collected from the bottom of the tube, made 0.3 N KOH and RNA was hydrolyzed by incubation for 1 h at 80°C. 100 μl aliquots from each fraction were applied to Whatman 3MM filter paper discs, the discs were washed four times in cold 1N HCl and twice in 95% ethanol, dried and counted in Permablend scintillation solution using a Nuclear Chicago Mark II liquid scintillation spectrometer. Correction has been made for ^{32}P counts overlapping into the ^{3}H channel. The ^{3}H peak on the heavy side of the ^{32}P bulk label represents rDNA and the ^{3}H on the light side of bulk DNA is presumably mitochondrial DNA (some mitochondria contaminate the crude nuclear preparations used in these experiments). ●——● ^{32}P cpm o——o ^{3}H cpm

tween 90 and 100 min after shift-up. This was demonstrated in the amicronucleate strain GL by DNA-RNA hybridization experiments (Engberg et al., 1972) and by buoyant density and electron microscope autoradiography experiments (Engberg et al., 1974b). This same pattern of preferential replication of rDNA and mitochondrial DNA is also observed in syngen 1, inbred family B micronucleate strains of *Tetrahymena* (Fig. 3). This is of interest because it is these micronucleate strains that must be utilized for any genetic experiments.

Ribosomal DNA in *Tetrahymena* is nucleolar, as is the case in most eukaryotes. In macronuclei, however, rDNA is autonomous and at least the majority is free and not integrated into chromosomal DNA. Its molecular weight is approximately 12.6 × 10^6 daltons depending on the strain from which it is isolated. It is isolated predominantly as linear molecules of defined length, although circular molecules have been observed in the electron microscope in some preparations (Gall, 1974; Engberg et al., 1974a). Partial denaturation mapping and restriction nuclease mapping has demonstrated that the linear molecules are unique and they do not arise from random breakage of circles. If rDNA does exist in vivo as a circular molecule, the circle is broken at a unique site to produce the linear molecules isolated (Gall, 1974; Engberg et al., 1976).

Of further interest is the unusual structure of the rDNA molecule. Electron microscopic visualization of the molecule as well as biochemical experiments utilizing the restriction endonuclease Eco R1 as well as the single strand specific nuclease S1 from *Aspergillus oryzae* have demonstrated that the molecule is a large reverse repeat containing a two-fold axis of rotational symmetry. In other words, the two strands of the molecule are not only complementary but are identical (Karrer and Gall, 1976; Engberg et al., 1976). A diagramatic representation of an rDNA molecule from amicronucleate strain GL together with a restriction map of this molecule is shown in Figure 4.

The preferential replication of rDNA in macronuclei observed following nutritional shift-up offers a unique opportunity to study the replication of a eukaryotic gene in its homologous nucleus. Our extensive knowledge of the structure of this gene will be of great value in these studies. Experiments both in vivo and in vitro to study rDNA replication in the shift-up condition are in progress. Preliminary observations (Andersson and Klenow, personal communication) suggest that DNA hybridizable to rRNA is synthesized in vitro in macronuclei isolated from shift-up *Tetrahymena*.

Continuing with DNA synthesis in the shift-up condition, we have investigated whether cells were capable of carrying out excision repair of ultraviolet (UV)-induced damage to DNA early after shift-up, at a time when unirradiated cells do not carry out any nucleoplasmic DNA synthesis. Our hope was to be able to study excision repair at a time when undamaged cells did not have the capacity for normal semiconservative replication. Details of the excision-repair of UV-induced damage to DNA or *Tetrahymena* following shift-up have been reported previously (Mowat and Pearlman, 1975). At this time, I will review only some of the results as they bear on the use of the shift-up condition.

When irradiation with sublethal doses of UV radiation (254 nm) is administered just prior to shift-up, excision repair begins 1 h later and appears to be essentially complete within 7 h. During this time, thymine dimers produced by irradiation are removed. Semiconservative DNA synthesis commences 2-3 h after the first appearance of excision repair as occurs in irradiated exponentially growing cells (Brunk,

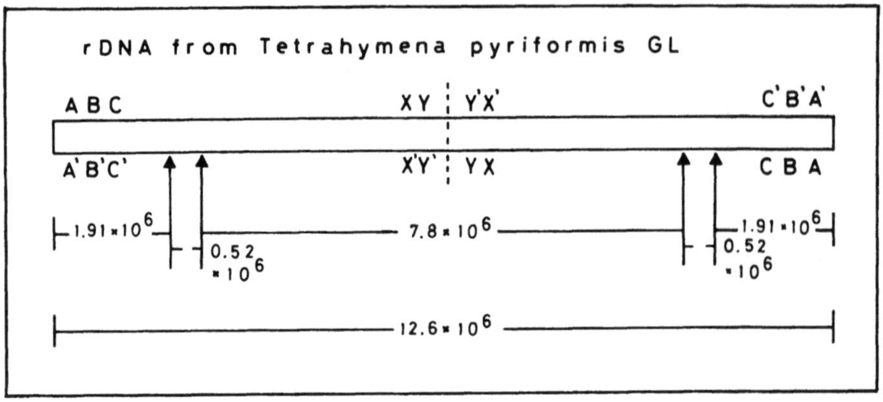

Fig. 4. Schematic representation of rDNA from *Tetrahymena pyriformis* amicronucleate strain GL showing the positions of cleavage by Eco Rl. rDNA was isolated as described by Engberg et al. (1974a). Restriction endonuclease mapping of the molecule is described by Engberg et al. (1976). rDNA isolated from syngen 1 strains of *Tetrahymena* has symmetry properties identical to those shown here but the Eco Rl restriction map differs from that for strain GL (Karrer and Gall, 1976). ↑ Sites of cleavage by Eco Rl restriction nuclease. ¦ axis of rotational symmetry. A-nucleotide or nucleotide sequence and its complement A'

1972). In addition, between 3 and 8 h after shift-up, the rate of DNA synthesis in irradiated cultures is much lower than the rate of DNA synthesis in unirradiated cultures. Also, the specific activity in vitro of DNA polymerase from irradiated shift-up cells is approximately 10-fold greater than that of polymerase from unirradiated shift-up cells (Mowat and Pearlman, 1975).

An ideal experimental situation would be to study the enzymology of excision repair at a time when no normal DNA synthesis occurs in non-irradiated cultures. Then, enzymatic activity, such as DNA polymerase activity, may be correlated with the repair process. This ideal situation does not occur in shift-up *Tetrahymena*. Normal nucleoplasmic DNA synthesis commences in unirradiated control cultures at the same time that excision repair commences in irradiated cultures. It has been observed previously, however, that DNA polymerase activity can be induced in *Tetrahymena* following treatment of exponentially growing cells with UV radiation or other agents causing excision-repairable damage to cells (Westergaard and Pearlman, 1969; Keiding and Westergaard, 1971). There is a large increase in DNA polymerase specific activity in lysates from irradiated shift-up cells, and these high levels of DNA polymerase activity are accompanied by low rates of DNA synthesis characterized as repair. In unirradiated control cultures, low levels of DNA polymerase are present, yet the rate of DNA synthesis is very high. This suggests that detailed studies of enzymes of DNA metabolism in irradiated shift-up *Tetrahymena* will be of significant interest in studying the mechanism of excision repair in vitro.

We have also studied ribosomal RNA (rRNA) synthesis at early times following shift-up. Synthesis of rRNA can be observed in nuclei by 30 min after shift-up, although essentially no label is seen in the cytoplasm at this time. By 50 min after shift-up howerer, label is found in 17 S and 25 S rRNA in the cytoplasm. Of interest also is that a radioactive peak migrating at 35 S, the presumptive primary rRNA transcript, is seen in nuclei 30 min after shift-up and persists until at least 50 min after shift-up (Fig. 5). The nuclear labelling pattern

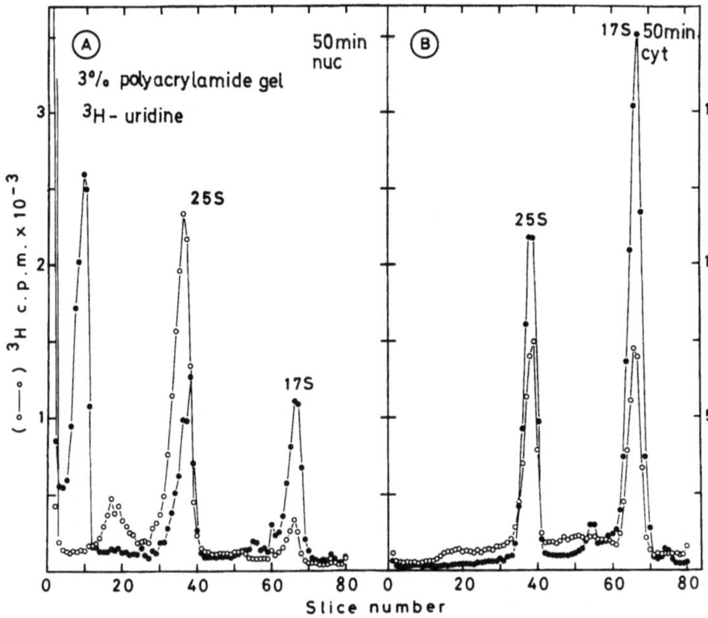

Fig. 5. Polyacrylamide gel electrophoresis of ^{3}H-uridine labelled RNA isolated from *T. pyriformis* amicronucleate strain GL 50 min following shift-up. Cells were grown for 8 generations in presence of 1 μCi/ml carrier free ^{32}P-orthophosphate (Risø, Denmark), starved, and then refed in presence of ^{3}H-uridine (0.5 μCi/ml, specific activity 29 Ci/mmole). Cells were harvested 50 min after shift-up, and nuclei and cytoplasm were isolated following lysis of cells in 0.5% Triton X-100 (Engberg and Pearlman, 1972). RNA was isolated by treatment with phenol-SDS as described by Leick and Plesner (1968). Electrophoresis was for 18 h in 3% polyacrylamide gels at 2.5 mA/tube in tubes 8 mm in diameter. Gels were sliced in 1-mm fractions and were counted using method of Hellung-Larsen (1971). •——• ^{32}P cpm; o——o ^{3}H cpm

also shows that the 25 S material labelled for 50 min following shift-up reproducibly migrates in polyacrylamide gels as if it were slightly larger than the steady-state labelled 25 S peak, again suggesting a possible precursor. Because in the shift-up condition we are observing the "turning on" of rRNA transcription, the system is extremely valuable in studying early events in the transcription of rRNA and the formation of ribosomes, particularly processing and transport from nucleus to cytoplasm. These in vivo studies can then be correlated with transcription studies in vitro that we are carrying out using the pure rDNA described earlier. Transcription in vivo has also been correlated with changes in nucleolar morphology from the existense of large fusion bodies in straved cells to the appearance of multiple peripheral nucleoli (characteristic of exponential cells), which occurs on shift-up (Nilsson and Leick, 1970; Leick et al., 1975).

The kinetics of the incorporation of radioactively labelled precursors into DNA, RNA, and protein following shift-up has previously been described by Cameron and co-workers (Cameron and Jeter, 1970; Cameron et al., 1971; Rudick and Cameron, 1972) and by Stocco and Zimmerman (1974). In these studies, the shift-up system was used as a means of obtaining a degree of cell synchrony and of studying events leading to and possibly involved in the control of cell division. In these studies, RNA and DNA synthesis following shift-up began at least one hour later than we observe. The reason for this difference is not clear although it is possible that the analytical methods used in their laboratory were not sensitive enough to detect the earliest events occurring. Strain differences are an unlikely explanation for the differences observed, because our experiments have been carried out with both the amicronucleate strain GL and with micronucleate strains of syngen 1, inbred family B. We (Mowat and Pearlman, 1975) also do not observe the increase in DNA polymerase specific activity on shift-up reported by Rudick and Cameron (1972). These differences may be important, because Cameron and co-workers have discussed events occurring between 120 and 140 min as being essential for DNA replication, which they describe as not beginning before 150 min after shift-up. However, because we have described nucleoplasmic DNA synthesis beginning by 120 min after shift-up and nucleolar and mitochondrial DNA synthesis beginning by 90 min after shift-up, the conclusions of Cameron and co-workers regarding the control of DNA replication may be subject to question. In any event, as long as care is taken not to damage the cells during centrifugation in preparation for starvation (Fig. 1), the shift-up system as described here is reproducible and is an extremely useful experimental situation for studying replication, repair, and transcription at the molecular level. Because cell division does not occur until about 240 min after division (Cameron and Jeter, 1970; Mowat et al., 1974) and because it has been shown that protein synthesis begins very early following shift-up (Cameron et al., 1971), this system should also be valuable in studying the synthesis of proteins that may be required for division. In particular, this system may be very useful for the study of the synthesis of ribosomal proteins.

Acknowledgements. Much of the work discussed in this review was done by and in collaboration with members of the Biochemical Institute B, University of Copenhagen, Medical School. In particular, I would like to thank Drs. J. Engberg, H. Klenow and V. Leick, Biochemical Institute B, Dr. J. Nilsson, Dept. of Zoology, University of Copenhagen, and Dr. J. Collins, Dept. of Microbiology, University of Copenhagen for their help in the preparation of this manuscript and for access to data, both published and unpublished. I also thank Drs. J. Gall and K. Karrer for making data available to me prior to publication.

References

Brunk, C.F.: The time sequence of events during excision repair in *Tetrahymena pyriformis*. Photochem. Photobiol. 15, 349-360 (1972)

Bruns, P., Brussard, T.: Positive selection for mating with functional heterokaryons in *Tetrahymena pyriformis*. Genetics 78, 831-841 (1974)

Cameron, I.L., Bols, N.C.: Effect of cell population density on G_2 arrest in *Tetrahymena*. J. Cell. Biol. 67, 518-522 (1975)

Cameron, I.L., Griffin, E.E., Rudick, M.J.: Macromolecular events following refeeding of starved *Tetrahymena*. Exptl. Cell Res. 65, 265-272 (1971)

Cameron, I.L., Jeter, J.R.: Synchronization of the cell cycle of *Tetrahymena* by starvation and refeeding. J. Protozool. 17, 429-431 (1970)

Elliott, A.M.: Biology of *Tetrahymena*. Dowden, Hutchinson and Ross, Stroudsburg, Pennsylvania 1973

Engberg, J., Christiansen, G., Leick, V.: Autonomous rDNA molecules containing single copies of the ribosomal RNA genes in the macronucleus of *Tetrahymena pyriformis*. Biochem. Biophys. Res. Comm. 59, 1356-1365 (1974a).

Engberg, J., Collins, J., Andersson, P., Leick, V.: The free rDNA molecules from *Tetrahymena pyriformis* GL are giant polindromes. J. Mol. Biol. (1976)

Engberg, J., Mowat, D., Pearlman, R.E.: Preferential replication of the ribosomal RNA genes during a nutritional shift-up in *Tetrahymena pyriformis*. Biochim. Biophys. Acta 272, 312-320 (1972)

Engberg, J., Nilsson, J.R., Pearlman, R.E., Leick, V.: Induction of nucleolar and mitochondrial DNA replication in *Tetrahymena pyriformis*. Proc. Nat. Acad. Sci. U.S.A. 71, 894-898 (1974b)

Engberg, J., Pearlman, R.E.: The amount of ribosomal RNA genes in *Tetrahymena pyriformis* in different physiological states. Eur. J. Biochem. 26, 393-400 (1972)

Gall, J.G.: Free ribosomal RNA genes in the macronucleus of *Tetrahymena*. Proc. Nat. Acad. Sci. U.S.A. 71, 3078-3081 (1974)

Gardonio, E., Crerar, M., Pearlman, R.E.: Manipulations with *Tetrahymena pyriformis* on solid medium. Methods in Cell Biology (ed. D.M. Prescott) Academic Press Vol. IX, 329-348 (1975)

Hellung-Larsen, P.: Liquid scintillation counting of ^{3}H and ^{32}P RNA in slices of polyacrylamide gels. Anal. Biochem. 39, 454-461 (1971)

Hill, D.L.: Biochemistry and Physiology of *Tetrahymena*. New York: Academic Press 1972

Jeter, J.R., Pavlat, W.A., Cameron, I.L.: Changes in the nuclear acidic proteins and chromatin structure in starved and refed *Tetrahymena*. Exptl. Cell Res. 93, 79-88 (1975)

Karrer, K.M., Gall, J.G.: The macronuclear ribosomal DNA of *Tetrahymena pyriformis* is a palindrome. J. Mol. Biol (1976)

Keiding, J., Westergaard, O.: Induction of DNA polymerase activity in irradiated *Tetrahymena* cells. Exptl. Cell Res. 64, 317-322 (1971)

Leick, V., Plesner, P.: Formation of ribosomes in *Tetrahymena pyriformis*. Biochim. Biophys. Acta 169, 398-408 (1968)

Leick, V., Tønnesen, T., Engberg, J., Nilsson, J.R.: Nucleolar reorganization in *Tetrahymena pyriformis* GL during a nutritional shift-up; timing of ribosomal RNA transcription and ribosomal DNA replication. Proceedings of the 9th Conference of the Nordic Society for Cell Biology, Odense University Press, Denmark 1975

Mowat, D., Pearlman, R.E.: Excision repair in *Tetrahymena:* Events following refeeding of starved UV-irradiated cells. Photochem. Photobiol. 21, 5-11 (1975)

Mowat, D., Pearlman, R.E., Engberg, J.: DNA synthesis following refeeding of starved *Tetrahymena pyriformis* GL. Exptl. Cell Res. 84, 282-286 (1974)

Nilsson, J.R., Leick, V.: Nucleolar organization and ribosome formation in *Tetrahymena pyriformis* GL. Exptl. Cell Res. 60, 361-372 (1970)

Orias, E., Bruns, P.: Induction and isolation of mutants in *Tetrahymena*. Methods in Cell Biology (ed. D.M. Prescott). New York: Academic Press 1975

Rasmussen, L., Modeweg-Hansen, L.: Cell multiplication in *Tetrahymena* cultures after addition of particulate material. J. Cell Sci. 12, 275-286 (1973)

Rudick, M.J., Cameron I.L.: Regulation of DNA synthesis and cell division in starved - refed synchronized *Tetrahymena pyriformis* HSM. Exptl. Cell Res. 70, 411-416 (1972)

Salamone, M., Pearlman, R.E.: Abstract to annual meeting of American Society for Cell Biology Puerto Rico 1975

Stocco, D.M., Zimmerman, A.M.: Metabolism of acid - soluble nucleotides in starved - refed synchronized *Tetrahymena pyriformis* GL. Can. J. Biochem. 52, 310-318 (1974)

Westergaard, O., Pearlman, R.E.: DNA polymerase activity in methotrexate plus uridine treated *Tetrahymena*. Exptl. Cell Res. 54, 309-313 (1969)

Wolfe, J.: Conjugation in *Tetrahymena:* The relationship between the division cycle and cell pairing. Dev. Biol. 35, 221-231 (1973)

7. Radiation-Induced Processes in Moss Cells: Short-Term and Long-Term Radiation Responses of Special Interest after Microbeam UV Irradiation of the Haploid Moss Cells of *Funaria hygrometrica*

Fr. R. von Döhren

A. Introduction

The response of a cell population to perturbation is a delicate one with a series of internal controls operating simultaneously. While this response is obvious for radiation, it also tends to operate in response to most potentially lethal influences in general (von Döhren, 1974). It would be desirable to have a complete model, which would help to interpret many complex cellular events. Perhaps the most interesting problem is the investigation of cellular components during the growth cycle. However, quantitative formulation of these components is limited by many prohibitive factors. Consequently the complexity of all established models is restricted by experimental data as well as by limitations in the generalization of data obtained from bacteria compared to those from yeast or other more developed eukaryotic systems, which in contrast are very complex. Furthermore, there are only a few accurate studies in the literature in which the growth kinetics of individual functional entities are analyzed in detail. Thus, it would be useful to have available another eukaryotic system that could be easily handled and that would survive under laboratory conditions.. Such a system is demonstrated in these experiments with the protonema of *Funaria hygrometrica*.

B. General Remarks

Growth at the cellular level is considered to be similar for all types of cells and is therefore a basic property of biologic species. *F. hygrometrica* exhibits the particularity that growing cells penetrate in only one direction. In order to comprehend basic cellular growth, it is essential that relationships between individual functional units should be analyzed as a function of time, since the growth process represents the changes of absolute concentrations of these entities.

The multiplicity of factors and functional entitities involved in such a system clearly suggests that only by quantitative analysis will it be possible to unravel the modes and kinetics of regulatory processes. Furthermore, multiple coupling between cells - as is suggested in this system (Bopp and Böhrs, 1965) - , and between various metabolic and physiologic processes suggests that a stepwise systems analysis approach should be used, whereby only a few parameters at a time are changed by the interfering experiment. Earlier experiments explored mainly the growth kinetics of the rapidly differentiating protonema including the cell division cycle, as far as is possible by microscopic observation. Experiments (in preparation) will deal with the relation of growth to DNA synthesis. The present results show short-term radiation effects after UV irradiation of selected parts of the cell as determined by UV microspectrophotometric measurements in the living cell, in a special case combined with long-term effects in the same cells, i.e., after some days, in relation to cell division and growth control.

C. Extrinsic and Intrinsic Conditions

I. Physiology

The growth characteristics of this rapidly differentiating system will be given elsewhere (von Döhren, in preparation) and are therefore mentioned only briefly. The moss plants are cultivated from spores under exactly controlled conditions on Knop-Agar under 6000 lux at 20°C. Under these conditions the cells grow horizontally in a filacious manner and show distinct apical growth (von Döhren, 1966). These cells are about 50-80 μ in length; the nucleus can be detected without phase contrast optics, and it has one nucleolus. Cell division takes place as soon as the apical cell reaches more than twice the length of the mother cell. For this reason cell division figures can be observed without difficulty. During cellular and nucleare division, which takes about 40 min, nuclear division figures can barely be observed. About 10 min after the disappearance of the nucleus, a barrel-shaped phragmoplast becomes visible; the primary membrane becomes visible 30 min later. When the nucleus reappears, the nucleolus can immediatly be recognized and within a few minutes it grows to its initial size. In the sampled state of "Caulonema Primanen" the apical cells of one protonema do not yet divide simultanously, and the apical growth pattern is therefore not as synchronized as it may become in a later differentiated state, the "Caulonema" (Bopp and Böhrs, 1965).

II. Experimental Conditions

Although the apical cells of the same protonema do not exhibit a synchronized growth pattern, the neighbor cells may act as controls in respect to laboratory conditions, because only one or two of the four to seven apical cells suffer from the UV dose, which has previously been applied to selected parts of the cell. Thus, the radiation response can easily be determined by microscopic examination of disordered growth characteristics, e.g., retardation of apical growth, delay of cell division or its deletion, and other factors, in comparison to the whole protonema. Here we have a unique possibility to exclude damages in the observed cell that are introduced by extrinsic factors not correlated to the experiment in question. Further, the observed trends in the extinction changes before irradiation in the living cell compared to those after irradiation at the same place only a few minutes later can thus be related to long-term effects occurring a few days later.

III. Cell Kinetics and Cell Division

The distinct apical growth pattern of apical cells and the occurrence of cell division when the daughter cell reaches more than twice the length of the mother cell will be used to classify the cell cycle in relation to the length of the apical cell to the length of its mother cell. It is well established that the velocity of growth in apical cells is a first order function with respect to time (only during light time), and the phases of the interphase can be correlated to special classes gained by dividing the length of the apical cells by the length of its mother cells. G2 and pro-, meta-, and telophase should have been completed at 240% as soon as the cell shows the phragmoplast. A few minutes later the primary membrane can be seen. After a while the new cell starts growing again and has to pass through G1, S, and G2 until it reaches 240% once again. Although in these experiments G1 is equal to more than 100%, S is assumed to remain in the middle, 140-160%, and G2 is placed just before 240%. One may fix the phases more precisely by determining the DNA contents in relation to this classification.

IV. Procedures and Apparatus

For UV irradiation three diaphragms are used: one comparable to the nucleolus, the next in the order of magnitude of the whole nucleus, and and one to irradiate most of the cell area. Details of the technique are given elsewhere (von Döhren, 1974, 1966) and shall not be discussed here. Two mercury lamps, one which emitts chiefly in the range of 254 nm and the other, in the range of 280 nm, are used for irradiation. Undesired wavelengths are suppressed by filters. Generally, irradiation in the range of 254 nm results in less detailed radiation effects in the nucleus and in the cytoplasm compared to that at 280 nm. In this range the radiation responses seem to be more specific. Therefore every specific effect requires careful discussion.

D. Radiation Response

I. Cell Suppression by Radiation

In the case of DNA-RNA specific irradiation by 254 nm the effects have not been very carefully observed because there are too many different factors contributing to the overall reaction. The sensitivities of most of these factors seem to be comparable, because they show in a remarkable manner a switch on-/switch off reaction. This might be related to overlapping repair effects, well known for years, which play a role in the excision of pyrimidine dimers and the action of DNA polymerase in filling postexcission single-strand gaps. Furthermore, irradiation administered for some minutes will interact with cell growth, i.e., the cell will change its position in this time interval. Irradiation in the range of 280 nm is administered only for some seconds, due to the higher intensity of the high-pressure mercury lamp. There was no emission at wavelengths less than 270 nm.

II. Specific Radiation Effects

Radiation effects specific for the delivered dose as well as for the irradiated area are the most desirable. Such an effect is observed with irradiation of the phragmoplast. In this case, the primary membrane is often not correctly inserted compared to controls, which suggests that the information for orientation of the membrane is impaired or that the building entities are closely connected to protein. The cell will react in either way if the applied dose is given next to the phragmoplast. Irradiation administered at different phases of the cell cycle may be useful. This is done with the DNA-RNA specific radiation source as well with the protein specific one. In both cases the diaphragm just covers the nucleus. Irradiation of the cytoplasm has not been so successful, because suppression of growth seems to be comparable for all phases as well as for the various areas of the cell as long as the irradiated area is small compared to the rest of the cell (von Döhren, 1966). This is to be assumed since only a little of the informational content or the basic entities necessary for cell growth will be damaged. An example is given (Fig. 1) that shows the radiation response to irradiation of the cytoplasm at different stages of the interphase. The protein specific irradiation of interphase nuclei is

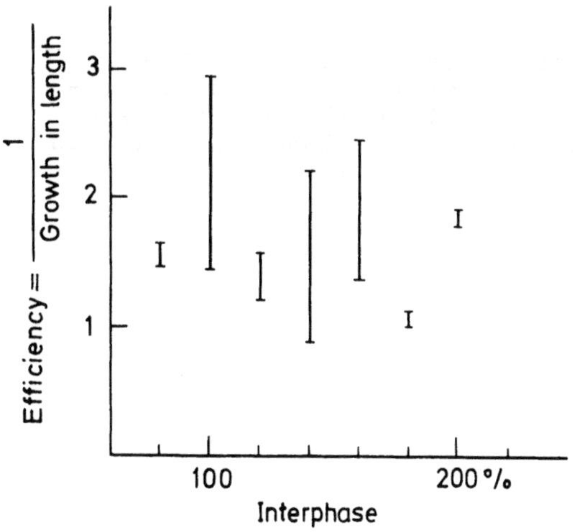

Fig. 1. Response to microbeam irradiation of cytoplasm at different stages of interphase measured by growth of controls in relation to growth of irradiated cell within one day; protein specific irradiation (λ ∿ 280 nm) 2.7×10^{-2} erg/μ^2

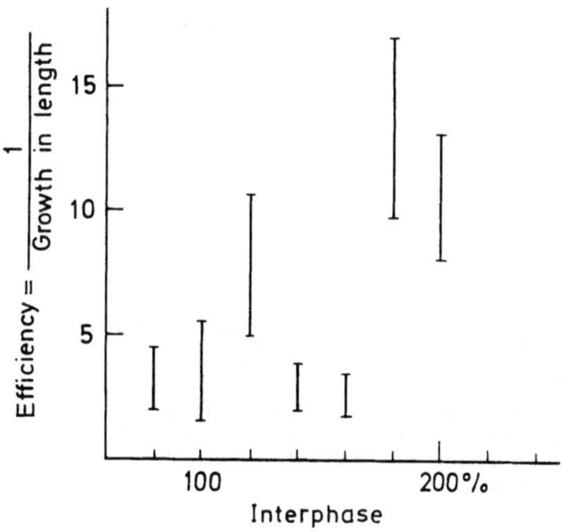

Fig. 2. Radiation response of UV microbeam-irradiated interphase nuclei. Efficiency is measured by apical growth of controls in relation to apical growth of irradiated cell. 1.3×10^{-2} erg/μ^2 protein specific irradiation (λ ∿ 280 nm)

much more effective and shows two peaks measured by the suppression of growth (Fig. 2).

The first relative maximum of radiation sensitivity is not only due to the reduction of growth. A larger number of cells in this state cannot be classified in this way because they stop growing completely. This proportion equals at the class of 100% half of the irradiated cells and at 120% just one third. The first maximum is placed in early interphase and is assumed to be correlated to G1 phase, because a relatively broad spectrum of possible radiation effects is observed. Furthermore, short-term measurements support this assumption. The second maximum of radiation efficiency in the late interphase or early prophase shows a remarkable and specific radiation response, because giants are found.

III. Short-Term Effects

Because early interphase is very sensitive to irradiation, UV microspectrophotometric measurements of the nuclei are performed just before the protein specific irradiation. Details of the technique are given elsewhere (von Döhren, 1964, 1966). The absorbance is increased in this case and drops after one hour (Fig. 3). The increasing amount

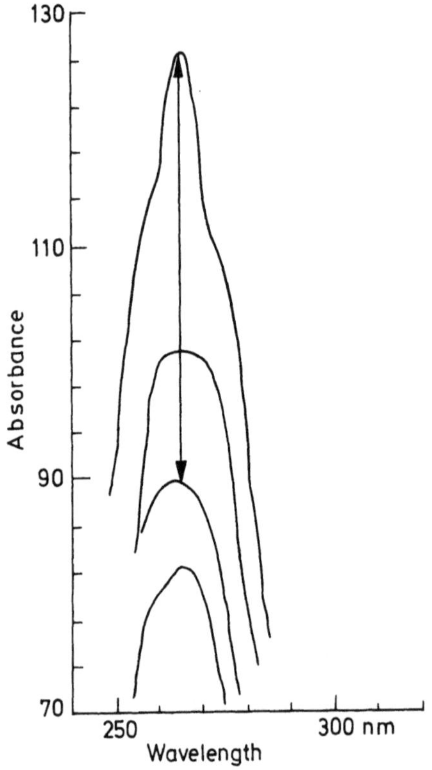

Fig. 3. Scheme of trend of absorbance in G1 phase nuclei after $4/3 \times 10^{-2}$ erg/μ^2 protein specific irradiation in living cell; increasing absorbance for 40 min with loss of absorbance after 2 h postirradiation

of absorbance after irradiation is in contrast to the well-known drop of absorbance after irradiation, which occurs if nuclei of this moss in other phases of the interphase are irradiated. This phenomenon requires discussion. It has been suggested that the increasing absorbance in this early prophase is due to irradiation of histones. Whether energy transfer to the DNA occurs with a breaking of the hydrogen bonds or whether there is a loss of histones from the DNA helix is not clear. In both cases the DNA has to be uncoiled and is not yet in S phase. This is in good agreement with the long-term effects, whereby some cells stop growing, others grow very slowly, and still others are only minimally affected.

IV. Giants

Another radiation response of some interest may be induced by UV irradiation of the nucleus just before cell division is initiated by forming the phragmoplast. In this case the cell grows to form giants. These giant cells will not divide and stay alive for some days (Fig. 4). After two or three days they are more difficult to observe since the rapidly growing neighboring cells of the protonema cover these cells. Therefore, cell growth in this case is not governed directly by cell division. Even untreated cells will not stop growing if the phragmoplast is formed.

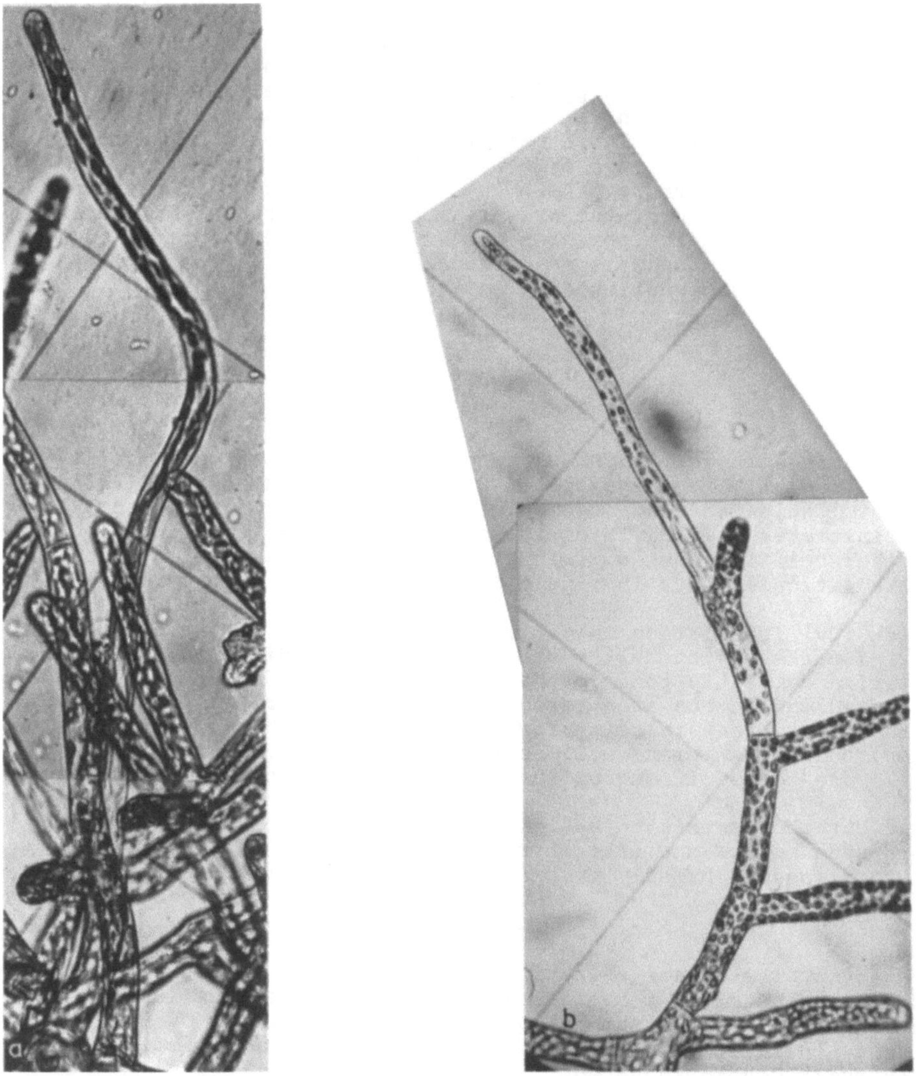

Fig. 4a and b. Giant cells formed after UV microbeam irradiation of G2 nuclei. (a) Stage of 240%, a 180 μ-long cell that grows to 360 μ and then stops growing for days (chloroplasts will bleach out after 4 days); (b) stage of 225%, cell grows to 246 μ

E. Discussion

Of special interest are the two maxima in radiation efficiency within the interphase, which were found by the protein specific irradiation of the nucleus. The first one is thought to be due to G1 phase, because the DNA seems to be uncoiled. This is supported by a broad variety of radiation responses as well as by the increase in extinction after the irradiation, which then drops after some time. At the end of interphase, a second maximum of radiation efficiency is found. In this case the cells are the stage just before cell division. The amazing and most interesting radiation response at this stage is the formation of giant cells. Evidently the nuclear division is perturbated by the delivered dose, although the purines and pyrimidines are not the primary absorbers. The basic assumption is that sufficient amino acid residues are present for absorption. These primary absorbers could be histones, DNA polymerases, or other enzymes as well whose action is related to nuclear division. Furthermore, there is strong evidence that protein participates in formation of the primary membrane. This is indicated by experiments in which the phragmoplast is irradiated by protein specific UV light. This leads to some perturbation, because the new membrane is often wrongly oriented.

F. Summary

The moss *Funaria hygrometrica* shows apical growth in protonema cells, which spread radially from the spore. Every apical daughter cell during the state of "Caulonema" and just before in the state of "Caulonema Primanen" initiates cell division as soon as more than twice the length of the mother cell is reached. This process lends itself well to the study of radiation effects in single cells.

Ultraviolet (UV) irradiation in the range of 254 nm - 280 nm delivered to different parts of the cell markedly delays cell division, may suppress it, and is able to stop the growth process, depending on the dose delivered and on the area irradiated. When the area adjacent to that where the membrane is being formed - i.e., the phragmoplast - is irradiated, the new membrane will be wrongly oriented. In particular, giant cells are formed if nuclei are irradiated during early prophase.

UV microspectrophotometric investigations of the cytoplasm and the nucleus were made before and after irradiation of living cells. These results are discussed briefly in terms of the long-term damages occurring in these cells.

Acknowledgement. The author regrets that he is unable to thank all the people who contributed by technical assistance, discussions, and financial support to this work. But all of them are represented by Prof. Dr. H. Glubrecht, Institut für Strahlenbiologie der T.H. Hannover, to whom the author feels deeply indebted.

References

Bopp, M., Böhrs, H.L.: Versuche zur Analyse der Protonema Entwicklung der Laubmoose. Planta 67, 357-374 (1965)

Döhren, Fr.R. von: Investigations of moss cells irradiated with UV-microbeam. 4th Int. Photobiol. Cong. Oxford, 1964, pp.
Döhren, Fr.R. von: Die UV-Partialbestrahlung von haploiden Protonemazellen des Mooses Funaria hygrometrica und ihre ultraviolettmikrospektrometrische Analyse. Dissertation. T.H. Hannover, 1966
Döhren, Fr.R. von: An approach to statistical factors in radiation damage through the analysis of the 3.5-day effect of whole-body x-rayed chick embryos. Radiat. Res. 57, 465-470 (1974)

8. Radiation Action on Lens Epithelial Cells

H. Rink and H.-R. Koch

The crystalline lens is a transparent, epithelial organ located in the anterior portion of the eye and forming part of its optical system. It is very sensitive to ionizing radiation and reacts by developing an opacity, called radiation cataract. Several peculiarities of this organ make it a suitable model not only for eye research but also in which to study differentiation behavior and control, cellular senescence, and radiobiology.

It seems useful to have a short look at the anatomy of the lens and to recall some of its peculiarities (Fig. 1). The lens is an avascular tissue composed of an outer, anterior monolayer of epithelial cells,

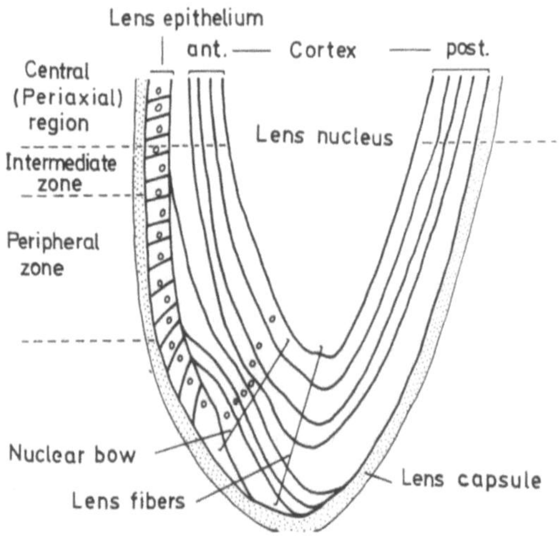

Fig. 1. Schematic diagram of meridional section through lens (redrawn from Wanko and Gavin, 1959)

a zone of cellular elongation (equatorial region) consisting of cells in the transition state from cuboidal epithelial cells to elongated fiber cells, and the inner fiber cells. Lens epithelial cells are the only cells able to divide. The preequatorial region, which is also called the germinative zone, is composed of cells having a high mitotic rate as compared to the cells in the central region, the anterior pole.

During the whole life-span of an individual the lens increases in cell mass without loss of cellular material. Cells grown in the embryonic phase are still found later on in the lens nucleus, whereas the newer cells are laid down layer by layer, the newest cells forming the periphery or cortical region. This growth pattern results in an adult tissue, the entire cellular history of which is preserved within the many layers of fiber cells. A unique feature of the lens is its transparency. This is given by the arrangement of the fiber cells and by specific proteins, the so-called crystallins (α-, β-, γ-crystallin). The synthesis of such specific proteins is often observed during cellular maturation, i.e., during the terminal stages of cellular differentiation. In muscle cells, erythrocytes, and lens cells, for example, the formation of myosin, hemoglobin, and γ-crystallin belongs to the biochemical differentiation of these cells. During the process of elongation to fiber cells, proliferative capacity stops, fiber cells lose their nuclei, DNA becomes inactive, mRNA is stabilized, and synthesis of the crystallins starts. These reactions are related to a change in the lactate dehydrogenase (LDH) isozyme pattern, going from LDH 5 to LDH 1 and consequently coupled with a change from oxidative metabolism to aerobic glycolysis (Stewart and Papaconstantinou, 1966). The synthesis of the 4 LDH subunits is genetically regulated (Shaw and Barto, 1963). Monitoring this step may be a good marker for the differentiation process per se and also for any disturbances (e.g., by irradiation) of the DNA program necessary for differentiation.

It has been shown by irradiation of partly shielded lenses (Pirie and Flanders, 1957) that the dividing cells of the preequatorial region are the primary targets of a radiation injury. A high radiation sensitivity is also known for other epithelial structures like skin, lining of the gastrointestinal tract, and sexual germinative tissues. All these tissues are characterized by continous cell proliferation. They have a resident population of young, rapidly growing cells, which are replenished as they mature. From these tissues the lens differs in one essential point. Whereas in other epithelial tissues postmature cells are sloughed off, maturing lens epithelial cells differentiate to fiber cells and are moved inward forming the so-called lens substance. In this way radiation-damaged cells (impaired survivors) are not eliminated by the physiologic process of tissue renovation but remain incorporated in the organ. These impaired cells disturb the organization and function of the surrounding unaffected fiber cells.

Investigations of lenses have been performed by protein chemists, metabolic researchers, and cell biologists. One may compile therefore three experimental approaches, all leading to a more or less sophisticated understanding of opacity development:

1. Lens damage caused by x-rays may be explained on a molecular level: lens proteins undergo changes in molecular weight by aggregation, oxidation of thiols and formation of disulphide bridges. This approach is supported by observations on physiologic aging of the lens and cataractogenesis due to noxae other than radiation (Kinoshita and Merola, 1973; van Kleff et al., 1974).

2. Metabolic changes of the lens including permeability factors may be causative. The lens is supplied with nutrients only by the aqueous humor and not directly by the blood. Alterations in metabolism (altered redoxstate, altered enzyme distribution pattern, altered permeability for substrates and ions) might consequently lead to changes in protein synthesis and/or structure and cause opacity (Hockwin, 1962; Kinsey and Reddy, 1964).

3. Irradiation acts mainly on the unique type of cells able to divide, i.e., the lens epithelial cells in the germinative zone. A disturbance of unknown kind in the dynamics of cellular proliferation, changes in differentiation behavior, DNA and protein synthesis, assembly of protein subunits or arrangement of fiber cells may trigger cataractogenesis.

Our investigations are concerned with the development of x-ray cataract in rat lenses. White Wistar rats with a mean body weight of 70 g and an age of about 6-7 weeks were irradiated with 440 R (head only).

Lens wet weight, protein content, and enzyme activities (LDH, MDH, G-6-PDH, HK, GK, PK, GOT, GPT, ALD)[1] were measured at 3, 13, 24, 80, 150, and 210 days after irradiation.

After 24 days the wet weight of the irradiated lenses diminishes as compared to the controls, reaching a significant difference of 10% on the 210th day (Fig. 2). Simultaneously the concentration of soluble proteins decreases (Fig. 3). The enzymes LDH and ALD were not influenced by irradiation. G-6-PDH decreased on the 24th day. But all 3 kinases (HK, GK, PK) were increased 3 days after irradiation. HK remained elevated until the 80th day, whereas GK and PK were temporarily depressed (20-50%). A striking finding was that GPT suffered a marked depression (Fig. 4), showing no activity on the 3rd day and a decrease to 25% on

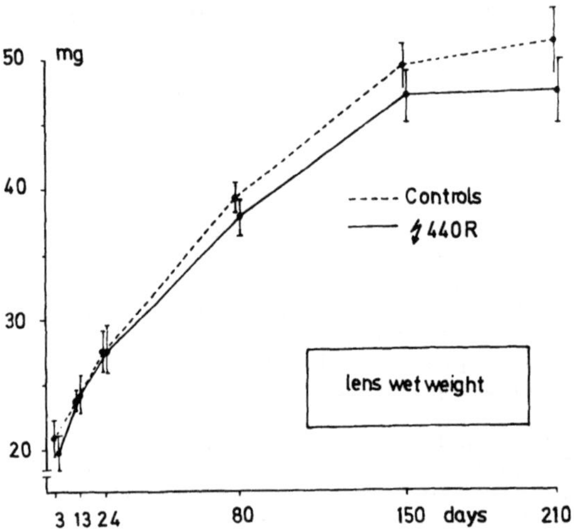

Fig. 2. Wet weight of rat lenses as function of time

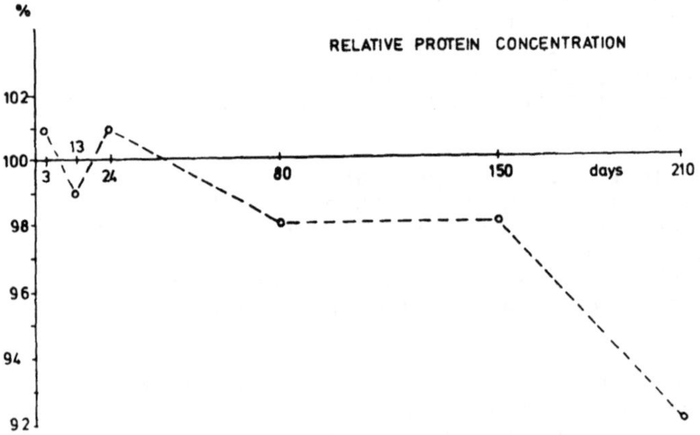

Fig. 3. Concentration of soluble proteins (%) dependent on time

[1]Abbreviations: LDH: lactate dehydrogenase; MDH: malate dehydrogenase; G-6-PDH: glucose-6-phosphate dehydrogenase; HK: hexokinase; GK: glycerokinase; PK pyruvate kinase; GOT: glutamate oxalacetate transaminase; GPT: glutamate pyruvate transaminase; ALD: aldolase.

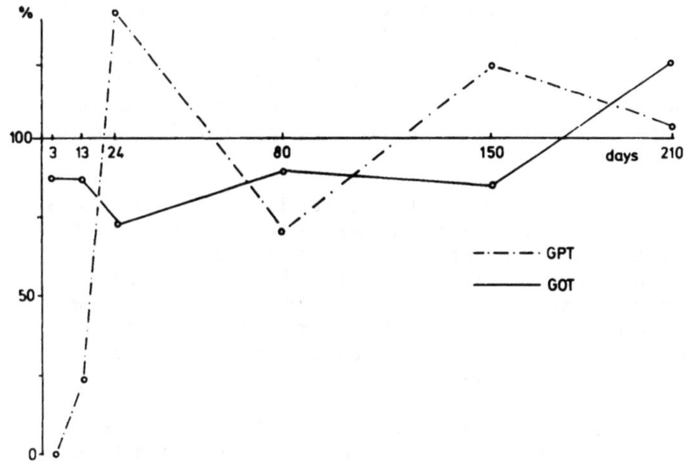

Fig. 4. Activities of the enyzmes GPT and GOT (%) as function of time

day 13. The activity of this enzyme, which is essential for protein synthesis, may be correlated with the interruption of mitotic activity, which occurs one day after irradiation and is reversible some days later.

The results indicate that irradiation primarily does not attack energy metabolism. The changes observed in enzyme activities of energy metabolism must be secondary or even tertiary effects of a general insult to lens metabolism. Protein metabolism however as an indicator for cell division capacity seems to be a main target of x-irradiation with respect to the development of x-ray cataract.

Further investigations into radiation action on rat lenses will be concerned with a more detailed analysis of the qualitative and quantitative changes of the crystallins. A study of this problem on a cellular level is also in progress.

Reviewing the current literature we should like to outline our experimental approach. Recent investigations of Worgul and Rothstein (1975) brought new light on these topics. They used several "kinds" of frogs, differing in mitotic activity: *Rana catesbeiana* with high and *Rana pipiens* with moderate and low mitotic activities. Furthermore mitosis was stimulated by hormone injections (T_3 = trijodothyronine, TSH = thyrotropine) or reduced by exposing the animals to low temperatures. The experiments demonstrated that the speed of development and the degree of severity of the radiation cataract formed were closely dependent on the mitotic activity of the germinative zone. Latency decreased and opacity increased with the absorbed dose. Stimulating mitotic activity by hormone administration (T_3/TSH) resulted in a sensitization of the lens, whereas a decrease in mitotic activity by exposure to low temperatures prior to irradiation and after irradiation lead to a suppression of radiocataractogenesis. For progression of cataractogenesis, mitotic activity or DNA synthesis is not required at the moment of exposure to radiant energy. Nonproliferating (G_0 and/ or G_1) cells sustain radiation damage, which is translated into opacity by subsequent division in the germinative zone. This means that postirradiation mitotic activity is absolutely essential for the development of a cataract. Even if division numbers are high at the time of irradiation no cataract will develop - if subsequent mitotic activity is held in check. Following these results an explanation of radiation-induced cataractogenesis could be: Kinetics of lens fiber genesis is affected, subsequently radiation-damaged cells give rise to fiber cells that will vary at least in cell size and maybe also in protein synthesis. This would lead to a disorganization in fiber orientation or to alterations in the specific protein pattern necessary for normal transparency.

From a methodologic point of view a paper of Bloemendal and co-workers (van Venrooij et al., 1974) is of interest. He was able to grow calf lens epithelial cells in vitro over about 40 cell passages (transfers) over a period of 12-18 months. In these long-term cultures monolayers of lens epithelial cells differentiated to fiber cells, with all features of morphologic differentiation, although the biochemical differentiation failed in the production of γ-crystallin and the change in LDH isozyme pattern from LDH 5 to LDH 1.

Using this system (lens epithelial cells cultured in vitro) it seems possible to clarify the action of radiation not only on cell proliferation capacity but also on differentiation behavior, DNA, and protein synthesis. Furthermore, the knowledge about aging mechanisms of diploid mammalian cells in vitro permits investigations of the radiosensitivity dependent on the cellular age in culture and on the age of the donor animals. This last point is of special interest because the severity of cataracts produced by x-rays clearly depends on age.

References

Hockwin, O.: Experimenteller Beitrag zur Wirkung von Röntgenstrahlen auf Stoffwechselabläufe der Augenlinse in Abhängigkeit vom Lebensalter. Fortschr. d. Med. 80, 659-666 (1962)

Kinoshita, J.H., Merola, L.O.: Oxidation of thiol groups in the human lens. In: The Human Lens in Relation to Cataract. Ciba Fnd. Symp. 19, 173-184 (1973)

Kinsey, V.E., Reddy, D.V.N.: Chemistry and dynamics of aqueous humour. In: The Rabbit in Eye Research. Prince, J.H. (ed.) Springfield: C.C Thomas 1964, Vol. IX, pp. 218-319

Kleff, F.S.M. van, Nijzink-Maas, M.J.C.M., Hoenders, H.J.: Intracellular degradation of γ-crystallin. Eur. J. Biochem. 48, 563-570 (1974)

Pirie, A., Flanders, H.: Effect of x-rays on partially shielded lens of the rabbit. Arch. Ophthal. 57, 849-854 (1957)

Shaw, C.R., Barto, E.: Genetic evidence for the subunit structure of LDH isozymes. Proc. Nat. Acad. Sci. USA 50, 211-214 (1963)

Stewart, J.A., Papaconstantinou, J.: Lactatedehydrogenase isozymes and their relationship to lens cell differentiation. Biochim. Biophys. Acta (Amst.) 121, 69-78 (1966)

Venrooij, W.J. van, Groeneveld, A.A., Bloemendal, H., Benedetti, E.L.: Cultured calf lens epithelium. I. Methods of cultivation and characteristics of the cultures. Exp. Eye Res. 18, 517-526 (1974)

Wanko, T., Gavin, M.A.: Electronmicroscopy of lens fibers. J. Biophys. Biochem. Cytol. 6, 97-102 (1959)

Worgul, B.V., Rothstein, H.: Radiation cataract and mitosis. Ophthal. Res. 7, 21-32 (1975)

9. Radiation and Endocellular Control of Cell Proliferation

K.-H. von Wangenheim

A. Introduction

There are two different levels at which mitotic activity can be controlled in higher organisms. The first level concerns the immature (undifferentiated) cells in which the steps of the mitotic cycle are controlled, whereas control at the other level concerns those processes that lead to terminal differentiation (cell maturation), and thereby to cessation of mitotic activity. For control at this level, studies in the plant endosperm and radiation experiments suggest the existence of a mechanism that enables the cell progenies, under certain environmental conditions, to proceed autonomously through a certain number of mitotic cycles and thereafter to differentiate. In normal cells of higher multicellular organisms, accordingly, the ability to restrain mitotic activity resides within the cells.

The mechanism is based on quantitative relations between the genetic systems of the nucleus and of the cytoplasm, although it is not consistent with the nucleus-cytoplasm relation theory. Radiation causes a shift in these quantitative relations, thereby causing the cell progenies either to differentiate and cease mitotic activity immediately or after a reduced number of mitotic cycles, provided that no specific conditions prevail that are needed in multicellular systems for maintenance and induction of mitotic activity. Since this radiation-induced "early differentiation" depends only on a quantitative shift, mutations are not a prerequisite for its induction. Under certain conditions the early differentiation is a major component of the radiation effect in higher organisms.

B. General Outline of the Control Mechanism

The mechanism is based on three principles:

I. Semi-Autonomy of Cytoplasmic Growth

Within certain limits, the amount of cytoplasm produced per mitotic cycle does not depend on nuclear ploidy, but on the pre-existing amount of cytoplasm. The phenomenon is most obviously explainable by dependence of cytoplasmic growth on the propagation of autoreduplicating cytoplasmic constituents such as mitochondria.

II. Induction of Differentiation by an Increase of the Quantitative Ratio of Cytoplasmic Amount (Number of Autoreduplicating Units): Nuclear Ploidy (Number of Chromosome Sets)

Unless the extracellular conditions do not enforce maintenance or induction of mitotic activity, an increase of this ratio to higher values induces successive steps that lead to terminal differentiation. It is not the cytoplasmic amount, however, that is important but an essential cytoplasmic component, which, under certain conditions, grows parallel to it. In the following, the cytoplasmic amount is still used as a parameter, because it is more easily determinable volumetrically, and the facts known are not yet sufficient for the use of another parameter (for example, number of mitochondrial genomes).

III. Suppression of Mitotic Activity by Differentiation

The end point of differentiation is reached gradually or stepwise. One of these steps includes cessation of mitotic activity.

From these principles it follows that in mitotically active tissues, under constant environmental conditions for which the rate of cytoplasmic growth is greater than that necessary for a doubling of the amount per mitotic cycle, the cell progenies have to pass through a certain number of cycles until terminal differentiation is initiated and mitotic activity ceases. Thus, the number of mitotic cycles passed through by the cell progenies, and, accordingly, the number of differentiated cells finally produced, is determined by the initial ratio of cytoplasmic amount:nuclear ploidy and by the increase in cytoplasmic amount per mitotic cycle (Fig. 1).

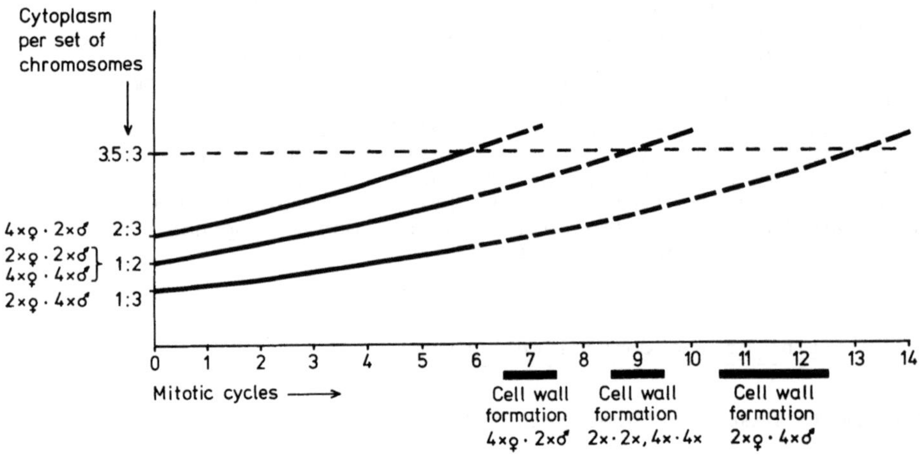

Fig. 1. Schematic representation of results obtained with *Oenothera hookeri*. Average volume of cytoplasm was determined from serial microtome sections of unfertilized embryo sacs and whole endosperms. Amount of cytoplasm per chromosome set was calculated accordingly. Increase in ratio of cytoplasmic amount:nuclear ploidy has been extrapolated beyond 6th synchronous mitotic cycle following fertilization, because determinations were not feasible thereafter (v. Wangenheim, 1967). For further explanation see text

C. Evidence from Studies in the Plant Endosperm and Other Observations

I. The Plant Endosperm

In the Angiosperms, as a rule, the endosperm develops abnormally if tetraploid plants are crossed with their original diploid lines (Müntzing, 1936). The phenomenon depends, according to a number of studies (von Wangenheim, 1962, 1967, 1969a; von Wangenheim et al., 1960), on a change of the initial ratio of cytoplasmic amount:nuclear ploidy, which is due to the fact that tetraploid females provide twice as much cytoplasm in their embryo sacs for endosperm development as diploid females, but the sperms transfer only relatively little cytoplasm. The initial ratio is, therefore, higher than normal in the cross tetraploid ♀·dipoid ♂ and lower than normal in the reciprocal cross. Although more cytoplasm is produced than is necessary for a doubling of the amount per mitotic cycle, the differences between the crosses could be shown to persist at least up to the 6th synchronous mitotic cycle following fertilization, thus demonstrating semiautonomy of cytoplasmic growth (see Fig. 1). A certain higher ratio of cytoplasm: ploidy is reached, therefore, significantly earlier in the cross tetraploid ♀·diploid ♂ than in the reciprocal cross. In accordance with this, cell wall formation is initiated two mitotic cycles earlier than normally in this cross, but at least two cycles later than normally in the other cross, and, similarly, all the other steps of differentiation that are discernible in the endosperm (cessation of growth substance production, RNA synthesis, and mitotic activity) are reached earlier or delayed, respectively. Finally, almost ten times more endosperm cells were counted in the cross diploid ♀·tetraploid ♂ than in the reciprocal cross. The attainment of certain thresholds of the quantitative ratio of cytoplasmic amount:nuclear ploidy, accordingly, induces the various steps of differentiation.

Interactions with other tissues and varying levels of ploidy per se have been excluded from being a possible cause of abnormal endosperm development by (1) regularities that are observable in the development of unreduced embryo sacs in certain crosses, (2) independence of endosperm development from ploidy level and presence of the embryo, and (3) by the use of *Oenothera hookeri*. This species is homozygous and the secondary embryo sac nucleus has the reduced chromosome number. Thus, the endosperms are triploid in both directions of the cross between diploid and tetraploid parents and do not differ genetically (see von Wangenheim, 1962, 1969a).

II. The Haploid Syndrome of Amphibia and Sex Determination in Insects

The haploid syndrome of Amphibia and sex determination by haploidy in insects point to the importance of the same mechanism in these organisms.

If a normal egg of Amphibia develops with only one chromosome set (male or female), the level of the ratio of cytoplasmic amount:ploidy is increased as is the case in the endosperm of the cross tetraploid ♀·diploid ♂. In some cell types this increased ratio has also been shown to persist in postgastrula stages (see Briggs and King, 1959), that is, when multiplication of mitochondria begins (Chase, 1971). Thus, as with the endosperms, those organs in which mitotic activity is not maintained permanently, as the nervous system, the heart, and visceral organs, remain too small. The development of the haploid animals is significantly improved if the initial ratio is decreased by production of the haploids from naturally occurring small eggs or from eggs whose size is decreased by surgery.

In *Mormoniella* (R.A. Whiting, 1955) and *Habrobracon* (P.W. Whiting and Keyhart, 1954), increasing x-ray exposures to eggs before fertilization increase the frequency of androgenetic males. Since the inactivation of the female chromosome set increases the ratio of cytoplasmic amount: nuclear ploidy, this suggests that sex is determined in these organisms, in the first place, by the level of this ratio.

D. Interference of Radiation with the Control Mechanism

In mitotically active tissues, x-rays induce "early differentiation." In essence, this means that the cell progenies, as is the case in the endosperm of the cross tetraploid ♀·diploid ♂, pass through a reduced number of mitotic cycles, or do not enter mitosis at all, before terminal differentiation. These differentiated cells are usually perfectly normal for at least a certain period of time (Haber and Foard, 1964; von Wangenheim, 1969b). The phenomenon is caused, as with the endosperm, by an increase of the ratio of cytoplasmic amount:nuclear ploidy. While the processes of cytoplasmic growth and differentiation are highly radioresistant, even relatively low exposures to ionizing radiations cause a transitory inhibition of mitosis (mitotic delay), which may last for hours or days. Therefore, cytoplasmic growth, and, according to recent investigations in the root tip meristem (von Wangenheim, 1975), the multiplication of mitochondria and proplastids, continue during the period of mitotic delay, thus increasing the amount of cytoplasm and the number and mass of the autoreduplicating cytoplasmic constituents after a LD_{37}-exposure to a level 51-123% above the corresponding unirradiated control cells. If the increase in the ratio of cytoplasmic amount:ploidy is sufficient, the cells differentiate even before the end of mitotic delay. Otherwise, the cells enter mitotic activity again. However, in accordance to the additional gain in cytoplasm and a natural gradient of cytoplasmic amount per cell in the normal root tip meristem, which increases from the stem cell region to the zone of differentiation, the cells pass through fewer mitotic cycles than normal. Only the cells in the stem cell region (quiescent centre, Clowes, 1963), which initially contain the lowest amount of cytoplasm, tend to counteract early differentiation and regenerate a new meristem.

There are various ways to demonstrate that the radiation-induced early differentiation depends on cytoplasmic growth during mitotic delay and not on mutations, although the latter may have an auxiliary effect (von Wangenheim, 1970, 1975; von Wangenheim et al., 1970).

1. Early differentiation proceeds with such regularity that a direct effect of mutations is excluded as a possible cause.

2. In diploid and tetraploid cells, the ratio of cytoplasm:ploidy is raised to approximately the same level if the same radiation exposure is applied, and, in accordance with this, the extent of early differentiation is the same at both levels of ploidy, although a higher level of ploidy protects significantly against genetic defects. Early differentiation means loss of cells that are capable of reproduction. Thus, as has been frequently observed with multicellular systems, a higher level of ploidy does not improve survival significantly, if mitotically active tissues are irradiated under usual conditions, because in this case early differentiation outweighs the effect of

mutations. If, however, dormant seeds are irradiated, in which case there is significantly less cytoplasmic growth during mitotic delay, or if anoxic hyperdiploid and hypertetraploid Ehrlich ascites tumor cells are irradiated (Revesz and Norman, 1963), not only is radiosensitivity lowered, but also the tetraploids survive significantly better than the diploids, as is expected if genetic damage is a major cause of nonsurvival.

3. Seed irradiation studies show that mutations per se, even in large numbers, neither produce differentiation nor easily inhibit the cells from entering at least the first mitotic cycle, and that a certain step of differentiation must be reached before mitotic activity is inhibited effectively. Conversely, spindle inhibition does not prevent early differentiation, although it prevents serious genetic damage, that is, loss and improper distribution of chromosome fragments, and thereby improves the survival of the differentiated cells. There are two ways, however, in which indirect effects of mutations may be auxiliary to the induction of early differentiation: (1) Various defects may increase the duration of the mitotic cycle, thus increasing the cytoplasmic production per cycle. (2) The level of ploidy determines the size of the differentiated cells. Loss of genetic material may, therefore, decrease the cell size and reduce the amount of cytoplasm needed for the induction of differentiation.

E. Maintenance and Induction of Mitotic Activity

During growth of the organism, as well as in cell renewal systems, mitotic activity has to be maintained. According to the control mechanism for cessation of mitotic activity, this is only possible by a reduced cytoplasmic growth rate that merely doubles the cytoplasmic amount per mitotic cycle and thereby prevents differentiation (the curves of Fig. 1 do not ascend). Thus, in the stem cell region of the root, the amount of cytoplasm and the number of mitochondria and proplastids are continuously kept at a low level (Clowes and Juniper, 1964; von Wangenheim, 1975). At the same time the mitotic cycle is very long (Clowes, 1963). Following irradiation, the amount of cytoplasm and the number of autoreduplicating units rise also in these cells. Instead of differentiating, however, the mitotic cycle shortens, and the cells tend to regenerate a new meristem with small cells. Thus, the same conditions that maintain mitotic activity counteract the radiation-induced early differentiation.

There are other examples that demonstrate the need of a low ratio between cytoplasmic amount (number of mitochondria) and nuclear ploidy for maintenance of mitotic activity. In regenerating rat liver, after partial hepatectomy, mitochondrial DNA synthesis begins with a delay, as compared with nuclear DNA synthesis (Baugnet-Mahieu et al., 1970), which, as a consequence, must decrease the number of mitochondria per cell during the first mitotic cycles. Similarly, in frog eggs, the huge number of mitochondria is brought to a low count per cell by a considerable delay in the onset of their multiplication during the early embryonic stages (Chase, 1971).

Thus, in multicellular systems special provisions exist for the purpose of counteracting the tendency of their cells to differentiate and cease mitotic activity. With these provisions, apparently, those substances are involved, which, in addition to pure nutrients, have to be added to cultures of normal cells in order to maintain mitotic activity, for example, auxins and cytokinins in plants and serum factors in mammals.

Under these conditions, mitotic control at the first level (see Introduction) is essential, and the assumption made by the nucleus-cytoplasm relation theory (Hertwig, 1903; Hartmann, 1928; Fantes et al., 1975), i.e., induction of mitosis by an increase of the quantitative ratio of cytoplasm:ploidy, appears to be valid. In the stem cell region of the root the long mitotic cycle is shortened following the radiation-induced increase in cytoplasmic amount (Clowes, 1963; von Wangenheim, 1975), and, similarly, the mitotic cycle in *Physarum plasmodia* is shortened following a relative increase in cytoplasmic amount due to selective destruction of some nuclei by ultraviolet (UV) treatment (Sachsenmaier et al., 1970). Apparently, a cytoplasmic component like an initiator protein is essential for the initiation of mitosis, whereas the autoreduplicating cytoplasmic units are involved with the induction of differentiation and cessation of mitotic activity.

F. Conclusions

Although most of the investigations cited above have been carried out with plants, the mechanism described appears to have a general importance for higher multicellular organisms. This is suggested by the fact that early differentiation can be induced by irradiation in various animal tissues and that, as with plants, a higher level of ploidy usually does not reduce radiosensitivity (see von Wangenheim, 1970, 1975). In mammals also, the prerequisites for early differentiation are known. Thus, in cells of some tumor types the cytoplasmic content increases following irradiation and the cells differentiate thereafter and cease mitotic activity (Klein and Forssberg, 1954; Glücksmann, 1941, 1974). In regenerating rat liver, mitochondrial DNA synthesis is significantly less radiosensitive than nuclear DNA-synthesis (Baugnet-Mahieu et al., 1970), from which it must be concluded that also in this organ the multiplication of mitochondria continues during mitotic delay. In the hematopoetic tissue cytoplasmic differentiation (maturation) is known to proceed independently of cell proliferation, so that with varying cell cycle and maturation times, the final output of cells can also be varied (Lord and Lajtha, 1974).

There are also other indications that may aid in further understanding of the control mechanism. According to a model of Barath and Küntzel (1972), mitochondrial DNA codes for one or several repressorlike proteins controlling those nuclear cistrons that code for mitochondrial proteins. If it is true that mitochondria evolved from an endosymbiotic bacterial cell, many genes from the original genome were lost, and this was only possible because their functions could be replaced by functions that were already present in the host cell genome. Thus, close metabolic interactions could develop, and the assumption follows that the postulated mitochondrial repressors control also the synthesis of enzymes that are essential for the host cell, e.g., those involved in DNA synthesis and differentiation. The demand of a low number of mitochondria per cell for maintenance of mitotic activity is also in agreement with the hypothesis of Welsh, according to which a low ATP level leads to the formation of open specific sites in the DNA strands, thus activating mRNA transcription (Welsh and Vyska, 1971), which would be expected to result in further mitotic activity.

For tumor therapy, it is of interest that the radiation-induced early differentiation is significantly more effective in inducing cessation of mitotic activity than pure genetic damage, and that it has been found to be inducible in various tumor types. The ability to restrain mitotic activity, accordingly, is not fully lost in at least these cases, and some support of cytoplasmic growth is sufficient for cessation of mitotic activity. For more detail see von Wangenheim (1976).

Acknowledgements. The author wishes to thank Miss H.M. Cichorius and Mrs. B. Kratz for careful technical assistance and Dr. R.S. Welsh for stimulating discussions and valuable aid in the preparation of the English text.

References

Barath, Z., Küntzel, H.: Cooperation of mitochondrial and nuclear genes specifying the mitochondrial genetic apparatus in Neurospora crassa. Proc. Nat. Acad. Sci. USA 69, 1371-1374 (1972)

Baugnet-Mahieu, L., Goutier, R., Baes, C.: Comparative sensitivity of the nuclear and mitochondrial DNA syntheses to X-irradiation and to the administration of a sulfhydryl radioprotector (AET) in normal and regenerating rat liver. Biophysik 6, 357-366 (1970)

Briggs, R., King, T.J.: Nucleocytoplasmic interactions in eggs and embryos. In: The Cell. Brachet, J., Mirsky, A.E. (ed.). New York, London: Academic Press,1959, Vol. I, pp. 537-617

Chase, J.W.: Mitochondrial RNAs in eggs and embryos of Xenopus. Carnegie Inst. Wash. Yearbook 69, 578-580 (1971)

Clowes, F.A.L.: X-irradiation of root meristems. Ann. Botany 27, 343-352 (1963)

Clowes, F.A.L., Juniper, B.E.: The fine structure of the quiescent centre and neighbouring tissues in root meristems. J. Exp. Botany 15, 622-630 (1964)

Fantes, P.A., Grant, W.D., Pritchard, R.H., Sudbery, P.E., Wheals, A.E.: The regulation of cell size and the control of mitosis. J. Theor. Biol. 50, 213-244 (1975)

Glücksmann, A.: Preliminary observations on the quantitative examination of human biopsy material taken from irradiated carcinomata. Brit. J. Radiol. 14, 187-198 (1941)

Glücksmann, A.: Histological features in the local radiocurability of carcinomas. In: The Biological and Clinical Basis of Radiosensitivity. Friedman, M. (ed.). Springfield, Ill.: C.C Thomas, 1974, pp. 203-219

Haber, A.H., Foard, D.E.: Further studies of gamma irradiated wheat and their relevance to use of mitotic inhibition for developmental studies. Amer. J. Botany 51, 151-159 (1964)

Hartmann, M.: Über experimentelle Unsterblichkeit von Protozoen-Individuen. Ersatz der Fortpflanzung von Amoeba proteus durch fortgesetzte Regeneration. Zool. Jahrb. Abtlg. Allg. Zool. u. Physiol. 45, 973-987 (1928)

Hertwig, R.: Über Kombination von Zell- und Kerngröße und ihre Bedeutung für die geschlechtliche Differenzierung und die Teilung der Zelle. Biol. Zentralbl. 23, 108-119 (1903)

Klein, G., Forssberg, A.: Studies on the effect of X-rays on the biochemistry and cellular composition of ascites tumors. Exp. Cell Res. 6, 211-220 (1954)

Lord, B.I., Lajtha, L.G.: Control of cell proliferation in haemopoetic tissue. Haematologia 39, 259-291 (1974)

Müntzing, A.: The evolutionary significance of autopolyploidy. Hereditas 21, 263-378 (1936)

Revesz, L., Norman, U.: Differential radiosensitivity in a neoplastic cell population. In: Radiation Effects in Physics, Chemistry and Biology. Ebert, M., Howard, A. (eds.). Amsterdam: North Holland, 1963, pp. 136-151

Sachsenmaier, W., Dönges, K.H., Rupff, H., Czihak, G.: Advanced initiation of synchronous mitoses in Physarum polycephalum following UV-irradiation. Z. Naturforsch. 25b, 866-871 (1970)

Wangenheim, K.-H. von: Zur Ursache der Abortion von Samenanlagen in Diploid-Polyploid-Kreuzungen. II. Unterschiedliche Differenzierung von Endospermen mit gleichem Genom. Z. Vererbungslehre 93, 319-334 (1962)

Wangenheim, K.-H. von: Entwicklungsphysiologische Untersuchungen über die Beteiligung nukleärer und extranukleärer Erbträger an der Phänogenese. Berlin Dtsch. Bot. Ges. 80, 228-236 (1967)

Wangenheim, K.-H. von: Zur Bedeutung extrachromosomaler Erbträger für die Steuerung von Nucleinsäurestoffwechsel, Zelldifferenzierung und Mitosetätigkeit. Berlin Dtsch. Bot. Ges. 82, 191-199 (1969a)

Wangenheim, K.-H. von: Wirkung von Röntgenbestrahlung auf Entwicklung und Ultrastruktur von Zellen des Weizenkeimlings. Radiat. Botany 9, 179-193 (1969b)

Wangenheim, K.-H. von: Genetische Defekte und Strahlenschädigung. Die Unterbindung der Mitose, ihre mutationsunabhängige Ursache und deren Bedeutung für Fragen der Zellproliferation, der Strahlenresistenz und der Strahlenstimulation. Radiat. Botany 10, 469-490 (1970)

Wangenheim, K.-H. von: A major component of the radiation effect: Interference with endocellular control of cell proliferation and differentiation. Int. J. Radiat. Biol. 27, 7-30 (1975)

Wangenheim, K.-H. von: A mechanism for the endocellular control of cell differentation and cell proliferation. J. Theor. Biol. 59, 205-222 (1976)

Wangenheim, K.-H. von, Peloquin, S.J., Hougas, R.W.: Embryological investigations on the formation of haploids in the potato (Solanum tuberosum). Z. Vererbungslehre 91, 391-399 (1960)

Wangenheim, K.-H. von, Walther, F., Kern, H.: DNA-Synthese nach Bestrahlung mit hohen Röntgendosen und Spindelhemmung. Z. f. Pflanzenzüchtung 64, 6-19 (1970)

Welsh, R.S., Vyska, K.: Properties of a new form of DNA from whole calf thymus nuclei: evidence for reactive, special sites in DNA. Arch. Biochem. Biophys. 142, 132-143 (1971)

Whiting, P.W., Kayhart, M.: A wasp for study of visible mutation rates. Radiat. Res. 1, 566-567 (1954)

Whiting, R.A.: Androgenesis as evidence for the nature of X-ray induced injury. Radiat. Res. 2, 71-78 (1955)

Closing Lecture - Modern Trends and Creeds in Radiobiology

T. ALPER

The title of this paper was chosen by Professor Kiefer, as a part of his very kind invitation to me to participate in this most interesting meeting. To be saved the trouble of thinking of a title for oneself has a peripheral advantage - one can examine the possible meanings of the words someone else has chosen, and, if one wishes, assign to them some significance that was perhaps not intended. It is a usual requirement of journals, in these days, that an author select key-words from his title and abstract, so as to keep computers happy. There are four key-words in my title, and I should like to consider the separate implications of three of them: What is "modern"? What are "trends"? and What do we mean by "creeds"?

Let us start by considering the phrase, "trends in radiobiology". We might think about these only in a scientific context - what problems are occupying the minds and hands of radiobiologists? Ideally, these should be dictated only by scientific interest: but none of us is free to pursue truth for its own sake, since research has to be funded, and we cannot divorce scientific trends from the tendencies of funding authorities (and therefore, usually, of Directors of research groups) to support, morally as well as financially, some lines or types of research rather than others. In my own country and, I am given to understand, also in others that have a tradition of radiobiological research, there is a distinct trend towards what is - in my opinion inaptly - called applied radiobiology: which means that, by and large, the research worker, in outlining his "project", must show that the results will be immediately, or almost immediately, applicable in the fields of either protection against radiation hazards or radiotherapy. Whether advances can be made in the former, without more information on fundamental processes, I do not know. I am better equipped to discuss the proposition that radiobiologists should be able to afford day-to-day assistance to the clinical radiotherapist, in a role analogous to that of the physicist (Windeyer, 1973). In my opinion, this state of affairs will not be achieved until much more has been learned about basic mechanisms of the action of radiation on cells. This is because cancerous tumours can be radiobiologically so different from each other, even when they are described as having the same histology. When a section from a tumour is examined microscopically we have no way of detecting features that would suggest an advantage in one method of radiotherapy over another. We cannot predict whether hypoxic cells will constitute a problem, because the tumour might fail to be adequately "reoxygenated" as a course of therapy proceeds; or whether the cells we see, either as individuals, or because of their contacts with others, will be able to sustain a great deal of damage in sublethal form, and then shed this damage in the intervals between treatments. If that happens, the tumour will effectively be "radio-resistant". The fraction of absorbed radiation that may be "wasted" by cells varies in many ways: for example, the way they are distributed in the cell cycle. The extent of sublethal damage may be increased if the cells are in contact with each other, but this happens only in some cell lines (Durand and Sutherland, 1975). The shape of cell survival curves was shown by Lehnert (1975) to depend also on the concentration of 3', 5'-cyclic adenosine monophosphate. These are just a few examples of the variability of "shoulders" to survival curves, observed with lower as well as with eukaryotic cells.

It might seem feasible to delineate the radiobiological properties of a given line of cancer cells by irradiating them in standard conditions in culture, but unfortunately such studies are only of limited value as a guide to what happens in situ. One example is afforded by comparing survival curves for freshly cultured normal mammalian cells with those obtained when the colonies of interest are formed in vivo. Puck et al. (1957) and, more recently, Cox and Masson (1975) found freshly cultured skin fibroblasts to be killed exponentially; whereas it has been shown that, in vivo, there is a large element of sublethal injury (Emery et al., 1970; Denekamp et al., 1971).

Surely it is the task of radiobiologists to discover the biological subtleties that determine such phenomena. I can see no reason for the assumption that the clues will come only from experiments with mammalian cells - on the contrary, the researcher should be free to test his hypothesis with whatever materials are the most suitable for his purpose. I would by no means suggest a moratorium on work aimed rather directly towards clinical aspects: indeed, there was some genuine "clinical radiobiology" in the programme of the Unit I used to direct; but if radiobiologists are constrained by funding authorities to engage *only* in applied radiobiology the source of profitable ideas and new concepts will dry up.

The topic and arrangement of this meeting have made me more optimistic on that score. It is unusual to attend a radiobiology meeting at which none of the words "protection", "hazards", "radiotherapy" or "cancer" appear in the programme! It therefore seems appropriate as a context for discussing "modern trends in radiobiology" in the scientific rather than the medical sense. My dictionary gives me a rather negative meaning of the word "modern" - things are "modern" that are "not antiquated or obsolete". But, in the pursuit of scientific truth, we are almost invariably building on what is antiquated, while (we hope) discarding the obsolete. It is of course almost a truism that some scientific topics, like fashions in dress, may be rather easily forgotten, and then recur as a novelty: so quite a lot of journal space is devoted to the rediscovery of results of research done a generation or so previously. However, a good deal of radiobiological effort is currently going into lines of research that have been going on for many years, but must be called "modern" because they are contemporary.

Judging by the proportion of papers devoted to the topic not only at this meeting, but at almost any other in the photo- or radiobiological fields, we must quote the phrase "repair and recovery" as *the* one that most readily elicits response from, I would guess, any institution engaged in research in those fields: and this is certainly no novelty. I find it regrettable that I have to report the failure of our attempt, made six years ago at this very place, to tidy up the nomenclature used in discussion on those topics. Muddled thinking is an inevitable consequence of the want of definition of those key words, which are applied to so many different phenomena frequently reported in those terms, although the evidence for positive "repair" is lacking, dare I say, more often than not. When the effects of irradiation differ, because different conditions have been used after irradiation, the results are almost invariably assumed to be "repair" of the DNA. This leads me directly to the third key-word of my title: "creeds".

Again referring to the Oxford English Dictionary, I find that the word has either a strictly religious significance - "a confession of faith" - or a very general definition - "a set of opinions on any subject".

The latter phrase is rather too weak to describe what I suppose Professor Kiefer to mean, and the former rather too strong - though it is true that the impression of religious fervour is sometimes gained from the tenacity with which certain radiobiological beliefs are defended!

The attachment of scientists to a belief that seems obviously justified is, of course, not confined to our own discipline. Only too often, there is a good deal of confusion between what is fact and what is supposition, in the supporting basis for a scientifically popular belief. Let me quote an example that is apposite, because it involves an important topic for us: the chemical nature of genes.

During the early years of this century research workers had amassed a great deal of accurate information about the location of genes on chromosomes, and it was generally recognized that the gene was a definable chemical entity, probably of molecular dimensions. Chromosomes were known to contain deoxyribose nucleic acid, the basic molecule of which contains sugar, phosphate, and four - *only* four - substituent bases. Both DNA and RNA were known to be very uniform in their chemical composition; so it was obvious that such components could not have the specificity necessary to account for the vast pool of different genes that between them determine the characteristics of all living organisms. Proteins, on the other hand, have the property of extreme diversity in structure and function (hence their name), so it was a reasonable and indeed obvious deduction that the "business" components of genes must be proteins - even if there were other components. Nucleic acid started to be mentioned in that role only after Avery et al. (1944) reported that the tranforming principle of *Pneumococcus* was DNA, to a high standard of purity. Presumably there could have been no *experimental* basis for assuming genes to be proteinaceous, or we have all been misled for thirty years! Yet, to demonstrate how this apparently obvious inference came to be accepted as fact, I quote from a book of the eminent scientist, Professor Julian Huxley, published in 1949.

"I would like to make a brief mention of another set of facts, concerning the chemical constitution and structure of the chromosomes. Of recent years, various elaborate physical methods such as spectrography under the microscope, x-ray analysis, etc. have enabled us to make a beginning in this difficult field. It is still only a beginning, but certain interesting facts are now established, and interesting probabilities opening up for investigation. These may be summed up for our purpose saying that a normal gene consists primarily of a particular kind of protein attached in some way to the neighbouring genes at its either end. The protein of genes is linked with nucleic acid, which becomes attached to and detached from the protein in a regular cycle. Both nucleic acid and protein are indispensable constituents of genes (including plasmagenes).

"The protein is in the form of an elongated chain of groups of atoms (polypeptide chain), and the molecules of nucleic acid seem to be fitted on to this chain rather as the steps of a spiral staircase are fitted on to the central column.

....

"Proteins are of course the most complicated of all the chemical constituents of living bodies, a single protein molecule sometimes containing many hundreds of atoms. Through slight variations in the position and number of the different kinds of atoms and atom-groupings, an almost infinite number of specifically different kinds of protein can be produced, each with its own particular chemical properties. The protein molecules of genes appear to be particularly large and complex, so that each gene presumably has its main framework composed of one particular kind of protein (or, possibly, a few particular kinds): it is this difference in protein structure which confers their specificity on genes, and makes every kind of gene different from every other.

"It seems probable that the combination of nucleic acid with a very complex kind of protein confers on the gene the capacity for self-reproduction or self-copying. The gene gets its specificity from the protein, while the nucleic acid seems to be responsible for various aspects of the behaviour of the chromosomes such as their spiral coiling at certain times. Mutations would then be due to minor rearrangements in the structure or composition of a particular kind of protein molecule.

"In any case, the facts, though still rudimentary, are important as taking our knowledge of the physical basis of heredity down below the microscopic level to the chemical level".

The book from which I have quoted was, of course, not intended as a discussion of the chemical nature of genes; nevertheless, those passages illustrate clearly how easy it is for unsupported beliefs to become accepted as facts, even by the most distinguished scientists. Once that has happened, it becomes difficult to assign a proper value to the emergence of *real* facts that reveal the false basis for the creed in question. So, unless we are very careful to examine the factual basis of our beliefs, we shall not be able to conduct ourselves as the grandfather of Julian Huxley, T.H. Huxley (1863), claimed that we do:

"Men of science do not pledge themselves to creeds; they are bound by articles of no sort; there is not a single belief that it is not a bounden duty with them to hold with a light hand and to part with it, cheerfully, the moment it is really proved to be contrary to any fact, great or small".

If, indeed, the *belief* that genes were proteins had been recognized as only a belief, with no true experimental basis, it might indeed have been parted with, in the face of certain "small facts", as early as 1931: there were reports that should have suggested nucleic acid, rather than protein, as the chemical compound responsible for carrying genetic information, at least in bacteria. Ultraviolet (UV) action spectra have long been used as a means for inferring the chemical nature of unidentified substances with biological activity, and Figure 1 presents results of Gates (1930) plotted as a conventional action spectrum to show the effectiveness of UV of various wavelengths in killing bacteria. It is very nicely equivalent to the UV absorption spectrum for nucleic acid. I once asked a prominent geneticist, who was an active worker already at that time, why the clue offered by Gates' work had been ignored. The answer was that the work had not been ignored, it had been discussed with interest, but, because the nucleic acids were so obviously incapable of affording the specificity required by genes, the action spectrum of Gates must be accounted for by some curiosity either of bacteria or of action spectra.

Well, we have gone through a U-turn, and it is now difficult for a biologist to entertain the idea that replication can occur without the involvement of either kind of nucleic acid. As it happens, certain researches in which I have been involved have suggested very strongly that this *does* indeed occur in the case of the transmitting agents of a small group of nervous diseases (Alper et al., 1967; Latarjet et al., 1970; Alper, Haig and Clarke, in preparation). In the corner of biology occupied by radiobiologists, the so called "central dogma" incorporates

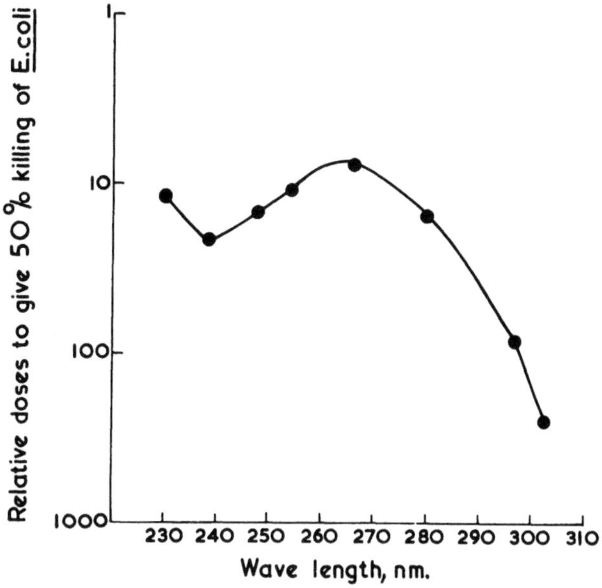

Fig. 1. Data of Gates (1930) plotted in form of an action spectrum, which corresponds with absorption spectrum for nucleic acids

an ancillary axiom - namely that the role of nuclear DNA is so overwhelmingly important that it is effectively the only structure in the cell in which the deposition of an ionization can be lethal. This appears to be as self-evident and obvious to many radiobiologists as the proteinaceous nature of genes was to geneticists, not so very long ago. If follows, of course, that the only type of "repair" with which most radiobiologists now concern themselves is repair of DNA.

I am as convinced as anybody that DNA is an important cellular target for lethal damage by radiation; I differ from the majority only in advocating that it is not the *only* important target. But what real experimental evidence have any of us that DNA *is* a target of major importance? We have all read, and heard, on numerous occasions that this is "a fact", but such evidence as is offered is inferential, and some of it rather weak. The assumption is sometimes alleged to gain support from the really good experimental evidence that lesions in DNA are overwhelmingly responsible for the lethal effects of UV - that is to say, UV in the range of wavelengths in which nucleic acids absorb strongly. To claim DNA as *the* target for x-ray damage on that account is palpably illogical. Surely we should expect to see differences in the effects of the two kinds of radiation precisely because UV in the range 250-270 nm *is* selectively absorbed by nucleic acid, whereas there is no selective absorption of ionizing radiation. It seems that nature has seen to it that several mechanisms exist by which UV-induced damage to DNA can be repaired - this could have been a necessary evolutionary development if free-living cells were to survive strong sunlight. Is it not logical, then, to expect that there might be other structures in the cell that would be less likely to be damaged by UV (because they do not specifically absorb at the wavelengths that reach the earth's surface) and for which repair mechanisms may not have evolved to the same extent?

I confess to adherence to an additional creed, namely that cell membranes constitute targets for damage by ionizing radiation at least

Fig. 2. To show "crystallization" of euchromatin around pores in nuclear membrane, soon after mitosis in mammalian cell. Traced from electron-micrograph by Everid et al. (1970)

as important as DNA. In this context, I refer to "targets" as the macromolecules or structures within which the ionizing event will have drastic consequences - although the pathway to lethality may well be via damage to the DNA. It is now well established that DNA is not a constituent floating freely in the cell; it is very firmly attached to membranes - nuclear membranes in eukaryotic cells (e.g., Fig. 2), or plasma membranes in prokaryotic ones; but I am conscious of the lack of direct evidence that cell membranes do constitute an important target for lethal or genetic effects of radiation. I wish I could say that investigation of effects of radiation on membranes, and how these interact with DNA lesions, was indeed a "modern trend in radiobiology": but I have hopes that it will be. I do not believe that radiobiologists can go on much longer ignoring all the current work that shows what important organelles cell membranes are.

Although such an "alternative" *biochemical* approach in radiobiology has yet to emerge, I can discern a trend towards the recognition of targets other than DNA by researchers currently concerned with events occurring too soon after the primary events for the involvement of cellular control processes. In recent years the techniques of flash photolysis and pulse radiolysis have revealed a good deal about occurrences within milli-, micro-, and even picoseconds after energy deposition in solids and liquids. Analogous techniques are being applied in experiments on living cells, although not yet on quite such a micro-time scale. With vegetative cells (i.e., other than dry spores or seeds) two techniques have been fruitful: (1) that in which a short pulse of radiation is followed within a very short space of time by an "explosion" that changes the gaseous environment; and (2) the "rapid-mix" technique, by which cells in suspension are mixed very soon after or before a pulse with a solution containing some modifying constituent.

As far as I know, the former technique has been used until now only to learn more about the oxygen effect - which continues to be a topic of great interest, despite its long history of recognition as a dominant factor in radiobiological effects. At the Gray Laboratory, Michael and his colleagues (1973) reported on the results of holding bacteria in an anoxic environment while they were irradiated by a 2 μsec pulse of electrons, after which oxygen was admitted within times that could be resolved to 0.1 msec. Sensitization by oxygen was found to decrease roughly exponentially within the interval between electron pulse and contact with oxygen, the half-life of the susceptible lesions being about 0.5 msec. This result provided support for the oxygen fixation hypothesis (Alper, 1956; Howard-Flanders, 1958), which, in its original form, postulated *one* sort of critical lesion that would be "fixed" by oxygen, or, in its absence, possess a probability equal to (m-1)/m of being restored to normal functioning, where m is the oxygen enhancement ratio. With those assumptions, the prediction from the hypothesis is indeed that the process by which restoration occurs should be exponential, provided first order kinetics are involved in the metionic reaction that results in restoration. However, subsequent exceedingly meticulous work by Maughan and Michael (as yet unpublished) has enabled them to look more closely at the time-course of the "radical decay" process. Their results are accurate enough for them to have found that a simple exponential function is not adequate as a description of the time course of the decay. They have therefore examined predictions from the assumption of a single decay or restoring process with second order kinetics: but these are likewise not fulfilled. On the other hand, on the assumption of two distinct types of damage contributing to the overall lethal effect, they have calculated that the time-related function would be compounded of two exponentials corresponding with two different rate constants for the different reactions, both obeying first order kinetics. This prediction is in good accord with the results obtained by Maughan and Michael. It might be postulated that those results suggest different types of damage to the same macromolecule, i.e., DNA, as suggested by Neary (1969). My own reasons for preferring the interpretation of *chemically different* sites for the potentially lethal primary lesions have been given in various publications, particularly in a paper at the Evian Congress, 1970.

The "fast-flow, rapid-mix" technique, likewise pioneered jointly by Adams and Michael (Adams et al., 1968) has been fruitful in elucidating the time scale, and therefore the mode, of the operation of several modifying agents, sensitizing as well as protective. In such experiments, the time resolution is of the order of milliseconds. In one important set of experiments (Shenoy et al., 1975), the sensitivity of mammalian cells was investigated as a function of the time *before* radiation at which oxygen was admitted to the suspension. Again, there is evidence of two distinctly resolvable components in the sensitizing action of oxygen. For one of these, the addition of quite low oxygen concentrations was sensitizing within the shortest measurable time (about 5 msec); for the other, however, full sensitization, even by 50% saturation of oxygen in the mix, was not achieved even when it was added 20-30 msec before the pulse of radiation. Subsequent experiments by Agnew, Stratford and Adams (reported by Adams et al., 1975) have demonstrated that the results cannot be accounted for by postulating delay before the oxygen gets into the nucleus. When identical techniques were used on the same cells, with the observation of DNA single-strand breaks as the end point, full sensitization by oxygen had been achieved within the shortest possible time of mixing before the pulse of irradiation. Furthermore, the same group found that in steady state conditions the relationship between extent of sensitization and oxygen concentration is quite different for results on cell killing and the induction of single-strand breaks.

The oxygen fixation hypothesis led also to the suggestion that molecular species other than oxygen might interact at the site of a primary event in such a way as to fix the damage (Alper, 1956). The existence of a small fraction of hypoxic cells in cancerous tumours has long been recognized as one possible reason for the failure of radiation to sterilize a tumour, so there could be a clinical advantage to be gained from the use of compounds that specifically sensitize anoxic cells, provided they are able to penetrate to the anoxic cells within the tumour. If oxygen acts within the physico-chemical time scale, it might be expected that this would be true also of non-gaseous sensitizers of anoxic cells. The current interest in this class of sensitizer extends through many fields of radiobiology - indeed, from their radiation chemical properties right up to their preliminary use in radiotherapy clinics.

The wide interest currently being shown in radiosensitizers of anoxic cells is the outcome of a series of investigations that very well illustrate my point about the indispensable nature of basic research. A proposed mechanism for the oxygen effect led to a carefully worked out rationale in the search for chemical radiosensitizers (cf. Adams, 1970), the investigation by radiation chemical techniques, with concomitant testing in living cells, and, within a remarkably short space of time, to their first tentative use in the clinic. Searches for sensitizers of tumour cells have, indeed, been conducted in other ways. I know of at least one project to "screen" for radiosensitizers of cells, by trying as many compounds as could be obtained, but with no guiding rationale that would confer a *therapeutic* advantage on the use of anything found to be effective in sensitizing cells in petri dishes. By contrast, the rational development I have described, invoking, as it does, ideas on basic mechanisms of radiation action, resulted also in the benefit of the feedback that can be obtained from the use of these compounds in learning still more about what happens within a very short time after energy is deposited in critical sites within cells (Chapman et al., 1975).

I have described work on the oxygen effect which provides more direct evidence of two chemically different sites of damage than I myself had presented - mine being inferential (e.g., Alper, 1963, 1969). It seems reasonable to predict that further work with chemical sensitizers will bear this out. The question then arises, do the different types of potentially lethal lesions act independently, or is there "interaction," either within the physico-chemical time scale or biochemically? The concept of interaction, although vague, is a part of classical radiobiology, though that word was not originally used. Lea and Catcheside (1942) calculated that breakage of a chromatid by an ionizing particle required the production of at least 17 ionizations within its volume. That calculation was thought to have been substantiated by further experimental work (Catcheside and Lea, 1943). Their basic concept of "cooperation" between ionizations was dominant for some time in the consideration of what we now refer to as "increase in RBE with increasing LET"; perhaps, conceptually, it is still dominant, despite Neary's challenge both of Catcheside and Lea's experimental results and of the calculation (17 ionizations to break a chromatid), which those results were thought to support. Neary (1965) proposed, instead, that the effectiveness of radiation in breaking chromosomes became greater as ionization density increased, because there was interaction between two single energy deposits. Neary applied his considerations to the induction of chromosome aberrations, and the interaction he envisaged was between two chromosome regions each carrying a primary lesion. While he did not discuss the time scale for these proposed interactions, his writing implies that he was thinking of processes of a biochemical rather than a physico-chemical nature.

Neary calculated the interaction distance, or site radius, he had posited as about 0.2 μ for chromosomal aberrations - too large to be accommodated within the concept that neighbouring radicals or ionizations within the track of a heavy particle co-operate to give a high biological effectiveness. Interaction distances of that order have been calculated by Kellerer and Rossi (1972) and others, on the basis of data from microdosimetry: that is, the measurement of the distribution of "event sizes" within a small volume, for radiation of a given quality. This branch of radiation physics has been developed only within the last decade and is potentially important in helping us to understand more about the initiating processes in radiation biology. At the recent biennial conference (5th Conference on Microdosimetry, 1975) authors of several papers on radiobiological topics included microdosimetry data on which they calculated interaction distances; but I could find no one who was prepared to define "interaction".

I think it important to remember that, when subcellular "targets" with biological activity are irradiated, the effectiveness of radiation *decreases* with increasing ionization density: this has been found with single- and double-strand DNA viruses (Schambra and Hutchinson, 1964), with enzymes (Brustad, 1961), and with lysosomal membranes (Watkins and Deacon, 1973). With cells, on the other hand, radiation is more effective lethally as the LET increases. It seems to me plausible that this happens because of interaction between sites of energy deposition, but not within the same macromolecule - and one thinks of DNA, of course - because of the evidence I have just quoted from the irradiation of viruses. The subcellular particles have no biochemistry, and this is why all the predictions of classical target theory are fulfilled when they are irradiated. Cells, however, contain structures of different composition in close proximity, and I refer particularly to the attachment of DNA to membranes. Figure 3 demonstrates in an oversimplified way how Relative Biological Effectiveness should de-

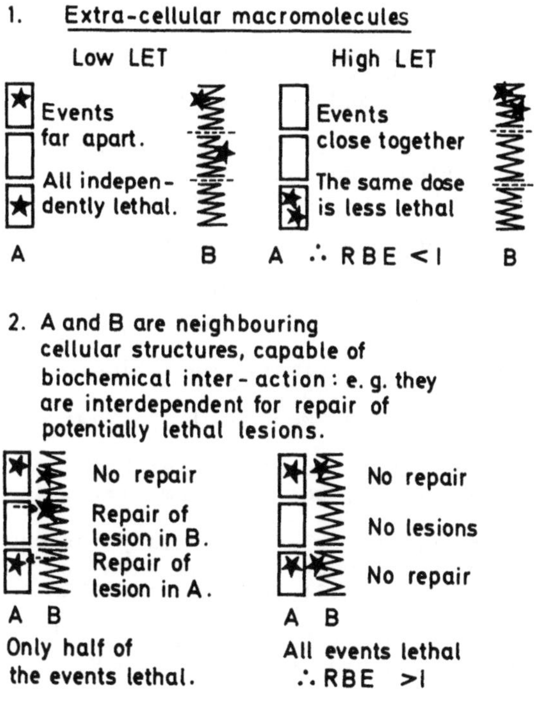

Fig. 3. Simple speculative rationale for increase in effectiveness of radiation for cellular effects, as ionization density increases. Upper panel shows why this is not true of subcellular "targets": target volume is that within which single energy deposit is sufficient for inactivation, so a second one would be wasted. Lower panel suggets that, with a pair of "targets" of different chemistry juxtaposed, a primary lesion in one member of a pair may be "repaired" if the other member is intact; events in both members would therefore be effective. Alternatively, a potentially lethal lesion in one member of the pair might be "fixed" as a consequence of an energy deposit in the other member

crease, when isolated targets are exposed to radiation of high LET, whereas, if there were biochemical interaction between lesions in adjoining units of structure, it could well increase. "Repair" of damage in one structure might be able to proceed, provided its partner were undamaged; alternatively, and perhaps this is more realistic, a potentially lethal lesion in DNA, say, might become fixed as a consequence of energy deposition in the neighbouring membrane. Many enzymes are, of course, tightly bound to membranes, and it was proposed many years ago that radiation damage might result from their release as a result of an ionization at the point of binding (Bacq and Alexander, 1955). Allison and Paton (1965) showed that the specific release of DNAse from lysosomes could cause chromosome breaks. How much more likely it is that DNA that has itself suffered a potentially damaging event should have the damage fixed by the untimely release of a degradative enzyme!

If there is anything in these speculations, the sort of "dual action" I have envisaged will certainly be reflected in its effect on cellular control processes. I look forward to a successor to this meeting, at which papers will be given that will elucidate the biochemistry of "interaction" and the two components of the oxygen effect, topics that are among the more exciting developments in that part of radiobiology that is concerned with the nature and crucial sites of primary energy deposition.

Acknowledgement. I am grateful to Drs. Adams, Agnew, Maughan, Michael and Stratford for allowing me to quote some of their unpublished results.

References

Adams, G.E.: Molecular mechanisms of cellular radiosensitization and protection. In: Radiation Protection and Sensitization. (Moroson, H., Quintiliana, M., eds.). London: Taylor and Francis, 1970, pp. 3-14

Adams, G.E., Agnew, D., Stratford, I., Wardman, P.: Applications of pulse radiolysis and cellular fast-mixing techniques to the study of radiation damage in cells. Proc. 5th Symp. Microdosimetry, Euratom. In press, 1975

Adams, G.E., Cooke, M.S., Michael, B.D.: Rapid mixing in radiobiology. Nature (Lond.) 219, 1368-1369 (1968)

Allison, A.C., Paton, G.R.: Chromosome damage to human diploid cells following activation of lysosomal enzymes. Nature (Lond.) 207, 1170-1173 (1965)

Alper, T.: The modification of damage caused by primary ionization of biological targets. Radiat. Res. 5, 573-586 (1956)

Alper, T.: Lethal mutations and cell death. Phys. Med. Biol. 8, 365-385 (1963)

Alper, T.: Mechanisms of lethal radiation damage to cells. In: Proc. 2nd Symp. Microdosimetry, 1969. (Ebert, H.G., ed.) Euratom, 1970, pp. 5-36

Alper, T.: Cell death and its modification: the roles of primary lesions in membranes and DNA. IAEA Symp. Biophys. Aspects of Radiation Quality. Lucas Heights, Australia, 1971, pp. 171-184. Vienna: IAEA Atomic Energy Agency, 1971, IAEA-SM 145/1

Alper, T., Cramp, V.A., Haig, D.A., Clarke, M.C.: Does the agent of scrapie replicate without nucleic acid? Nature (Lond.) 214, 5090, 764-766 (1967)

Avery, O.T., MacLeod, C.M., McCarty, M.: Studies on the chemical nature of the substance inducing transformation of pneumococcal types. Induction of transformation by a desoxyribonucleic acid fraction isolated from Pneumococcus type III. J. Exp. Med. 79, 137-157 (1944)

Bacq, Z.M., Alexander, P.: Fundamentals of Radiobiology. 1st Ed. London: Butterworth's, 1955, pp. 185-188

Brustad, T.: Molecular and cellular effect of fast charged particles. Radiat. Res. 15, 139-158 (1961)

Catcheside, D.G., Lea, D.: The effect of ionizing distribution on chromosome breakage by X-rays. J. Genet. 45, 186-196 (1943)

Chapman, J.D., Reuvers, A.P., Doern, S.D., Gillespie, C.J., Dugle, D.L.: Radiation chemical probes in the study of mammalian cell investigation and their influence on radiobiological effectiveness. Proc. 5th Symp. Microdosimetry, Euratom. In press 1975

Cox, R., Masson, W.K.: X-ray survival curves of cultured human diploid fibroblasts. In: Cell Survival after Low doses of Radiation: Theoretical and Clinical Implications. Alper, T. (ed.) United Kingdom: The Institute of Physics and John Wiley & Sons, 1975, pp. 217-222

Denekamp, J., Emery, E.W., Field, S.B.: Response of mouse epidermal cells to single and divided doses of fast neutrons. Radiat. Res. 45, 80-84 (1971)

Durand, R.E., Sutherland, R.M.: Intercellular contact, its influence on the D_q of mammalian survival curves. In: Cell Survival after Low Doses of Radiation: Theoretical and Clinical Implications. Alper, T. (ed.) United Kingdom: The Institute of Physics and John Wiley & Sons, 1975, pp. 237-247

Emery, E.W., Denekamp, J., Ball, M.M., Field, S.B.: Survival of mouse skin epithelial cells following single and divided doses of x-rays. Radiat. Res. 41, 450-466 (1970)

Everid, A.C., Small, J.V., Davies, H.G.: Electron-microscope observations on the structure of condensed chromatin: Evidence for orderly arrays of unit threads on the surface of chicken erythrocyte nuclei. J. Cell Sci. 7, 35-48 (1970)

Gates, F.L.: A study of the bactericidal actions of ultraviolet light. VII. The absorption of ultraviolet light by bacteria. J. Gen. Physiol. 14, 31-42 (1930)

Howard-Flanders, P.: Physical and chemical mechanisms in the injury of cells by ionizing radiations. Adv. Biol. Med. Phys. 6, 553-603 (1958)

Huxley, J.: Soviet Genetics and World Science. London: Chatto and Windus, 1949

Huxley, T.H.: On our knowledge of the causes of the phenomena of organic nature. London: Robert Hardwicke, 1863, p. 150

Kellerer, A.N., Rossi, H.H.: The theory of dual radiation action. Current Topics Radiat. Res. Qtly. 8, 85-158 (1972)

Latarjet, R., Muel, B., Haig, D.A., Clarke, M.C., Alper, T.: Inactivation of the Scrapie agent by near-monochromatic ultraviolet light. Nature (Lond.) 227, 1341-1343 (1970)

Lea, D.E., Catcheside, D.G.: The mechanism of the induction by radiation of chromosome aberrations in tradescantia. J. Genet. 44, 216-245 (1942)

Lehnert, S.: Relation of intracellular cyclic AMP to the shape of mammalian survival curves. In: Cell Survival after Low Doses of Radiation: Theoretical and Clinical Implications. Alper, T. (ed.) United Kingdom: The Institute of Physics and John Wiley & Sons, 1975, pp. 226-236

Michael, B.D., Adams, G.E., Hewitt, H.B., Jones, W.B.G., Watts, M.E.: A post-effect of oxygen in irradiated bacteria: a submillisecond fast mixing study. Rad. Res. 54, 239-251 (1973)

Neary, G.J.: Chromosome aberrations and the theory of R.B.E. I. General considerations. Int. J. Rad. Biol., 9, 477-502 (1965)

Neary, G.J.: The interpretation of survival curves in relation to radiation quality and cellular repair mechanisms. In: 2nd Symp. Microdosimetry, Stresa (Italy), Ebert, H.G. (ed.). Euratom, 4452 dfe, 1969, p. 51

Puck, T.T., Morkovin, D., Marcus, P.I., Cieciura, S.J.: Action of X-rays on mammalian cells II. Survival curves of cells from normal human tissues. J. Exp. Med. 106, 483-500 (1957)

Schambra, P.E., Hutchinson, F.: The action of fast heavy ions on biological material. II Effects on T_1 and Øx-174 bacteriophage and double-strand and single-strand DNA. Radiat. Res. 23, 514-526 (1964)

Shenoy, M.A., Asquith, J.C., Adams, G.E., Michael, B.D., Watts, M.E.: Time-resolved oxygen effects in irradiated bacteria and mammalian cells. A rapid-mix study. Radiat. Res. 62, 498-512 (1975)

Watkins, D.K., Deacon, S.: Comparative effects of electron and neutron irradiation on the release of enzymes from isolated rat-spleen lysosomes. Int. J. Radiat. Biol. 23, No. 1, 41-50 (1973)

Windeyer, B.: Biological Basis of Radiotherapy. Introduction. Brit. Med. Bulletin 29, No. 1, 1-2 (1973)

Subject Index

Note: Only the starting pages of the papers are indicated

MIX
Papier aus verantwortungsvollen Quellen
Paper from responsible sources
FSC® C105338

If you have any concerns about our products,
you can contact us on
ProductSafety@springernature.com

In case Publisher is established outside the EU,
the EU authorized representative is:
Springer Nature Customer Service Center GmbH
Europaplatz 3, 69115 Heidelberg, Germany

Printed by Libri Plureos GmbH
in Hamburg, Germany